BUT IS IT SCIENCE?

UPDATED EDITION

BUT IS IT SCIENCE?

THE PHILOSOPHICAL QUESTION IN THE CREATION/EVOLUTION CONTROVERSY

EDITED BY
ROBERT T. PENNOCK
AND MICHAEL RUSE

Prometheus Books
59 John Glenn Drive
Amherst, New York 14228-2119

Published 2009 by Prometheus Books

But Is It Science? The Philosophical Question in the Creation/Evolution Controversy—Updated Edition.

Inquiries should be addressed to
Prometheus Books
59 John Glenn Drive
Amherst, New York 14228–2119
VOICE: 716–691–0133, ext. 210
FAX: 716–691–0137
WWW.PROMETHEUSBOOKS.COM

13 12 11 10 09 5 4 3 2 1

Library of Congress Cataloging-in-Publication Data

But is it science? : the philosophical question in the creation/evolution controversy / edited by Robert T. Pennock and Michael Ruse. — Updated ed.

p. cm.

ISBN 978–1–59102–582–5 (pbk. : alk. paper)

1. Evolution (biology)—Philosophy. 2. Creationism—Philosophy. I. Pennock, Robert T. II. Ruse, Michael.

QH360.5.B87 2008
576.801—dc22

2008048840

Printed in the United States of America on acid-free paper

CONTENTS

Preface 9

PART I: RELIGIOUS, SCIENTIFIC, AND PHILOSOPHICAL BACKGROUND

Introduction to Part I 19

1. The Bible (Genesis 1 and 2; John 1:1–18) 27

2. *Natural Theology* 33
   William Paley

3. *On the Origin of Species* 38
   Charles Darwin

4. Objections to Mr. Darwin's Theory of the Origin of Species 65
   Adam Sedgwick

5. The Origin of Species 73
Thomas H. Huxley

6. *What Is Darwinism?* 82
Charles Hodge

7. Darwinism as a Metaphysical Research Program 105
Karl Popper

8. Karl Popper's Philosophy of Biology 116
Michael Ruse

9. Human Nature: One Evolutionist's View 136
Francisco Ayala

10. Universal Darwinism 158
Richard Dawkins

PART II: CREATION SCIENCE AND THE *McLEAN* CASE

Introduction to Part II 187

11. The Creationists 192
Ronald L. Numbers

12. Creation, Evolution, and the Historical Evidence 231
Duane T. Gish

13. Witness Testimony Sheet: *McLean v. Arkansas Board of Education* 253
Michael Ruse

14. United States District Court Opinion:
McLean v. Arkansas Board of Education 279
Judge William R. Overton

15. The Demise of the Demarcation Problem 312
Larry Laudan

16. Science at the Bar—Causes for Concern 331
Larry Laudan

17. *Pro Judice* 337
Michael Ruse

18. More on Creationism 345
Larry Laudan

19. Commentary: Philosophers at the Bar—Some Reasons for Restraint 350
Barry R. Gross

PART III: INTELLIGENT DESIGN CREATIONISM AND THE *KITZMILLER* CASE

Introduction to Part III 369

20. But Isn't It Creationism? The Beginnings of "Intelligent Design" in the Midst of the *Arkansas* and *Louisiana* Litigation 377
Nick Matzke

21. What Is Darwinism? 414
Phillip E. Johnson

22. Is It Science Yet? Intelligent Design, Creationism, and the Constitution 426
Matthew J. Brauer, Barbara Forrest, and Steven G. Gey

23. *Kitzmiller v. Dover Area School District* Expert Witness Testimony 434
Michael Behe

24. *Kitzmiller v. Dover Area School District* Expert Report 456
Robert T. Pennock

25. A Step toward the Legalization of Science Studies 485
Steve Fuller

26. What Is Wrong with Intelligent Design? 495
Elliott Sober

27. United States District Court Memorandum Opinion:
Kitzmiller, et al. v. Dover Area School District, et al. 506
Judge John E. Jones II

28. Can't Philosophers Tell the Difference between
Science and Religion? Demarcation Revisited 536
Robert T. Pennock

PREFACE

Since at least the time of Saint Augustine in the fourth and fifth centuries of the Common Era, it has been part of Christian tradition that parts of the Bible—Old Testament and New—have to be interpreted metaphorically. A literal reading just will not do. The great Reformers, Luther and Calvin, urged a return to a more scripture-based religion, but they too recognized that sometimes one must go beyond the literal words of the text. God "accommodates" his language to the common people. For various reasons, however, in the nineteenth century, American Protestantism developed into a religion that urged a much more direct and uninterpreted approach to the sacred text. In part, this was because a people building a new land needed direct guides to action and everyday living, and the preachers told them that the answers lay close at hand, in the Bible. In part, particularly in the South, this was because a literal reading of the Bible confirmed opinions and prejudices already held. Above all, a literalist reading of scripture justified slavery for its supporters. Saint Paul did not free the runaway slave but told him to return to his master and to obey!

The Civil War hardened positions and divisions that had grown in the land. In the North, the last part of the nineteenth century was an age of progress, of science and technology and industry, of advances in medicine and other human-related endeavors, of new German-style universities, and above all in religion of the embrace of "modernism"—a belief that much in Christianity is metaphorical at best and that actions should be directed toward social improvements. In the South, and increasingly in the West as the national moved out across the plains, a Bible-based, evangelical Christianity ruled supreme. Far from seeing defeat in the war as a refutation of such a religion, sermon after sermon stressed that God most oppresses those whom he loves most. Frequent analogies were drawn between the American losers and the Israelites in captivity.

Evolution was always part of this divide, perhaps less in its own right—one doubts that many actually lay awake at night worrying about gaps in the fossil record—and more as a kind of litmus test for other, deeper concerns. To the Northern modernist, evolution was a mark of science's triumph over old prejudices, and people almost took pride in pointing to simian aspects of their own natures. To the Southern literalist, evolution was the epitome of a false doctrine that went against scripture and that stood for lax living and godlessness. North and South agreed that evolution is incompatible with a literalist Bible-based Christianity, although North and South differed over the consequences to be drawn.

Things continued this way right through the first half of the twentieth century. Showing that what troubled people was the symbolism of evolution rather than the actual science, we find indeed considerable flexibility in its Christian opponents. The most famous public confrontation between creationism and evolution was the trial in Dayton, Tennessee, in 1925 when a young school teacher, John Thomas Scopes, was accused of teaching evolution to the young charges in his classroom. He was defended by the noted free-thinking lawyer Clarence Darrow who, on being denied the opportunity to present his own scientific witnesses, put on the stand the prosecuting attorney, William Jennings Bryan, the Great Commoner, three-time presidential candidate for the Democratic party, and sometime cabinet minister for Woodrow Wilson. Bryan, who was certainly considered by himself and by others to be an orthodox Bible Christian, calmly denied that he thought the six days of creation were six days of twenty-four-hour limits. As he said, a thousand years are as but a day in the eye of the Lord, and a day is as a thousand years.

Although Scopes was found guilty, America laughed so much at the case that the literalist movement—now known as fundamentalism, after a series of pamphlets attacking modernism (but surprisingly, though it was soon forgotten by fundamentalists, finding no necessary opposition between scripture and evolution) that had been published between 1910 and 1915—rather faded from sight. But it was sleeping, not dead. Textbook manufacturers took note of the opposition to evolution, and it practically disappeared from the curriculum. The paradox is that this was precisely the time (the 1930s and '40s) when evolutionary theory itself was undergoing massive changes, as the new Mendelian genetics was blended with the achievements of Charles Darwin, author in 1859 of *On the Origin of Species* and discoverer of the mechanism of natural selection (or the survival of the fittest). Just at the time when evolution finally had a fully functioning paradigm, it was disappearing from the classrooms!

Things changed in the late 1950s, thanks primarily to the Cold War and Russia's success in launching a satellite, *Sputnik*. Rightly or wrongly, America saw itself as losing the battle with Communism and started to pour money into science and technology, including science education. New textbooks were written that made it clear that evolution was one of the truly fundamental theories of science. Evolution was back in public schools, and fundamentalists awoke to challenge what they took to be the teaching of godlessness. After the 1968 *Epperson v. Arkansas* decision in which the US Supreme Court ruled that Scopes-era laws banning the teaching of evolution were unconstitutional, supporters of creationism attempted at first to mandate that evolution not be presented as a scientific fact but just a theory and that texts also give equal emphasis to the alternative creationist account in the Bible. But such laws were immediately struck down, particularly in the 1975 *Daniel v. Waters* case, for the preferential treatment they gave to the biblical religious view over the scientific view and over the creation beliefs of other religions. Fundamentalists' next move was inspired by a 1961 book, *The Genesis Flood*, co-written by a biblical scholar and a hydraulic engineer, that was to become the bedrock work of the invigorated literalist movement.

John C. Whitcomb and Henry M. Morris's *The Genesis Flood* was no mere restatement of past ideas and beliefs. Deeply influenced by the thinking of Ellen G. White, the mid-nineteenth-century founder of the Seventh-day Adventists, Whitcomb and Morris endorsed a "young-earth" creationism.

Going beyond the more flexible creationism of the 1920s that Bryan exemplified and that allowed for an old earth, they stood firm for a literal six-day creation about six thousand years ago (as calculated from the genealogies given in the Bible). They were also premillennial dispensationalists, believing that earth history is divided into periods, each ended by some awful catastrophe. The end will come in the near future, when the forces of evil clash with the forces of good in the battle of Armageddon. Because Noah's Flood marked the end of the first dispensation, its proof became a key point in the creationist scenario, more so even than such things as the expulsion from Eden or the Jews' travails in Egypt. Most important, they claimed that these views were supported by science without the need for any reference to the Bible. Thus was born a new form of creationism, what they called creation science or scientific creationism. Morris built a grassroots ministry upon this idea—the Institute for Creation Research—but more politically active fundamentalists pushed it as a way to get creationism back in the schools by portraying it as science.

Matters came to a head in 1981 in Arkansas. The state's Republican governor, Frank D. White, signed into law a bill that demanded "balanced treatment," meaning that if evolution was to be taught in biology classes, then the same courtesy had to be extended to creation science. Immediately, seeing a gross violation of the United States Constitution's separation of church and state, the American Civil Liberties Union sprang into action with the 1982 *McLean v. Arkansas* case, opposing the law on the grounds that creation science, despite its name, was not science but disguised religion. The federal judge trying the case agreed with the ACLU, and it was thrown out. Subsequently, in the state of Louisiana, a similar law was declared unconstitutional, and, having been appealed unsuccessfully right up to the US Supreme Court in the 1987 *Edwards v. Aguillard* case, that seemed to be the end of creation science in the public domain.

However, like the Hydra that grows two heads for each one struck off, evangelical biblicalism was far from finished. At the beginning of the 1990s, a new version of creationism appeared that was renamed and tailored to get around the 1987 *Edwards* decision. Much more user-friendly, it avoided specific statements designed to bring science into harmony with Genesis read literally—some of its supporters were not interested in such harmony, others most certainly were but realized that for the moment a policy of silence was golden. So-called intelligent design theory was, however, continuous with older thinking in many respects, but especially in its keystone belief that a purely naturalistic—lawbound—evolutionary theory is simply impossible and that some intelligence beyond nature had to have been responsible for the design of creation. Arguing that aspects of the organic world exhibit "irreducible complexity," that simply cannot be explained by regular science, its supporters claimed that at certain periods of life history it is necessary to suppose that an intelligence—a nonnatural designing agent—got involved to move along the development and unfurling of life here on earth. In such works as *Darwin on Trial* by the Berkeley lawyer Phillip Johnson, *Darwin's Black Box* by the Lehigh University biochemist Michael Behe, and *The Design Inference* by William Dembski, who now teaches philosophy and theology at Southwestern Baptist Theological Seminary, the case was made for the incorporation of infusions of design by a nonnatural intelligent being in the history of life. Although, like the scientific creationists, the intelligent design enthusiasts did not want publicly to identify their thinking too closely with Christian theology—there is the separation of church and state to be evaded—in truth they made little secret that they believed the designer was the God of the Bible. A particularly favorite identification was with God as described at the beginning of Saint John's Gospel.

Once again the drama moved to the law courts. In the town of Dover, Pennsylvania, the school board declared that intelligent design must be introduced into the biology classrooms of the district. Once again, the ACLU, together with Americans United for Separation of Church and State, sprang into action, arguing that such a move is unconstitutional. Once again, a federal judge agreed, dismissing the case of the school board in even more withering terms than the judge had used more than twenty years earlier in Arkansas. In *Kitzmiller et al. v. Dover Area School District et al.*, intelligent design theory was thrown on the same garbage dump as was creation science. However, it would be a person optimistic to the point of naiveté who argued that we shall never again see the rise and force of some form of creationism—whether it be called literalism, fundamentalism, creation science, intelligent design theory, or what have you.

It is with this last point in mind that this collection has been developed. It is the second edition of a work with the same title that was put together twenty years ago by one of the present editors, Michael Ruse, who was an expert witness for the ACLU in Arkansas. He is now joined by Robert T. Pennock, who was an expert witness for the ACLU in Pennsylvania. We are both professional philosophers (Pennock also does scientific research in experimental evolution) and it was this that got us involved in the two trials, for both times it became obvious to all that although these events demanded knowledge of science and of theology, another significant part of the division between evolutionists and their opponents was philosophical. There were questions of methodology and of evaluating evidence, for example. Most importantly, given that the United States Constitution bars the teaching of religion and not of bad science, there was the task of showing that while evolutionary theorizing and experimentation is scientific, creation science or intelligent design theory are not. They are religious.

It seemed important twenty years ago, and it seems equally important now, given that we surely have not seen the end of this debate, that the philosophical elements in the creationism wars be given a full airing. Accordingly, this volume has several aims. First, at a historical and conceptual level, to see how philosophy got involved and what contributions it could make. Second, to bring the material up to date and to deal with not only the new bottles into which creationism has put its old wine but also the ways that science and philosophy of science have evolved. Third, to learn from the errors of the past so that we might try to avoid or at least mitigate such controversies in the future. If the job was not done as well as it might have been the first time, then let us see why and consider how arguments and policies may be improved to maintain the integrity of science and science education. Finally, at a more pedagogical level, to introduce students to these issues through actual cases, so that they can grasp philosophical points of great interest and importance in the rough, as it were—not sanitized in conventional textbooks, but out in the real world where decisions matter. For these reasons, we offer a new edition of *But Is It Science?*

We have significantly revised the earlier edition. The aims are the same, but the content has been updated. We eliminated some outdated or redundant pieces and added newer ones. For pedagogical reasons, we reduced the number of general secondary sources and expanded the length of the primary sources. We have kept portions of the earlier sections that provide key background information on the religious and scientific perspectives that led to the conflict over evolution and that show how philosophical issues have been important right from the start. We have also, in a somewhat revised form, retained the most significant material on the Arkansas trial, both the trial itself and the philosophical discussions about the nature of science to which it led. To this we have now added much new material on the *Kitzmiller* trial, trying to give enough information that the reader can understand the history linking the two cases, and can grasp how the philosophical issues were no less important in Pennsylvania than they were in Arkansas for answering the salient questions about the relationship between creationism and science.

We hope that you enjoy this collection and learn from it. We hope sincerely that in twenty years it will not be necessary to bring out a third edition.

Robert T. Pennock
Michael Ruse

PART I

RELIGIOUS, SCIENTIFIC, AND PHILOSOPHICAL BACKGROUND

INTRODUCTION TO PART I

Part I aims to spell out using mostly primary sources some of the historical facts alluded to in the preface and to provide some background information that is needed to begin to evaluate our philosophical question as it figures in the contemporary creationism controversy: *But is it science?* We will see that right from the coming of Darwin's theory—before, even—there was concern and debate around the status of the various claims made about origins. Many of the issues that are discussed in the creation science and intelligent design creationism debate—the status of the evidence for evolution, evolution's relation to religious belief and nonbelief, whether science can countenance notions of supernatural design, and more—will be seen to have clear antecedents in earlier periods. The section can best be thought of as falling into four subsections.

The first begins "in the beginning" with some of the biblical passages that many creationists take to be incompatible with evolution. The book of Genesis is the most problematic, of course, in that on a literalist reading it gives a story of creation (or is it two stories of creation?) that differs not only from evolutionary biology but also from astronomy, geology, and other sciences. Here we reproduce the first two chapters with their two accounts of creation. We might note, parenthetically, that modern biblical scholars date the second creation story before the first. The second appears at the time of King David (he reigned 1000–961 BCE), and with its twin emphases on God's having created man, yet also having ceded man's power over the earth, represents an attempt to define the proper role of a leader to a population unused to (and somewhat dubious about) having a king. The first story appears during the Babylonian exile (post-587 BCE), picks up on other Near Eastern creation stories, and tries to reconvince the Jews of both the power of their God and of their own special place in his heart.

Of course, this is to focus on the creation stories for their historical and religious significance, rather than treating them as though they were meant to be a science textbook, as creationists are wont to do. The difficulty of reconciling the two accounts on a literal interpretation, even on points as simple as the order of events, was well known by scholars of biblical exegesis. Such internal interpretive problems arise completely independently of the relationship between the passages and scientific findings, and it was such issues that led the early Church fathers such as Augustine to opt for a figurative interpretation. Fundamentalists, however, insist upon a plain reading. Fundamentalists also tend to ignore problems of translation by taking the King James Version as divinely guided and thus authoritative, which is why we use that translation here. We also include a selection from the Gospel of John. While creation science focuses its attention mostly on Genesis, intelligent design creationism especially brings up the Gospel of John, claiming that that the phrase, "In the beginning was the Word," indicates the priority of a designing intellect and of information over matter. The key issue for the question we are considering in this anthology is whether it makes sense to treat the Bible in these and other such passages as though it were offering scientific hypotheses.

***

Next we present the most famous version of what is known as the argument from design, that given by Archdeacon William Paley in his book *Natural Theology*. Although he was not the first to make the argument, his watchmaker analogy—as the functional arrangement of parts of a watch required the designing skill of a watchmaker to craft, so does the purposeful arrangement of parts found in the biological world require a designing creator—made his version of the design argument the most influential. In his book, Paley gives example after example to illustrate his point. What plainer manifestation of design is there, he says, than to see the difference in roundness between the crystalline lens of the fish and that of terrestrial animals; surely such examples of purposeful differences could not arise but by the action of a supreme intelligent being.

In the second part of the section, we move to Charles Darwin's contribution, particularly that given in his *Origin of Species*. Darwin came into an intellectual community that, although much concerned with the question of organic origins because of its religious implications, was becoming more and more aware that naive, literalistic readings of the Bible would not do. Such readings put far too much strain on the facts as then known. This is not to say that the pre-Darwinian community had felt at liberty to theorize and conclude precisely as their fancies took them. Constraints had to be met. First, it was demanded that one stay true to the principles of science. One may speak of two factions, "liberals" and "conservatives," although in a way they might more readily be characterized as "empiricists" and "rationalists," for the former were ardent to argue from experience (to beliefs about God and the like) whereas the latter were ardent to argue to experience (from beliefs about God and the like). But, either way, both factions agreed that the best kinds of explanations are causal, exhibiting so-called true causes, *vera causae*. Second, in dealing with origins, in line with the natural theology of which Paley was the classic expositor, one had to respect the teleological nature of organisms. One had to see organisms not as randomly thrown together, but as integrated functioning entities, marked above all else by having features that are adapted to life's needs.

For all of their differences, almost everyone in Britain agreed that the earliest attempts at evolutionary theorizing failed these two tests. This applied particularly to the most notorious speculator of them all, the French biologist Jean Baptiste de Lamarck. At the beginning of the nineteenth century, he had argued that organisms "transform" in a chain or ladder of development, from primitive blobs up to humans. This transformation was supposedly aided, in part, by a mechanism that assumed the inheritance of acquired characters. However, in the opinion of just about all, these "Lamarckian" speculations were a methodological failure from beginning to end. There was neither a *vera causa* nor an adequate understanding of design.

It was Darwin's genius to respond to both of the demands that his elders and teachers put on solutions to the problem of origins. Darwin did not so much break with the past as mold it to his own ends. Thus, on the one hand, he strove to put his theorizing on a *vera causa* basis. Darwin succeeded in his work (we believe) to such an extent that he satisfied both of the chief interpretations that others put on the *vera causa* notion.

On the other hand, through his mechanism of natural selection, Darwin hoped to speak not only to evolution but also to the adaptedness of organisms. This latter point cannot be overstressed. What really distinguishes Darwin and his true supporters is the conviction, first, that adaptation is the key aspect of organic nature and, second, that natural selection is a mechanism adequate to an understanding of this adaptation. In one fundamental way, Darwin differed from Paley. After the *Origin*, there was no need for the Master Craftsman. In another equally fundamental way, there was no difference at all. Darwin was at one with Paley in seeing the world *as if* designed.

The next part of this subsection takes up three of the contemporary responses of the *Origin*. In the decade after Darwin published, there was a huge amount written, both in favor of and against evolutionism. Many of the criticisms were well taken. Darwin himself was much troubled by his inability to produce a coherent theory of heredity—of how the new variations (the "raw stuff" of evolution) might be produced and transmitted—and the critics were quick to magnify and exploit these troubles. Other criticisms worried Darwin less. Many critics harped constantly on the implications of evolution for our own species, but Darwin was always convinced that one must go the whole hog (or, rather, the whole primate). In 1871, in *The Descent of Man*, he focused explicit attention on humankind, but essentially his thinking was an expansion rather than an innovation.

What is interesting is the extent to which one finds thoughts and worries about methodology running through the responses of the ideas of the *Origin*. Is evolution the kind of approach that could, in principle, be scientific? Is natural selection an adequate cause? We have chosen three such responses where methodological concerns take a prominent role. Two responses are critical (though in different ways), and one response is favorable, although when you read them, you will see there is slightly more to the story than simple agreement or disagreement.

Adam Sedgwick, Darwin's old friend and teacher, vehemently opposes evolution. Darwin has failed to be truly "inductive," he thinks. One might, perhaps, be forgiven for wondering what Sedgwick means by this, but whatever it is, Darwin has not got it. Note, however, that Sedgwick is at one with Darwin on the significance of adaptation. It is just that he does not think selection is adequate to the task. This, we suspect, is the real crux of Sedgwick's opposition.

Thomas Henry Huxley, Darwin's great popular spokesperson, has a much more sophisticated grasp of methodology. He picks right up on the three key aspects of Darwin's argument. Note, however, that although Huxley clearly accepts common descent, he is much more reticent about the gradual action of natural selection, supposing even that evolution might go in jumps. Huxley's agnosticism (his own word!) about selection becomes more understandable when you remember what we just said above. Huxley, unlike Darwin and Sedgwick, was never that moved by adaptation. He had never been under the spell of natural theology. Hence, he never accepted as a problem that which selection was intended to solve. Paradoxical though it may seem, we are led to conclude that although Huxley was a Darwinian emotionally and socially, intellectually there was always a chasm between the two men.

Chapter 6, the last piece in the historical subsection, is by theologian Charles Hodge, who was a professor at Princeton Theological Seminary (PTS) at the time of the publication of Darwin's *Origin*. "What Is Darwinism?" is an excerpt from Hodge's book of the same title. The answer he gives to his question is that Darwinism is atheism. Although Hodge's old-school theological views quickly faded in influence (to the degree that they were largely ignored even at PTS), he has more recently been resurrected in the creationism debate. As we will see in part III, Hodge's views about the relationship of evolution and Christianity are especially influential for intelligent design creationists. In the excerpts included here, we see not only Hodge's objections to Darwinism but also some of his thoughts on what he took to be some overlap between science and religion. There can be no harmony between science and religion, he opined, until scientists come to recognize that there are other kinds of evidence of truth than the testimony of the senses. Comparing scientists to the Pharisees who will not listen to the man whose sight was restored by Christ, he writes that "[m]en of science must not speak thus. They must not say to every objector, Thou art not scientific, and therefore hast no right to speak." We will see that attitude expressed again later by creationists, who want their views to be taken as science. Intelligent design leader William Dembski, for instance, writes:

> If we take seriously the word-flesh Christology of Chalcedon (i.e., the doctrine that Christ is fully human and fully divine) and view Christ as the telos toward which God is drawing the whole of creation, then any view of the sciences that leaves Christ out of the picture must be seen as fundamentally deficient. (W. A. Dembski, *Intelligent Design: The Bridge between Science and Theology* [Downers Grove, IL: InterVarsity Press, 1999], p. 206)

One will need to ask what such a theistic science would look like and how it could function.

***

Chapters 7 and 8 are a philosophical pair. The first is by Karl Popper, one of the most influential philosophers of science of the twentieth century. It was Popper who focused philosophical interest on the question of how to demarcate science from pseudoscience and whose own proposed falsificationist criterion continues to be echoed by many scientists today. For this reason alone, Popper's work has been a regular element of the creationism debate. But, ironically, some creationists have appealed to Popper to try to discredit evolution, arguing that Popper himself said that evolution by natural selection was unfalsifiable. If creationism is not science, then neither is evolution, they say. Here we reproduce the Popper article they had in mind, in which he criticizes some aspects of the orthodox modern theory of evolution (generally called "neo-Darwinism" or the "synthetic theory"). Among other charges he levels one of the most familiar that one hears from creationists, namely, that the theory is not genuine science because selection is not a genuine mechanism—Darwin's argument purportedly reduces to the shallow tautology that those that survive are those that survive.

Ruse's critique of Popper does not challenge his "criterion of demarcation" between science and nonscience; he does not deny that a mark of genuine science is that it be, in principle, falsifiable—that is to say, open to test and refutation. However, he does take issue with Popper's understanding of evolution. The reader can decide whether Ruse answers Popper adequately.

It is only fair to point out that Popper later changed his view, admitting that he had been mistaken about natural selection, writing that

> [t]he fact that the theory of natural selection is difficult to test has led some people... to claim that it is tautology.... I mention this problem because I too belong among the culprits.... The theory of natural selection may be so formulated that it is far from tautological. In this case it is not only testable, but it turns out to be not strictly universally true. There seem to be exceptions... and considering the random character of the variations on which natural selection operates, the occurrence of exceptions is not surprising. (Popper, "Natural Selection and the Emergence of Mind," *Dialectica* 32 [1978]: 344, 346)

***

The last two articles in part I are by two distinguished contemporary evolutionary biologists, Francisco Ayala and Richard Dawkins. Francisco Ayala is a professor of both biology and philosophy at University of California, Irvine. He is a member of the National Academy of Sciences and has received numerous other scientific distinctions. Like Ruse, he was an expert witness in the *McLean* trial. Richard Dawkins is Charles Simonyi Chair of Public Understanding of Science and a member of the zoology department at Oxford University. The gene-centered analysis of evolution he articulated in his book *The Selfish Gene* has been one of the most provocative and stimulating recent ideas in the field, and he is known internationally for his work explaining evolution to the general public. Both Ayala and Dawkins have written books that criticize creationism—*Darwin and Intelligent Design* and *The Blind Watchmaker*, respectively. However, they have very different views about religion and its relationship to evolution. Dawkins is an atheist who, rather like the creationists, sees Darwinian evolution as fundamentally at odds with the idea of a divine creator, while Ayala is a theist who, like most mainstream Christians, sees no incompatibility between evolution and his religious belief.

Ayala's article takes on one of the most worrisome aspects of evolution for creationists, namely, human evolution. Human evolution has always been the major sticking point for creationists, who take human beings to be distinct from animals and unique in being created in the image of God. But the scientific evidence of human continuity with other animals is overwhelming. Ayala reviews some of this evidence from both paleontology and molecular biology in his article. He goes on to examine human nature, giving an evolutionary account of many of our distinctive features, including our moral nature. Human nature is biological nature, he concludes, but also much more in that it has given rise to cultural evolution, which can transcend it and make sense of human uniqueness. Such a picture, he argues, takes science seriously and abides by St. Augustine's injunction against "maintaining foolish opinions about the Scriptures" of the sort that creationists advance.

The selection from Richard Dawkins is a thought-provoking look at evolution in a way that helps reveal its power and generality. The Darwinian law, he concludes, may be as universal as the great laws of physics. On the way to this conclusion, Dawkins reviews six distinct theories of evolution, showing how only Darwinian evolution has the explanatory power to explain adaptive complexity. For the purposes of the question of this book, another point to note is how Dawkins dismisses the first of the proposed theories—namely, that evolution works by means of a built-in capacity for, or drive toward, increasing perfection—for being essentially mystical and not a real scientific theory at all. Thus, although the article does not address our question directly, it implicitly assumes a conception of science of what is and what is not science. Theories that do not qualify as scientific are simply not taken seriously. Dawkins's article was somewhat controversial when it was published, having been dismissed by one eminent evolutionist of my (Ruse's) acquaintance as "banal." Perhaps, taken as science, it did not say much new; but, this is to read it the wrong way. Dawkins is doing philosophy, trying to set up the very conditions for an adequate evolutionary theory. This study is far from banal. Dawkins's arguments also suggest what I (Ruse) hint at in my critique of Popper, namely, that selection is deeply embedded in and connected to beliefs about the general lawlike nature of the world. Denial of selection may not be contradictory, but given all of selection's implications about its systematic effect, such denial may push you close to a disregard for the general uniformity of biological nature.

And that issue, as we will see in the second and third parts, is closely connected with what it means for something to count as science.

CHAPTER 1
THE BIBLE

GENESIS: CHAPTER 1

[1] In the beginning God created the heaven and the earth.

[2] And the earth was without form, and void; and darkness was upon the face of the deep. And the Spirit of God moved upon the face of the waters.

[3] And God said, Let there be light: and there was light.

[4] And God saw the light, that it was good: and God divided the light from the darkness.

[5] And God called the light Day, and the darkness he called Night. And the evening and the morning were the first day.

[6] And God said, Let there be a firmament in the midst of the waters, and let it divide the waters from the waters.

[7] And God made the firmament, and divided the waters which were under the firmament from the waters which were above the firmament: and it was so.

[8] And God called the firmament Heaven. And the evening and the morning were the second day.

From King James Version (Genesis 1 and 2; John 1:1–18).

[9] And God said, Let the waters under the heaven be gathered together unto one place, and let the dry land appear: and it was so.

[10] And God called the dry land Earth; and the gathering together of the waters called he Seas: and God saw that it was good.

[11] And God said, Let the earth bring forth grass, the herb yielding seed, and the fruit tree yielding fruit after his kind, whose seed is in itself, upon the earth: and it was so.

[12] And the earth brought forth grass, and herb yielding seed after his kind, and the tree yielding fruit, whose seed was in itself, after his kind: and God saw that it was good.

[13] And the evening and the morning were the third day.

[14] And God said, Let there be lights in the firmament of the heaven to divide the day from the night; and let them be for signs, and for seasons, and for days, and years:

[15] And let them be for lights in the firmament of the heaven to give light upon the earth: and it was so.

[16] And God made two great lights; the greater light to rule the day, and the lesser light to rule the night: he made the stars also.

[17] And God set them in the firmament of the heaven to give light upon the earth,

[18] And to rule over the day and over the night, and to divide the light from the darkness: and God saw that it was good.

[19] And the evening and the morning were the fourth day.

[20] And God said, Let the waters bring forth abundantly the moving creature that hath life, and fowl that may fly above the earth in the open firmament of heaven.

[21] And God created great whales, and every living creature that moveth, which the waters brought forth abundantly, after their kind, and every winged fowl after his kind: and God saw that it was good.

[22] And God blessed them, saying, Be fruitful, and multiply, and fill the waters in the seas, and let fowl multiply in the earth.

[23] And the evening and the morning were the fifth day.

[24] And God said, Let the earth bring forth the living creature after his kind, cattle, and creeping thing, and beast of the earth after his kind: and it was so.

[25] And God made the beast of the earth after his kind, and cattle after their kind, and every thing that creepeth upon the earth after his kind: and God saw that it was good.

[26] And God said, Let us make man in our image, after our likeness: and let them have dominion over the fish of the sea, and over the fowl of the air, and over the cattle, and over all the earth, and over every creeping thing that creepeth upon the earth.

[27] So God created man in his own image, in the image of God created he him; male and female created he them.

[28] And God blessed them, and God said unto them, Be fruitful, and multiply, and replenish the earth, and subdue it: and have dominion over the fish of the sea, and over the fowl of the air, and over every living thing that moveth upon the earth.

[29] And God said, Behold, I have given you every herb bearing seed, which is upon the face of all the earth, and every tree, in the which is the fruit of a tree yielding seed; to you it shall be for meat.

[30] And to every beast of the earth, and to every fowl of the air, and to every thing that creepeth upon the earth, wherein there is life, I have given every green herb for meat: and it was so.

[31] And God saw every thing that he had made, and, behold, it was very good. And the evening and the morning were the sixth day.

GENESIS: CHAPTER 2

[1] Thus the heavens and the earth were finished, and all the host of them.

[2] And on the seventh day God ended his work which he had made; and he rested on the seventh day from all his work which he had made.

[3] And God blessed the seventh day, and sanctified it: because that in it he had rested from all his work which God created and made.

[4] These are the generations of the heavens and of the earth when they were created, in the day that the LORD God made the earth and the heavens,

[5] And every plant of the field before it was in the earth, and every herb of the field before it grew: for the LORD God had not caused it to rain upon the earth, and there was not a man to till the ground.

[6] But there went up a mist from the earth, and watered the whole face of the ground.

[7] And the LORD God formed man of the dust of the ground, and breathed into his nostrils the breath of life; and man became a living soul.

[8] And the LORD God planted a garden eastward in Eden; and there he put the man whom he had formed.

[9] And out of the ground made the LORD God to grow every tree that is pleasant to the sight, and good for food; the tree of life also in the midst of the garden, and the tree of knowledge of good and evil.

[10] And a river went out of Eden to water the garden; and from thence it was parted, and became into four heads.

[11] The name of the first is Pison: that is it which compasseth the whole land of Havilah, where there is gold;

[12] And the gold of that land is good: there is bdellium and the onyx stone.

[13] And the name of the second river is Gihon: the same is it that compasseth the whole land of Ethiopia.

[14] And the name of the third river is Hiddekel: that is it which goeth toward the east of Assyria. And the fourth river is Euphrates.

[15] And the LORD God took the man, and put him into the garden of Eden to dress it and to keep it.

[16] And the LORD God commanded the man, saying, Of every tree of the garden thou mayest freely eat:

[17] But of the tree of the knowledge of good and evil, thou shalt not eat of it: for in the day that thou eatest thereof thou shalt surely die.

[18] And the LORD God said, It is not good that the man should be alone; I will make him an help meet for him.

[19] And out of the ground the LORD God formed every beast of the field, and every fowl of the air; and brought them unto Adam to see what he would call them: and whatsoever Adam called every living creature, that was the name thereof.

[20] And Adam gave names to all cattle, and to the fowl of the air, and to every beast of the field; but for Adam there was not found an help meet for him.

[21] And the LORD God caused a deep sleep to fall upon Adam and he slept: and he took one of his ribs, and closed up the flesh instead thereof;

[22] And the rib, which the LORD God had taken from man, made he a woman, and brought her unto the man.

[23] And Adam said, This is now bone of my bones, and flesh of my flesh: she shall be called Woman, because she was taken out of Man.

[24] Therefore shall a man leave his father and his mother, and shall cleave unto his wife: and they shall be one flesh.

[25] And they were both naked, the man and his wife, and were not ashamed.

GOSPEL OF JOHN: CHAPTER 1:1–18

[1] In the beginning was the Word, and the Word was with God, and the Word was God.

[2] The same was in the beginning with God.

[3] All things were made by him; and without him was not any thing made that was made.

[4] In him was life; and the life was the light of men.

[5] And the light shineth in darkness; and the darkness comprehended it not.

[6] There was a man sent from God, whose name was John.

[7] The same came for a witness, to bear witness of the Light, that all men through him might believe.

[8] He was not that Light, but was sent to bear witness of that Light.

[9] That was the true Light, which lighteth every man that cometh into the world.

[10] He was in the world, and the world was made by him, and the world knew him not.

[11] He came unto his own, and his own received him not.

[12] But as many as received him, to them gave he power to become the sons of God, even to them that believe on his name:

[13] Which were born, not of blood, nor of the will of the flesh, nor of the will of man, but of God.

[14] And the Word was made flesh, and dwelt among us, (and we beheld his glory, the glory as of the only begotten of the Father,) full of grace and truth.

[15] John bare witness of him, and cried, saying, This was he of whom I spake, He that cometh after me is preferred before me: for he was before me.

[16] And of his fulness have all we received, and grace for grace.

[17] For the law was given by Moses, but grace and truth came by Jesus Christ.

[18] No man hath seen God at any time, the only begotten Son, which is in the bosom of the Father, he hath declared him.

CHAPTER 2
NATURAL THEOLOGY
WILLIAM PALEY

I. STATE OF THE ARGUMENT

In crossing a heath, suppose I pitched my foot against a *stone*, and were asked how the stone came to be there; I might possibly answer, that, for any thing I knew to the contrary, it had lain there for ever: nor would it perhaps be very easy to show the absurdity of this answer. But suppose I had found a *watch* upon the ground, and it should be inquired how the watch happened to be in that place; I should hardly think of the answer which I had before given, that, for any thing I knew, the watch might have always been there. Yet why should not this answer serve for the watch as well as for the stone? Why is it not as admissible in the second case, as in the first? For this reason, and for no other, viz. that, when we come to inspect the watch, we perceive (what we could not discover in the stone) that its several parts are framed and put together for a purpose, e.g. that they are so formed and adjusted as to produce motion, and that motion so regulated as to point out the hour of the day; that, if the different parts had been differently shaped from what they are, of a different size from what they are, or placed after any other manner, or in

---

From William Paley, *Natural Theology; or Evidences of the Existence and Attributes of the Deity collected from the Appearance of Nature*, 12th edition (1809).

any other order, than that in which they are placed, either no motion at all would have been carried on in the machine, or none which would have answered the use that is now served by it. To reckon up a few of the plainest of these parts, and of their offices, all tending to one result: We see a cylindrical box containing a coiled elastic spring, which, by its endeavor to relax itself, turns round the box. We next observe a flexible chain (artificially wrought for the sake of flexure), communicating the action of the spring from the box to the fusee. We then find a series of wheels, the teeth of which catch in, and apply to, each other, conducting the motion from the fusee to the balance, and from the balance to the pointer; and at the same time, by the size and shape of those wheels, so regulating that motion, as to terminate in causing an index, by an equable and measured progression, to pass over a given space in a given time. We take notice that the wheels are made of brass in order to keep them from rust; the springs of steel, no other metal being so elastic; that over the face of the watch there is placed a glass, a material employed in no other part of the work, but in the room of which, if there had been any other than a transparent substance, the hour could not be seen without opening the case. This mechanism being observed (it requires indeed an examination of the instrument, and perhaps some previous knowledge of the subject, to perceive and understand it; but being once, as we have said, observed and understood), the inference, we think, is inevitable, that the watch must have had a maker: that there must have existed, at some time, and at some place or other, an artificer or artificers who formed it for the purpose which we find it actually to answer; who comprehended its construction, and designed its use....

II. STATE OF THE ARGUMENT CONTINUED

Suppose, in the next place, that the person who found the watch, should, after some time, discover that, in addition to all the properties which he had hitherto observed in it, it possessed the unexpected property of producing, in the course of its movement, another watch like itself (the thing is conceivable); that it contained within it a mechanism, a system of parts, a mold for instance, or a complex adjustment of lathes, files, and other tools, evidently and separately calculated for this purpose; let us inquire, what effect ought such a discovery to have upon his former conclusion.

I. The first effect would be to increase his admiration of the contrivance, and his conviction of the consummate skill of the contriver. Whether he regarded the object of the contrivance, the distinct apparatus, the intricate, yet in many parts intelligible mechanism, by which it was carried on, he would perceive, in this new observation, nothing but an additional reason for doing what he had already done—for referring the construction of the watch to design, and to supreme art. If that construction *without* this property, or which is the same thing, before this property had been noticed, proved intention and art to have been employed about it; still more strong would the proof appear, when he came to the knowledge of this further property, the crown and perfection of all the rest.

II. He would reflect, that though the watch before him were, in some sense, the maker of the watch, which was fabricated in the course of its movements, yet it was in a very different sense from that, in which a carpenter, for instance, is the maker of a chair; the author of its contrivance, the cause of the relation of its parts to their use. With respect to these, the first watch was no cause at all to the second: in no such sense as this was it the author of the constitution and order, either of the parts which the new watch contained, or of the parts by the aid and instrumentality of which it was produced. We might possibly say, but with great latitude of expression, that a stream of water ground corn: but no latitude of expression would allow us to say, no stretch of conjecture could lead us to think, that the stream of water built the mill, though it were too ancient for us to know who the builder was. What the stream of water does in the affair, is neither more nor less than this; by the application of an unintelligent impulse to a mechanism previously arranged, arranged independently of it, and arranged by intelligence, an effect is produced, viz. the corn is ground. But the effect results from the arrangement. The force of the stream cannot be said to be the cause or author of the effect, still less of the arrangement. Understanding and plan in the formation of the mill were not the less necessary, for any share which the water has in grinding the corn: yet is this share the same, as that which the watch would have contributed to the production of the new watch, upon the supposition assumed in the last section. Therefore,

III. Though it be now no longer probable, that the individual watch, which our observer had found, was made immediately by the hand of an artificer, yet doth not this alteration in anywise affect the inference, that an artificer had been originally employed and concerned in the production. The argument from design remains as it was. Marks of design and contrivance are no more accounted for now, than they were before. In the same thing, we may ask for the cause of different properties. We may ask for the cause of the color of a body, of its hardness, of its head; and these causes may be all different. We are now asking for the cause of that subserviency to a use, that relation to an end, which we have remarked in the watch before us. No answer is given to this question, by telling us that a preceding watch produced it. There cannot be design without a designer; contrivance without a contriver; order without choice; arrangement, without any thing capable of arranging; subserviency and relation to a purpose, without that which could intend a purpose; means suitable to an end, and executing their office, in accomplishing that end, without the end ever having been contemplated, or the means accommodated to it. Arrangement, disposition of parts, subserviency of means to an end, relation of instruments to a use, imply the presence of intelligence and mind. No one, therefore, can rationally believe, that the insensible, inanimate watch, from which the watch before us issued, was the proper cause of the mechanism we so much admire in it—could be truly said to have constructed the instrument, disposed its parts, assigned their office, determined their order, action, and mutual dependency, combined their several motions into one result, and that also a result connected with the utilities of other beings. All these properties, therefore, are as much unaccounted for, as they were before....

The conclusion of which the first examination of the watch, of its works, construction, and movement, suggested, was, that it must have had, for the cause and author of that construction, an artificer, who understood its mechanism, and designed its use. This conclusion is invincible. A second examination presents us with a new discovery. The watch is found, in the course of its movement, to produce another watch, similar to itself; and not only so, but we perceive in it a system or organization, separately calculated for that purpose. What effect would this discovery have, or ought it to have, upon our former inference? What, as hath already been said, but to increase, beyond measure, our admiration of the skill, which had been employed in the formation of such a machine? Or shall it, instead of this, all at once turn us round to an opposite conclusion, viz. that no art or skill whatever has been concerned in the business, although all other evidences of art and skill remain as they were, and this last and supreme piece of art be now added to the rest? Can this be maintained without absurdity? Yet this is atheism.

III. APPLICATION OF THE ARGUMENT

This is atheism: for every indication of contrivance, every manifestation of design, which existed in the watch, exists in the works of nature; with the difference, on the side of nature, of being greater and more, and that in a degree which exceeds all computation. I mean that the contrivances of nature surpass the contrivances of art, in the complexity, subtility, and curiosity of the mechanism; and still more, if possible, do they go beyond them in number and variety; yet, in a multitude of cases, are not less evidently mechanical, not less evidently contrivances, not less evidently accommodated to their end, or suited to their office, than are the most perfect productions of human ingenuity.

I know no better method of introducing so large a subject, than that of comparing a single thing with a single thing; an eye, for example, with a telescope. As far as the examination of the instrument goes, there is precisely the same proof that the eye was made for vision, as there is that the telescope was made for assisting it. They are made upon the same principles; both being adjusted to the laws by which the transmission and refraction of rays of light are regulated. I speak not of the origin of the laws themselves; but such laws being fixed, the construction, in both cases, is adapted to them. For instance; these laws require, in order to produce the same effect, that the rays of light, in passing from water into the eye, should be refracted by a more convex surface, than when it passes out of air into the eye. Accordingly we find that the eye of a fish, in that part of it called the crystalline lens, is much rounder than the eye of terrestrial animals. What plainer manifestation of design can there be than this difference? What could a mathematical-instrument-maker have done more, to show his knowledge of his principle, his application of that knowledge, his suiting of his means to his end; I will not say to display the compass or excellence of his skill and art, for in these all comparison is indecorous, but to testify counsel, choice, consideration, purpose?

CHAPTER 3
ON THE ORIGIN OF SPECIES
CHARLES DARWIN

CHAPTER XV: RECAPITULATION AND CONCLUSION

As this whole volume is one long argument, it may be convenient to the reader to have the leading facts and inferences briefly recapitulated.

That many and serious objections may be advanced against the theory of descent with modification through variation and natural selection, I do not deny. I have endeavored to give to them their full force. Nothing at first can appear more difficult to believe than that the more complex organs and instincts have been perfected, not by means superior to, though analogous with, human reason, but by the accumulation of innumerable slight variations, each good for the individual possessor. Nevertheless, this difficulty, though appearing to our imagination insuperably great, cannot be considered real if we admit the following propositions, namely, that all parts of the organization and instincts offer, at least individual differences—that there is a struggle for existence leading to the preservation of profitable deviations of structure or instinct—and, lastly, that gradations in the state of perfection of each organ

---

From Charles Darwin, *On the Origin of Species*, 6th edition (London: John Murray, 1872).

may have existed, each good of its kind. The truth of these propositions cannot, I think, be disputed.

It is, no doubt, extremely difficult even to conjecture by what gradations many structures have been perfected, more especially among broken and failing groups of organic beings, which have suffered much extinction; but we see so many strange gradations in nature, that we ought to be extremely cautious in saying that any organ or instinct, or any whole structure, could not have arrived at its present state by many graduated steps. There are, it must be admitted, cases of special difficulty opposed to the theory of natural selection; and one of the most curious of these is the existence in the same community of two or three defined castes of workers or sterile female ants; but I have attempted to show how these difficulties can be mastered.

With respect to the almost universal sterility of species when first crossed, which forms so remarkable a contrast with the almost universal fertility of varieties when crossed, I must refer the reader to the recapitulation of the facts given at the end of the ninth chapter, which seem to me conclusively to show that this sterility is no more a special endowment than is the incapacity of two distinct kinds of trees to be grafted together; but that it is incidental on differences confined to the reproductive systems of the intercrossed species. We see the truth of this conclusion in the vast difference in the results of crossing the same two species reciprocally—that is, when one species is first used as the father and then as the mother. Analogy from the consideration of dimorphic and trimorphic plants clearly leads to the same conclusion, for when the forms are illegitimately united, they yield few or no seed, and their offspring are more or less sterile; and these forms belong to the same undoubted species, and differ from each other in no respect except in their reproductive organs and functions.

Although the fertility of varieties when intercrossed, and of their mongrel offspring, has been asserted by so many authors to be universal, this cannot be considered as quite correct after the facts given on the high authority of Gartner and Kolreuter. Most of the varieties which have been experimented on have been produced under domestication; and as domestication (I do not mean mere confinement) almost certainly tends to eliminate that sterility which, judging from analogy, would have affected the parent-species if intercrossed, we ought not to expect that domestication would likewise induce sterility in their modified descendants when crossed. This elimination of sterility apparently follows from the same cause which allows our domestic animals to breed freely under diversified circumstances; and this again apparently follows from their having been gradually accustomed to frequent changes in their conditions of life.

A double and parallel series of facts seems to throw much light on the sterility of species, when first crossed, and of their hybrid offspring. On the one side, there is good reason to believe that slight changes in the conditions of life give vigor and fertility to all organic beings. We know also that a cross between the distinct individuals of the same variety, and between distinct varieties, increases the number of their offspring, and certainly gives to them increased size and vigor. This is chiefly owing to the forms which are crossed having been exposed to somewhat different conditions of life; for I have ascertained by a laborious series of experiments that if all the individuals of the same variety be subjected during several generations to the same conditions, the good derived from crossing is often much diminished or wholly disappears. This is one side of the case. On the other side, we know that species which have long been exposed to nearly uniform conditions, when they are subjected under confinement to new and greatly changed conditions, either perish, or if they survive, are rendered sterile, though retaining perfect health. This does not occur, or only in a very slight degree, with our domesticated productions, which have long been exposed to fluctuating conditions. Hence when we find that hybrids produced by a cross between two distinct species are few in number, owing to their perishing soon after conception or at a very early age, or if surviving that they are rendered more or less sterile, it seems highly probable that this result is due to their having been in fact subjected to a great change in their conditions of life, from being compounded of two distinct organizations. He who will explain in a definite manner why, for instance, an elephant or a fox will not breed under confinement in its native country, whilst the domestic pig or dog will breed freely under the most diversified conditions, will at the same time be able to give a definite answer to the question why two distinct species, when crossed, as well as their hybrid offspring, are generally rendered more or less sterile, while two domesticated varieties when crossed and their mongrel offspring are perfectly fertile.

Turning to geographical distribution, the difficulties encountered on the theory of descent with modification are serious enough. All the individuals of the same species, and all the species of the same genus, or even higher group, are descended from common parents; and therefore, in however distant and isolated parts of the world they may now be found, they must in the course of successive generations have traveled from some one point to all the others. We are often wholly unable even to conjecture how this could have been effected. Yet, as we have reason to believe that some species have retained the same specific form for very long periods of time, immensely long as measured by years, too much stress ought not to be laid on the occasional wide diffusion of the same species; for during very long periods there will always have been a good chance for wide migration by many means. A broken or interrupted range may often be accounted for by the extinction of the species in the intermediate regions. It cannot be denied that we are as yet very ignorant as to the full extent of the various climatical and geographical changes which have affected the earth during modern periods; and such changes will often have facilitated migration. As an example, I have attempted to show how potent has been the influence of the Glacial period on the distribution of the same and of allied species throughout the world. We are as yet profoundly ignorant of the many occasional means of transport. With respect to distinct species of the same genus, inhabiting distant and isolated regions, as the process of modification has necessarily been slow, all the means of migration will have been possible during a very long period; and consequently the difficulty of the wide diffusion of the species of the same genus is in some degree lessened.

As according to the theory of natural selection an interminable number of intermediate forms must have existed, linking together all the species in each group by gradations as fine as our existing varieties, it may be asked, Why do we not see these linking forms all around us? Why are not all organic beings blended together in an inextricable chaos? With respect to existing forms, we should remember that we have no right to expect (excepting in rare cases) to discover DIRECTLY connecting links between them, but only between each and some extinct and supplanted form. Even on a wide area, which has during a long period remained continuous, and of which the climatic and other conditions of life change insensibly in proceeding from a district occupied by one species into another district occupied by a closely allied species, we have no just right to expect often to find intermediate varieties in the intermediate zones. For we have reason to believe that only a few species of a genus ever undergo change; the other species becoming utterly extinct and leaving no modified progeny. Of the species which do change, only a few within the same country change at the same time; and all modifications are slowly effected. I have also shown that the intermediate varieties which probably at first existed in the intermediate zones, would be liable to be supplanted by the allied forms on either hand; for the latter, from existing in greater numbers, would generally be modified and improved at a quicker rate than the intermediate varieties, which existed in lesser numbers; so that the intermediate varieties would, in the long run, be supplanted and exterminated.

On this doctrine of the extermination of an infinitude of connecting links, between the living and extinct inhabitants of the world, and at each successive period between the extinct and still older species, why is not every geological formation charged with such links? Why does not every collection of fossil remains afford plain evidence of the gradation and mutation of the forms of life? Although geological research has undoubtedly revealed the former existence of many links, bringing numerous forms of life much closer together, it does not yield the infinitely many fine gradations between past and present species required on the theory, and this is the most obvious of the many objections which may be urged against it. Why, again, do whole groups of allied species appear, though this appearance is often false, to have come in suddenly on the successive geological stages? Although we now know that organic beings appeared on this globe, at a period incalculably remote, long before the lowest bed of the Cambrian system was deposited, why do we not find beneath this system great piles of strata stored with the remains of the progenitors of the Cambrian fossils? For on the theory, such strata must somewhere have been deposited at these ancient and utterly unknown epochs of the world's history.

I can answer these questions and objections only on the supposition that the geological record is far more imperfect than most geologists believe. The number of specimens in all our museums is absolutely as nothing compared with the countless generations of countless species which have certainly existed. The parent form of any two or more species would not be in all its characters directly intermediate between its modified offspring, any more than the rock-pigeon is directly intermediate in crop and tail between its descendants, the pouter and fantail pigeons. We should not be able to recognize a species as the parent of another and modified species, if we were to examine the two ever so closely, unless we possessed most of the intermediate links; and owing to the imperfection of the geological record, we have no just right to expect to find so many links. If two or three, or even more linking forms were discovered, they would simply be ranked by many naturalists as so many new species, more especially if found in different geological substages, let their differences be ever so slight. Numerous existing doubtful forms could be named which are probably varieties; but who will pretend that in future ages so many fossil links will be discovered, that naturalists will be able to decide whether or not these doubtful forms ought to be called varieties? Only a small portion of the world has been geologically explored. Only organic beings of certain classes can be preserved in a fossil condition, at least in any great number. Many species when once formed never undergo any further change but become extinct without leaving modified descendants; and the periods during which species have undergone modification, though long as measured by years, have probably been short in comparison with the periods during which they retained the same form. It is the dominant and widely ranging species which vary most frequently and vary most, and varieties are often at first local—both causes rendering the discovery of intermediate links in any one formation less likely. Local varieties will not spread into other and distant regions until they are considerably modified and improved; and when they have spread, and are discovered in a geological formation, they appear as if suddenly created there, and will be simply classed as new species. Most formations have been intermittent in their accumulation; and their duration has probably been shorter than the average duration of specific forms. Successive formations are in most cases separated from each other by blank intervals of time of great length, for fossiliferous formations thick enough to resist future degradation can, as a general rule, be accumulated only where much sediment is deposited on the subsiding bed of the sea. During the alternate periods of elevation and of stationary level the record will generally be blank. During these latter periods there will probably be more variability in the forms of life; during periods of subsidence, more extinction.

With respect to the absence of strata rich in fossils beneath the Cambrian formation, I can recur only to the hypothesis given in the tenth chapter; namely, that though our continents and oceans have endured for an enormous period in nearly their present relative positions, we have no reason to assume that this has always been the case; consequently formations much older than any now known may lie buried beneath the great oceans. With respect to the lapse of time not having been sufficient since our planet was consolidated for the assumed amount of organic change, and this objection, as urged by Sir William Thompson, is probably one of the gravest as yet advanced, I can only say, firstly, that we do not know at what rate species change, as measured by years, and secondly, that many philosophers are not as yet willing to admit that we know enough of the constitution of the universe and of the interior of our globe to speculate with safety on its past duration.

That the geological record is imperfect all will admit; but that it is imperfect to the degree required by our theory, few will be inclined to admit. If we look to long enough intervals of time, geology plainly declares that species have all changed; and they have changed in the manner required by the theory, for they have changed slowly and in a graduated manner. We clearly see this in the fossil remains from consecutive formations invariably being much more closely related to each other than are the fossils from widely separated formations.

Such is the sum of the several chief objections and difficulties which may justly be urged against the theory; and I have now briefly recapitulated the answers and explanations which, as far as I can see, may be given. I have felt these difficulties far too heavily during many years to doubt their weight. But it deserves especial notice that the more important objections relate to questions on which we are confessedly ignorant; nor do we know how ignorant we are. We do not know all the possible transitional gradations between the simplest and the most perfect organs; it cannot be pretended that we know all the varied means of Distribution during the long lapse of years, or that we know how imperfect is the Geological Record. Serious as these several objections are, in my judgment they are by no means sufficient to overthrow the theory of descent with subsequent modification.

Now let us turn to the other side of the argument. Under domestication we see much variability, caused, or at least excited, by changed conditions of life; but often in so obscure a manner, that we are tempted to consider the variations as spontaneous. Variability is governed by many complex laws, by correlated growth, compensation, the increased use and disuse of parts, and the definite action of the surrounding conditions. There is much difficulty in ascertaining how largely our domestic productions have been modified; but we may safely infer that the amount has been large, and that modifications can be inherited for long periods. As long as the conditions of life remain the same, we have reason to believe that a modification, which has already been inherited for many generations, may continue to be inherited for an almost infinite number of generations. On the other hand we have evidence that variability, when it has once come into play, does not cease under domestication for a very long period; nor do we know that it ever ceases, for new varieties are still occasionally produced by our oldest domesticated productions.

Variability is not actually caused by man; he only unintentionally exposes organic beings to new conditions of life and then nature acts on the organization and causes it to vary. But man can and does select the variations given to him by nature, and thus accumulates them in any desired manner. He thus adapts animals and plants for his own benefit or pleasure. He may do this methodically, or he may do it unconsciously by preserving the individuals most useful or pleasing to him without any intention of altering the breed. It is certain that he can largely influence the character of a breed by selecting, in each successive generation, individual differences so slight as to be inappreciable except by an educated eye. This unconscious process of selection has been the great agency in the formation of the most distinct and useful domestic breeds. That many breeds produced by man have to a large extent the character of natural species, is shown by the inextricable doubts whether many of them are varieties or aboriginally distinct species.

There is no reason why the principles which have acted so efficiently under domestication should not have acted under nature. In the survival of favored individuals and races, during the constantly recurrent Struggle for Existence, we see a powerful and ever-acting form of Selection. The struggle for existence inevitably follows from the high geometrical ratio of increase which is common to all organic beings. This high rate of increase is proved by calculation—by the rapid increase of many animals and plants during a succession of peculiar seasons, and when naturalized in new countries. More individuals are born than can possibly survive. A grain in the balance may determine which individuals shall live and which shall die—which variety or species shall increase in number, and which shall decrease, or finally become extinct. As the individuals of the same species come in all respects into the closest competition with each other, the struggle will generally be most severe between them; it will be almost equally severe between the varieties of the same species, and next in severity between the species of the same genus. On the other hand the struggle will often be severe between beings remote in the scale of nature. The slightest advantage in certain individuals, at any age or during any season, over those with which they come into competition, or better adaptation in however slight a degree to the surrounding physical conditions, will, in the long run, turn the balance.

With animals having separated sexes, there will be in most cases a struggle between the males for the possession of the females. The most vigorous males, or those which have most successfully struggled with their conditions of life, will generally leave most progeny. But success will often depend on the males having special weapons or means of defense or charms; and a slight advantage will lead to victory.

As geology plainly proclaims that each land has undergone great physical changes, we might have expected to find that organic beings have varied under nature, in the same way as they have varied under domestication. And if there has been any variability under nature, it would be an unaccountable fact if natural selection had not come into play. It has often been asserted, but the assertion is incapable of proof, that the amount of variation under nature is a strictly limited quantity. Man, though acting on external characters alone and often capriciously, can produce within a short period a great result by adding up mere individual differences in his domestic productions; and every one admits that species present individual differences. But, besides such differences, all naturalists admit that natural varieties exist, which are considered sufficiently distinct to be worthy of record in systematic works. No one has drawn any clear distinction between individual differences and slight varieties; or between more plainly marked varieties and subspecies and species. On separate continents, and on different parts of the same continent, when divided by barriers of any kind, and on outlying islands, what a multitude of forms exist, which some experienced naturalists rank as varieties, others as geographical races or sub species, and others as distinct, though closely allied species!

If, then, animals and plants do vary, let it be ever so slightly or slowly, why should not variations or individual differences, which are in any way beneficial, be preserved and accumulated through natural selection, or the survival of the fittest? If man can by patience select variations useful to him, why, under changing and complex conditions of life, should not variations useful to nature's living products often arise, and be preserved or selected? What limit can be put to this power, acting during long ages and rigidly scrutinizing the whole constitution, structure, and habits of each creature, favoring the good and rejecting the bad? I can see no limit to this power, in slowly and beautifully adapting each form to the most complex relations of life. The theory of natural selection, even if we look no further than this, seems to be in the highest degree probable. I have already recapitulated, as fairly as I could, the opposed difficulties and objections: now let us turn to the special facts and arguments in favor of the theory.

On the view that species are only strongly marked and permanent varieties, and that each species first existed as a variety, we can see why it is that no line of demarcation can be drawn between species, commonly supposed to have been produced by special acts of creation, and varieties which are acknowledged to have been produced by secondary laws. On this same view we can understand how it is that in a region where many species of a genus have been produced, and where they now flourish, these same species should present many varieties; for where the manufactory of species has been active, we might expect, as a general rule, to find it still in action; and this is the case if varieties be incipient species. Moreover, the species of the larger genera, which afford the greater number of varieties or incipient species, retain to a certain degree the character of varieties; for they differ from each other by a less amount of difference than do the species of smaller genera. The closely allied species also of a larger genera apparently have restricted ranges, and in their affinities they are clustered in little groups round other species—in both respects resembling varieties. These are strange relations on the view that each species was independently created, but are intelligible if each existed first as a variety.

As each species tends by its geometrical rate of reproduction to increase inordinately in number; and as the modified descendants of each species will be enabled to increase by as much as they become more diversified in habits and structure, so as to be able to seize on many and widely different places in the economy of nature, there will be a constant tendency in natural selection to preserve the most divergent offspring of any one species. Hence during a long-continued course of modification, the slight differences characteristic of varieties of the same species, tend to be augmented into the greater differences characteristic of the species of the same genus. New and improved varieties will inevitably supplant and exterminate the older, less improved and intermediate varieties; and thus species are rendered to a large extent defined and distinct objects. Dominant species belonging to the larger groups within each class tend to give birth to new and dominant forms; so that each large group tends to become still larger, and at the same time more divergent in character. But as all groups cannot thus go on increasing in size, for the world would not hold them, the more dominant groups beat the less dominant. This tendency in the large groups to go on increasing in size and diverging in character, together with the inevitable contingency of much extinction, explains the arrangement of all the forms of life in groups subordinate to groups, all within a few great classes, which has prevailed throughout all time. This grand fact of the grouping of all organic beings under what is called the Natural System, is utterly inexplicable on the theory of creation.

As natural selection acts solely by accumulating slight, successive, favorable variations, it can produce no great or sudden modifications; it can act only by short and slow steps. Hence, the canon of "Natura non facit saltum," which every fresh addition to our knowledge tends to confirm, is on this theory intelligible. We can see why throughout nature the same general end is gained by an almost infinite diversity of means, for every peculiarity when once acquired is long inherited, and structures already modified in many different ways have to be adapted for the same general purpose. We can, in short, see why nature is prodigal in variety, though niggard in innovation. But why this should be a law of nature if each species has been independently created no man can explain.

Many other facts are, as it seems to me, explicable on this theory. How strange it is that a bird, under the form of a woodpecker, should prey on insects on the ground; that upland geese, which rarely or never swim, would possess webbed feet; that a thrush-like bird should dive and feed on sub-aquatic insects; and that a petrel should have the habits and structure fitting it for the life of an auk and so in endless other cases. But on the view of each species constantly trying to increase in number, with natural selection always ready to adapt the slowly varying descendants of each to any unoccupied or ill-occupied place in nature, these facts cease to be strange, or might even have been anticipated.

We can to a certain extent understand how it is that there is so much beauty throughout nature; for this may be largely attributed to the agency of selection. That beauty, according to our sense of it, is not universal, must be admitted by every one who will look at some venomous snakes, at some fishes, and at certain hideous bats with a distorted resemblance to the human face. Sexual selection has given the most brilliant colors, elegant patterns, and other ornaments to the males, and sometimes to both sexes of many birds, butterflies and other animals. With birds it has often rendered the voice of the male musical to the female, as well as to our ears. Flowers and fruit have been rendered conspicuous by brilliant colors in contrast with the green foliage, in order that the flowers may be easily seen, visited and fertilized by insects, and the seeds disseminated by birds. How it comes that certain colors, sounds and forms should give pleasure to man and the lower animals, that is, how the sense of beauty in its simplest form was first acquired, we do not know any more than how certain odors and flavors were first rendered agreeable.

As natural selection acts by competition, it adapts and improves the inhabitants of each country only in relation to their co-inhabitants; so that we need feel no surprise at the species of any one country, although on the ordinary view supposed to have been created and specially adapted for that country, being beaten and supplanted by the naturalized productions from another land. Nor ought we to marvel if all the contrivances in nature be not, as far as we can judge, absolutely perfect; as in the case even of the human eye; or if some of them be abhorrent to our ideas of fitness. We need not marvel at the sting of the bee, when used against the enemy, causing the bee's own death; at drones being produced in such great numbers for one single act, and being then slaughtered by their sterile sisters; at the astonishing waste of pollen by our fir-trees; at the instinctive hatred of the queen-bee for her own fertile daughters; at ichneumonidae feeding within the living bodies of caterpillars; and at other such cases. The wonder, indeed, is, on the theory of natural selection, that more cases of the want of absolute perfection have not been detected.

The complex and little known laws governing the production of varieties are the same, as far as we can judge, with the laws which have governed the production of distinct species. In both cases physical conditions seem to have produced some direct and definite effect, but how much we cannot say. Thus, when varieties enter any new station, they occasionally assume some of the characters proper to the species of that station. With both varieties and species, use and disuse seem to have produced a considerable effect; for it is impossible to resist this conclusion when we look, for instance, at the logger-headed duck, which has wings incapable of flight, in nearly the same condition as in the domestic duck; or when we look at the burrowing tucu-tucu, which is occasionally blind, and then at certain moles, which are habitually blind and have their eyes covered with skin; or when we look at the blind animals inhabiting the dark caves of America and Europe. With varieties and species, correlated variation seems to have played an important part, so that when one part has been modified other parts have been necessarily modified. With both varieties and species, reversions to long-lost characters occasionally occur. How inexplicable on the theory of creation is the occasional appearance of stripes on the shoulders and legs of the several species of the horse genus and of their hybrids! How simply is this fact explained if we believe that these species are all descended from a striped progenitor, in the same manner as the several domestic breeds of the pigeon are descended from the blue and barred rock-pigeon!

On the ordinary view of each species having been independently created, why should specific characters, or those by which the species of the same genus differ from each other, be more variable than the generic characters in which they all agree? Why, for instance, should the color of a flower be more likely to vary in any one species of a genus, if the other species possess differently colored flowers, than if all possessed the same colored flowers? If species are only well-marked varieties, of which the characters have become in a high degree permanent, we can understand this fact; for they have already varied since they branched off from a common progenitor in certain characters, by which they have come to be specifically distinct from each other; therefore these same characters would be more likely again to vary than the generic characters which have been inherited without change for an immense period. It is inexplicable on the theory of creation why a part developed in a very unusual manner in one species alone of a genus, and therefore, as we may naturally infer, of great importance to that species, should be eminently liable to variation; but, on our view, this part has undergone, since the several species branched off from a common progenitor, an unusual amount of variability and modification, and therefore we might expect the part generally to be still variable. But a part may be developed in the most unusual manner, like the wing of a bat, and yet not be more variable than any other structure, if the part be common to many subordinate forms, that is, if it has been inherited for a very long period; for in this case it will have been rendered constant by long-continued natural selection.

Glancing at instincts, marvelous as some are, they offer no greater difficulty than do corporeal structures on the theory of the natural selection of successive, slight, but profitable modifications. We can thus understand why nature moves by graduated steps in endowing different animals of the same class with their several instincts. I have attempted to show how much light the principle of gradation throws on the admirable architectural powers of the hive-bee. Habit no doubt often comes into play in modifying instincts; but it certainly is not indispensable, as we see in the case of neuter insects, which leave no progeny to inherit the effects of long-continued habit. On the view of all the species of the same genus having descended from a common parent, and having inherited much in common, we can understand how it is that allied species, when placed under widely different conditions of life, yet follow nearly the same instincts; why the thrushes of tropical and temperate South America, for instance, line their nests with mud like our British species. On the view of instincts having been slowly acquired through natural selection, we need not marvel at some instincts being not perfect and liable to mistakes, and at many instincts causing other animals to suffer.

If species be only well-marked and permanent varieties, we can at once see why their crossed offspring should follow the same complex laws in their degrees and kinds of resemblance to their parents—in being absorbed into each other by successive crosses, and in other such points—as do the crossed offspring of acknowledged varieties. This similarity would be a strange fact, if species had been independently created and varieties had been produced through secondary laws.

If we admit that the geological record is imperfect to an extreme degree, then the facts, which the record does give, strongly support the theory of descent with modification. New species have come on the stage slowly and at successive intervals; and the amount of change after equal intervals of time, is widely different in different groups. The extinction of species and of whole groups of species, which has played so conspicuous a part in the history of the organic world, almost inevitably follows from the principle of natural selection; for old forms are supplanted by new and improved forms. Neither single species nor groups of species reappear when the chain of ordinary generation is once broken. The gradual diffusion of dominant forms, with the slow modification of their descendants, causes the forms of life, after long intervals of time, to appear as if they had changed simultaneously throughout the world. The fact of the fossil remains of each formation being in some degree intermediate in character between the fossils in the formations above and below, is simply explained by their intermediate position in the chain of descent. The grand fact that all extinct beings can be classed with all recent beings, naturally follows from the living and the extinct being the offspring of common parents. As species have generally diverged in character during their long course of descent and modification, we can understand why it is that the more ancient forms, or early progenitors of each group, so often occupy a position in some degree intermediate between existing groups. Recent forms are generally looked upon as being, on the whole, higher in the scale of organization than ancient forms; and they must be higher, in so far as the later and more improved forms have conquered the older and less improved forms in the struggle for life; they have also generally had their organs more specialized for different functions. This fact is perfectly compatible with numerous beings still retaining simple and but little improved structures, fitted for simple conditions of life; it is likewise compatible with some forms having retrograded in organization, by having become at each stage of descent better fitted for new and degraded habits of life. Lastly, the wonderful law of the long endurance of allied forms on the same continent—of marsupials in Australia, of edentata in America, and other such cases—is intelligible, for within the same country the existing and the extinct will be closely allied by descent.

Looking to geographical distribution, if we admit that there has been during the long course of ages much migration from one part of the world to another, owing to former climatical and geographical changes and to the many occasional and unknown means of dispersal, then we can understand, on the theory of descent with modification, most of the great leading facts in distribution. We can see why there should be so striking a parallelism in the distribution of organic beings throughout space, and in their geological succession throughout time; for in both cases the beings have been connected by the bond of ordinary generation, and the means of modification have been the same. We see the full meaning of the wonderful fact, which has struck every traveler, namely, that on the same continent, under the most diverse conditions, under heat and cold, on mountain and lowland, on deserts and marshes, most of the inhabitants within each great class are plainly related; for they are the descendants of the same progenitors and early colonists. On this same principle of former migration, combined in most cases with modification, we can understand, by the aid of the Glacial period, the identity of some few plants, and the close alliance of many others, on the most distant mountains, and in the northern and southern temperate zones; and likewise the close alliance of some of the inhabitants of the sea in the northern and southern temperate latitudes, though separated by the whole intertropical ocean. Although two countries may present physical conditions as closely similar as the same species ever require, we need feel no surprise at their inhabitants being widely different, if they have been for a long period completely sundered from each other; for as the relation of organism to organism is the most important of all relations, and as the two countries will have received colonists at various periods and in different proportions, from some other country or from each other, the course of modification in the two areas will inevitably have been different.

On this view of migration, with subsequent modification, we see why oceanic islands are inhabited by only few species, but of these, why many are peculiar or endemic forms. We clearly see why species belonging to those groups of animals which cannot cross wide spaces of the ocean, as frogs and terrestrial mammals, do not inhabit oceanic islands; and why, on the other hand, new and peculiar species of bats, animals which can traverse the ocean, are often found on islands far distant from any continent. Such cases as the presence of peculiar species of bats on oceanic islands and the absence of all other terrestrial mammals, are facts utterly inexplicable on the theory of independent acts of creation.

The existence of closely allied representative species in any two areas, implies, on the theory of descent with modification, that the same parent-forms formerly inhabited both areas; and we almost invariably find that wherever many closely allied species inhabit two areas, some identical species are still common to both. Wherever many closely allied yet distinct species occur, doubtful forms and varieties belonging to the same groups likewise occur. It is a rule of high generality that the inhabitants of each area are related to the inhabitants of the nearest source whence immigrants might have been derived. We see this in the striking relation of nearly all the plants and animals of the Galapagos Archipelago, of Juan Fernandez, and of the other American islands, to the plants and animals of the neighboring American mainland; and of those of the Cape de Verde Archipelago, and of the other African islands to the African mainland. It must be admitted that these facts receive no explanation on the theory of creation.

The fact, as we have seen, that all past and present organic beings can be arranged within a few great classes, in groups subordinate to groups, and with the extinct groups often falling in between the recent groups, is intelligible on the theory of natural selection with its contingencies of extinction and divergence of character. On these same principles we see how it is that the mutual affinities of the forms within each class are so complex and circuitous. We see why certain characters are far more serviceable than others for classification; why adaptive characters, though of paramount importance to the beings, are of hardly any importance in classification; why characters derived from rudimentary parts, though of no service to the beings, are often of high classificatory value; and why embryological characters are often the most valuable of all. The real affinities of all organic beings, in contradistinction to their adaptive resemblances, are due to inheritance or community of descent. The Natural System is a genealogical arrangement, with the acquired grades of difference, marked by the terms, varieties, species, genera, families, etc.; and we have to discover the lines of descent by the most permanent characters, whatever they may be, and of however slight vital importance.

The similar framework of bones in the hand of a man, wing of a bat, fin of the porpoise, and leg of the horse—the same number of vertebrae forming the neck of the giraffe and of the elephant—and innumerable other such facts, at once explain themselves on the theory of descent with slow and slight successive modifications. The similarity of pattern in the wing and in the leg of a bat, though used for such different purpose—in the jaws and legs of a crab—in the petals, stamens, and pistils of a flower, is likewise, to a large extent, intelligible on the view of the gradual modification of parts or organs, which were aboriginally alike in an early progenitor in each of these classes. On the principle of successive variations not always supervening at an early age, and being inherited at a corresponding not early period of life, we clearly see why the embryos of mammals, birds, reptiles, and fishes should be so closely similar, and so unlike the adult forms. We may cease marveling at the embryo of an air-breathing mammal or bird having branchial slits and arteries running in loops, like those of a fish which has to breathe the air dissolved in water by the aid of well-developed branchiae.

Disuse, aided sometimes by natural selection, will often have reduced organs when rendered useless under changed habits or conditions of life; and we can understand on this view the meaning of rudimentary organs. But disuse and selection will generally act on each creature, when it has come to maturity and has to play its full part in the struggle for existence, and will thus have little power on an organ during early life; hence the organ will not be reduced or rendered rudimentary at this early age. The calf, for instance, has inherited teeth, which never cut through the gums of the upper jaw, from an early progenitor having well-developed teeth; and we may believe, that the teeth in the mature animal were formerly reduced by disuse owing to the tongue and palate, or lips, having become excellently fitted through natural selection to browse without their aid; whereas in the calf, the teeth have been left unaffected, and on the principle of inheritance at corresponding ages have been inherited from a remote period to the present day. On the view of each organism with all its separate parts having been specially created, how utterly inexplicable is it that organs bearing the plain stamp of inutility, such as the teeth in the embryonic calf or the shriveled wings under the soldered wing-covers of many beetles, should so frequently occur. Nature may be said to have taken pains to reveal her scheme of modification, by means of rudimentary organs, of embryological and homologous structures, but we are too blind to understand her meaning.

I have now recapitulated the facts and considerations which have thoroughly convinced me that species have been modified, during a long course of descent. This has been effected chiefly through the natural selection of numerous successive, slight, favorable variations; aided in an important manner by the inherited effects of the use and disuse of parts; and in an unimportant manner, that is, in relation to adaptive structures, whether past or present, by the direct action of external conditions, and by variations which seem to us in our ignorance to arise spontaneously. It appears that I formerly underrated the frequency and value of these latter forms of variation, as leading to permanent modifications of structure independently of natural selection. But as my conclusions have lately been much misrepresented, and it has been stated that I attribute the modification of species exclusively to natural selection, I may be permitted to remark that in the first edition of this work, and subsequently, I placed in a most conspicuous position—namely, at the close of the Introduction—the following words: "I am convinced that natural selection has been the main but not the exclusive means of modification." This has been of no avail. Great is the power of steady misrepresentation; but the history of science shows that fortunately this power does not long endure.

It can hardly be supposed that a false theory would explain, in so satisfactory a manner as does the theory of natural selection, the several large classes of facts above specified. It has recently been objected that this is an unsafe method of arguing; but it is a method used in judging of the common events of life, and has often been used by the greatest natural philosophers. The undulatory theory of light has thus been arrived at; and the belief in the revolution of the earth on its own axis was until lately supported by hardly any direct evidence. It is no valid objection that science as yet throws no light on the far higher problem of the essence or origin of life. Who can explain what is the essence of the attraction of gravity? No one now objects to following out the results consequent on this unknown element of attraction; notwithstanding that Leibnitz formerly accused Newton of introducing "occult qualities and miracles into philosophy."

I see no good reasons why the views given in this volume should shock the religious feelings of any one. It is satisfactory, as showing how transient such impressions are, to remember that the greatest discovery ever made by man, namely, the law of the attraction of gravity, was also attacked by Leibnitz, "as subversive of natural, and inferentially of revealed, religion." A celebrated author and divine has written to me that "he has gradually learned to see that it is just as noble a conception of the Deity to believe that He created a few original forms capable of self-development into other and needful forms, as to believe that He required a fresh act of creation to supply the voids caused by the action of His laws."

Why, it may be asked, until recently did nearly all the most eminent living naturalists and geologists disbelieve in the mutability of species? It cannot be asserted that organic beings in a state of nature are subject to no variation; it cannot be proved that the amount of variation in the course of long ages is a limited quantity; no clear distinction has been, or can be, drawn between species and well-marked varieties. It cannot be maintained that species when intercrossed are invariably sterile and varieties invariably fertile; or that sterility is a special endowment and sign of creation. The belief that species were immutable productions was almost unavoidable as long as the history of the world was thought to be of short duration; and now that we have acquired some idea of the lapse of time, we are too apt to assume, without proof, that the geological record is so perfect that it would have afforded us plain evidence of the mutation of species, if they had undergone mutation.

But the chief cause of our natural unwillingness to admit that one species has given birth to other and distinct species, is that we are always slow in admitting any great changes of which we do not see the steps. The difficulty is the same as that felt by so many geologists, when Lyell first insisted that long lines of inland cliffs had been formed, and great valleys excavated, by the agencies which we still see at work. The mind cannot possibly grasp the full meaning of the term of even a million years; it cannot add up and perceive the full effects of many slight variations, accumulated during an almost infinite number of generations.

Although I am fully convinced of the truth of the views given in this volume under the form of an abstract, I by no means expect to convince experienced naturalists whose minds are stocked with a multitude of facts all viewed, during a long course of years, from a point of view directly opposite to mine. It is so easy to hide our ignorance under such expressions as the "plan of creation," "unity of design," etc., and to think that we give an explanation when we only restate a fact. Any one whose disposition leads him to attach more weight to unexplained difficulties than to the explanation of a certain number of facts will certainly reject the theory. A few naturalists, endowed with much flexibility of mind, and who have already begun to doubt the immutability of species, may be influenced by this volume; but I look with confidence to the future, to young and rising naturalists, who will be able to view both sides of the question with impartiality. Whoever is led to believe that species are mutable will do good service by conscientiously expressing his conviction; for thus only can the load of prejudice by which this subject is overwhelmed be removed.

Several eminent naturalists have of late published their belief that a multitude of reputed species in each genus are not real species; but that other species are real, that is, have been independently created. This seems to me a strange conclusion to arrive at. They admit that a multitude of forms, which till lately they themselves thought were special creations, and which are still thus looked at by the majority of naturalists, and which consequently have all the external characteristic features of true species—they admit that these have been produced by variation, but they refuse to extend the same view to other and slightly different forms. Nevertheless, they do not pretend that they can define, or even conjecture, which are the created forms of life, and which are those produced by secondary laws. They admit variation as a *vera causa* in one case, they arbitrarily reject it in another, without assigning any distinction in the two cases. The day will come when this will be given as a curious illustration of the blindness of preconceived opinion. These authors seem no more startled at a miraculous act of creation than at an ordinary birth. But do they really believe that at innumerable periods in the earth's history certain elemental atoms have been commanded suddenly to flash into living tissues? Do they believe that at each supposed act of creation one individual or many were produced? Were all the infinitely numerous kinds of animals and plants created as eggs or seed, or as full grown? And in the case of mammals, were they created bearing the false marks of nourishment from the mother's womb? Undoubtedly some of these same questions cannot be answered by those who believe in the appearance or creation of only a few forms of life or of some one form alone. It has been maintained by several authors that it is as easy to believe in the creation of a million beings as of one; but Maupertuis' philosophical axiom "of least action" leads the mind more willingly to admit the smaller number; and certainly we ought not to believe that innumerable beings within each great class have been created with plain, but deceptive, marks of descent from a single parent.

As a record of a former state of things, I have retained in the foregoing paragraphs, and elsewhere, several sentences which imply that naturalists believe in the separate creation of each species; and I have been much censured for having thus expressed myself. But undoubtedly this was the general belief when the first edition of the present work appeared. I formerly spoke to very many naturalists on the subject of evolution, and never once met with any sympathetic agreement. It is probable that some did then believe in evolution, but they were either silent or expressed themselves so ambiguously that it was not easy to understand their meaning. Now, things are wholly changed, and almost every naturalist admits the great principle of evolution. There are, however, some who still think that species have suddenly given birth, through quite unexplained means, to new and totally different forms. But, as I have attempted to show, weighty evidence can be opposed to the admission of great and abrupt modifications. Under a scientific point of view, and as leading to further investigation, but little advantage is gained by believing that new forms are suddenly developed in an inexplicable manner from old and widely different forms, over the old belief in the creation of species from the dust of the earth.

It may be asked how far I extend the doctrine of the modification of species. The question is difficult to answer, because the more distinct the forms are which we consider, by so much the arguments in favor of community of descent become fewer in number and less in force. But some arguments of the greatest weight extend very far. All the members of whole classes are connected together by a chain of affinities, and all can be classed on the same principle, in groups subordinate to groups. Fossil remains sometimes tend to fill up very wide intervals between existing orders.

Organs in a rudimentary condition plainly show that an early progenitor had the organ in a fully developed condition, and this in some cases implies an enormous amount of modification in the descendants. Throughout whole classes various structures are formed on the same pattern, and at a very early age the embryos closely resemble each other. Therefore I cannot doubt that the theory of descent with modification embraces all the members of the same great class or kingdom. I believe that animals are descended from at most only four or five progenitors, and plants from an equal or lesser number.

Analogy would lead me one step further, namely, to the belief that all animals and plants are descended from some one prototype. But analogy may be a deceitful guide.

Nevertheless all living things have much in common, in their chemical composition, their cellular structure, their laws of growth, and their liability to injurious influences. We see this even in so trifling a fact as that the same poison often similarly affects plants and animals; or that the poison secreted by the gall-fly produces monstrous growths on the wild rose or oak-tree. With all organic beings, excepting perhaps some of the very lowest, sexual reproduction seems to be essentially similar. With all, as far as is at present known, the germinal vesicle is the same; so that all organisms start from a common origin. If we look even to the two main divisions—namely, to the animal and vegetable kingdoms—certain low forms are so far intermediate in character that naturalists have disputed to which kingdom they should be referred. As Professor Asa Gray has remarked, "the spores and other reproductive bodies of many of the lower algae may claim to have first a characteristically animal, and then an unequivocally vegetable existence." Therefore, on the principle of natural selection with divergence of character, it does not seem incredible that, from some such low and intermediate form, both animals and plants may have been developed; and, if we admit this, we must likewise admit that all the organic beings which have ever lived on this earth may be descended from some one primordial form. But this inference is chiefly grounded on analogy, and it is immaterial whether or not it be accepted. No doubt it is possible, as Mr. G. H. Lewes has urged, that at the first commencement of life many different forms were evolved; but if so, we may conclude that only a very few have left modified descendants. For, as I have recently remarked in regard to the members of each great kingdom, such as the Vertebrata, Articulata, etc., we have distinct evidence in their embryological, homologous, and rudimentary structures, that within each kingdom all the members are descended from a single progenitor.

When the views advanced by me in this volume, and by Mr. Wallace or when analogous views on the origin of species are generally admitted, we can dimly foresee that there will be a considerable revolution in natural history. Systematists will be able to pursue their labors as at present; but they will not be incessantly haunted by the shadowy doubt whether this or that form be a true species. This, I feel sure and I speak after experience, will be no slight relief. The endless disputes whether or not some fifty species of British brambles are good species will cease. Systematists will have only to decide (not that this will be easy) whether any form be sufficiently constant and distinct from other forms, to be capable of definition; and if definable, whether the differences be sufficiently important to deserve a specific name. This latter point will become a far more essential consideration than it is at present; for differences, however slight, between any two forms, if not blended by intermediate gradations, are looked at by most naturalists as sufficient to raise both forms to the rank of species.

Hereafter we shall be compelled to acknowledge that the only distinction between species and well-marked varieties is, that the latter are known, or believed to be connected at the present day by intermediate gradations, whereas species were formerly thus connected. Hence, without rejecting the consideration of the present existence of intermediate gradations between any two forms, we shall be led to weigh more carefully and to value higher the actual amount of difference between them. It is quite possible that forms now generally acknowledged to be merely varieties may hereafter be thought worthy of specific names; and in this case scientific and common language will come into accordance. In short, we shall have to treat species in the same manner as those naturalists treat genera, who admit that genera are merely artificial combinations made for convenience. This may not be a cheering prospect; but we shall at least be freed from the vain search for the undiscovered and undiscoverable essence of the term species.

The other and more general departments of natural history will rise greatly in interest. The terms used by naturalists, of affinity, relationship, community of type, paternity, morphology, adaptive characters, rudimentary and aborted organs, etc., will cease to be metaphorical and will have a plain signification. When we no longer look at an organic being as a savage looks at a ship, as something wholly beyond his comprehension; when we regard every production of nature as one which has had a long history; when we contemplate every complex structure and instinct as the summing up of many contrivances, each useful to the possessor, in the same way as any great mechanical invention is the summing up of the labor, the experience, the reason, and even the blunders of numerous workmen; when we thus view each organic being, how far more interesting—I speak from experience—does the study of natural history become!

A grand and almost untrodden field of inquiry will be opened, on the causes and laws of variation, on correlation, on the effects of use and disuse, on the direct action of external conditions, and so forth. The study of domestic productions will rise immensely in value. A new variety raised by man will be a far more important and interesting subject for study than one more species added to the infinitude of already recorded species. Our classifications will come to be, as far as they can be so made, genealogies; and will then truly give what may be called the plan of creation. The rules for classifying will no doubt become simpler when we have a definite object in view. We possess no pedigree or armorial bearings; and we have to discover and trace the many diverging lines of descent in our natural genealogies, by characters of any kind which have long been inherited. Rudimentary organs will speak infallibly with respect to the nature of long-lost structures. Species and groups of species which are called aberrant, and which may fancifully be called living fossils, will aid us in forming a picture of the ancient forms of life. Embryology will often reveal to us the structure, in some degree obscured, of the prototypes of each great class.

When we can feel assured that all the individuals of the same species, and all the closely allied species of most genera, have, within a not very remote period descended from one parent, and have migrated from some one birthplace; and when we better know the many means of migration, then, by the light which geology now throws, and will continue to throw, on former changes of climate and of the level of the land, we shall surely be enabled to trace in an admirable manner the former migrations of the inhabitants of the whole world. Even at present, by comparing the differences between the inhabitants of the sea on the opposite sides of a continent, and the nature of the various inhabitants of that continent in relation to their apparent means of immigration, some light can be thrown on ancient geography.

The noble science of geology loses glory from the extreme imperfection of the record. The crust of the earth, with its embedded remains, must not be looked at as a well-filled museum, but as a poor collection made at hazard and at rare intervals. The accumulation of each great fossiliferous formation will be recognized as having depended on an unusual occurrence of favorable circumstances, and the blank intervals between the successive stages as having been of vast duration. But we shall be able to gauge with some security the duration of these intervals by a comparison of the preceding and succeeding organic forms. We must be cautious in attempting to correlate as strictly contemporaneous two formations, which do not include many identical species, by the general succession of the forms of life. As species are produced and exterminated by slowly acting and still existing causes, and not by miraculous acts of creation; and as the most important of all causes of organic change is one which is almost independent of altered and perhaps suddenly altered physical conditions, namely, the mutual relation of organism to organism—the improvement of one organism entailing the improvement or the extermination of others; it follows, that the amount of organic change in the fossils of consecutive formations probably serves as a fair measure of the relative, though not actual lapse of time. A number of species, however, keeping in a body might remain for a long period unchanged, whilst within the same period, several of these species, by migrating into new countries and coming into competition with foreign associates, might become modified; so that we must not overrate the accuracy of organic change as a measure of time.

In the future I see open fields for far more important researches. Psychology will be securely based on the foundation already well laid by Mr. Herbert Spencer, that of the necessary acquirement of each mental power and capacity by gradation. Much light will be thrown on the origin of man and his history.

Authors of the highest eminence seem to be fully satisfied with the view that each species has been independently created. To my mind it accords better with what we know of the laws impressed on matter by the Creator, that the production and extinction of the past and present inhabitants of the world should have been due to secondary causes, like those determining the birth and death of the individual. When I view all beings not as special creations, but as the lineal descendants of some few beings which lived long before the first bed of the Cambrian system was deposited, they seem to me to become ennobled. Judging from the past, we may safely infer that not one living species will transmit its unaltered likeness to a distinct futurity. And of the species now living very few will transmit progeny of any kind to a far distant futurity; for the manner in which all organic beings are grouped, shows that the greater number of species in each genus, and all the species in many genera, have left no descendants, but have become utterly extinct. We can so far take a prophetic glance into futurity as to foretell that it will be the common and widely spread species, belonging to the larger and dominant groups within each class, which will ultimately prevail and procreate new and dominant species. As all the living forms of life are the lineal descendants of those which lived long before the Cambrian epoch, we may feel certain that the ordinary succession by generation has never once been broken, and that no cataclysm has desolated the whole world. Hence, we may look with some confidence to a secure future of great length. And as natural selection works solely by and for the good of each being, all corporeal and mental endowments will tend to progress towards perfection.

It is interesting to contemplate a tangled bank, clothed with many plants of many kinds, with birds singing on the bushes, with various insects flitting about, and with worms crawling through the damp earth, and to reflect that these elaborately constructed forms, so different from each other, and dependent upon each other in so complex a manner, have all been produced by laws acting around us. These laws, taken in the largest sense, being Growth with reproduction; Inheritance which is almost implied by reproduction; Variability from the indirect and direct action of the conditions of life, and from use and disuse; a Ratio of Increase so high as to lead to a Struggle for Life, and as a consequence to Natural Selection, entailing Divergence of Character and the Extinction of less improved forms. Thus, from the war of nature, from famine and death, the most exalted object which we are capable of conceiving, namely, the production of the higher animals, directly follows. There is grandeur in this view of life, with its several powers, having been originally breathed by the Creator into a few forms or into one; and that, whilst this planet has gone circling on according to the fixed law of gravity, from so simple a beginning endless forms most beautiful and most wonderful have been, and are being evolved.

CHAPTER 4

OBJECTIONS TO MR. DARWIN'S THEORY OF THE ORIGIN OF SPECIES

ADAM SEDGWICK

Before writing about the transmutation theory, I must give you a skeleton of what the theory is: 1st. *Species* are *not permanent*; *varieties* are the beginning of new species.

2nd. Nature began from the simplest forms—probably from one form—the primeval monad, the parent of all organic life.

3rd. There has been a continual ascent on the organic scale, till organic nature became what it is, by one continued and unbroken stream of onward movement.

4th. The organic ascent is secured by a Malthusian principle through nature—by a battle of life, in which the best in organization (the best varieties of plants and animals) encroach upon and drive off the less perfect. This is called the theory of *natural selection*. It is admirably worked up, and contains a great body of important truth; and it is eminently amusing. But it gives no element of strength to the fundamental theory of transmutation; and without specific transmutations natural selection can do nothing for the general theory.[1] The flora and fauna of North America are very different from what they were when the Pilgrim Fathers were driven out from old England; but

---

From Adam Sedgwick, "Objections to Mr. Darwin's Theory of the Origin of Species," *Spectator* (April 7, 1860).

changed as they are, they do not one jot change the collective fauna and flora of the actual world.

5th. We do not mark any great organic changes *now*, because they are so slow that even a few thousand years may produce no changes that have fixed the notice of naturalists.

6th. But *time is the agent*, and we can mark the effects of time by the organic changes on the great geological scale. And on every part of that scale, where the organic changes are great in two contiguous deposits of the scale, there must have been a corresponding lapse of time between the periods of their deposition—perhaps millions of years.

I think the foregoing heads give the substance of Darwin's theory; and I think that the great broad facts of geology are directly opposed to it.

Some of these facts I shall presently refer to. But I must in the first place observe that Darwin's theory is not *inductive*—not based on a series of acknowledged facts pointing to a *general conclusion*—not a proposition evolved out of the facts, logically, and of course including them. To use an old figure, I look on the theory as a vast pyramid resting on its apex, and that apex a mathematical point. The only facts he pretends to adduce, as true elements of proof, are the *varieties* produced by domestication, or the *human artifice* of cross-breeding. We all admit the varieties, and the very wide limits of variation, among domestic animals. How very unlike are poodles and greyhounds! Yet they are of one species. And how nearly alike are many animals—allowed to be of distinct species, on any acknowledged views of species. Hence there may have been very many blunders among naturalists in the discrimination and enumeration of species. But this does not undermine the grand truth of nature and the continuity of true species.

Again, the varieties, built upon by Mr. Darwin, are varieties of domestication and human *design*. Such varieties could have no existence in the old world. Something may be done by crossbreeding; but mules are generally sterile, or the progeny (in some rare instances) passes into one of the original crossed forms. The Author of Nature will not permit His work to be spoiled by the wanton curiosity of Man. And in a state of nature (such as that of the old world before Man came upon it) wild animals of different species do not desire to cross and unite.

Species have been constant for thousands of years; and time (so far as I see my way) though multiplied by millions and billions would never change them, so long as the conditions remained constant. Change the conditions, and old species would disappear; and new species *might* have room to come in and flourish. But how, and by what causation? I say by *creation*. But, what do I mean by creation? I reply, the operation of a power quite beyond the powers of a pigeon-fancier, a cross-breeder, or hybridizer; a power I cannot imitate or comprehend; but in which I can believe, by a legitimate conclusion of sound reason draw from the laws and harmonies of Nature. For I can see in all around me a design and purpose, and a mutual adaptation of parts which I *can* comprehend—and which prove that there is exterior to, and above, the mere phenomena of Nature a great prescient and designing cause. Believing this, I have no difficulty in the repetition of new species during successive epochs in the history of the earth.

But Darwin would say I am introducing a miracle by the supposition. In one sense, I am; in another, I am not. The hypothesis does not suspend or interrupt an established law of Nature. It does suppose the introduction of a new phenomenon unaccounted for by the operation of any *known* law of Nature; and it appeals to a power above established laws, and yet acting in harmony and conformity with them.

The pretended physical philosophy of modern days strips Man of all his moral attributes, or holds them of no account in the estimate of his origin and place in the created world. A cold atheistical materialism is the tendency of the so-called material philosophy of the present day. Not that I believe that Darwin is an atheist; though I cannot but regard his materialism as atheistical; because it ignores all rational conception of a final cause. I think it untrue because opposed to the obvious course of Nature, and the very opposite of inductive truth. I therefore think it intensely mischievous.

Let no one say that it is held together by a cumulative argument. Each series of facts is laced together by a series of assumptions, which are mere repetitions of the one false principle. You cannot make a good rope out of a string of air-bubbles.

I proceed now to notice the manner in which Darwin tries to fit his principles to the facts of geology.

I will take for granted that the known series of fossil-bearing rocks or deposits may be divided into the Palaeozoic; the Mesozoic; the Tertiary or

Neozoic; and the Modern, with the spoils of the actual flora and fauna of the world, and with wrecks of the works of Man.

To begin then, with the Palaeozoic rocks. Surely we ought on the transmutation theory, to find near their base great deposits with *none but the lowest forms of organic life.* I know of no such deposits. Oken contends that life began with the infusorial forms. They are at any rate well fitted for fossil preservation; but we do not find them. Neither do we find beds exclusively of hard corals and other humble organisms, which ought, on the theory, to mark a period of vast duration while the primeval monads were working up into the higher types of life. Our evidence is, no doubt, very scanty; but let not our opponents dare to say that it makes *for them.* So far as it is positive, it seems to me pointblank against them. If we build upon imperfect evidence, they commence without any evidence whatsoever, and against the evidence of actual nature. As we ascend in the great stages of the Palaeozoic series (through Cambrian, Silurian, Devonian, and Carboniferous rocks) we have in each a *characteristic* fauna; we have no wavering of species, we have the noblest cephalopods and brachiopods that ever existed; and they preserve their typical forms till they disappear. And a few of the types have endured, with specific modifications, through all succeeding ages of the earth, it is during these old periods that we have some of the noblest ichthyic forms that ever were created. The same may be said, I think, of the carboniferous flora. As a whole, indeed, it is lower than the living flora of our own period; but many of the old types were grander and of higher organization than the corresponding families of the living flora; and there is no wavering, no wanting of organic definition, in the old types. We have some land reptiles (batrachians), in the higher Palaeozoic periods, but not of a very low type; and the reptiles of the permian groups (at the very top of the Palaezoic rocks), are of a high type. If all this be true (and I think it is), it gives but a sturdy grist for the transmutation-mill, and may soon break its cogs.[2]

We know the complicated organic phenomena of the Mesozoic (or Oolitic) period. It defies the transmutationist at every step. Oh! but the document, says Darwin, is a fragment. I will interpolate long periods to account for all the changes. I say, in reply, if you deny my conclusion grounded on positive evidence, I toss back your conclusions, derived from negative evidence—the inflated cushion on which you try to bolster up the defects of your hypothesis.

The reptile fauna of the Mesozoic period is the grandest and highest that ever lived. How came these reptiles to die off, or to degenerate? And how came the dinosaurs to disappear from the face of Nature, and leave no descendants like themselves, or of a corresponding nobility? By what process of *natural selection* did they disappear? Did they tire of the land, and become whales, casting off their hind-legs? And, after they had lasted millions of years as whales, did they tire of the water, and leap out again as pachyderms? I have heard of both hypotheses; and I cannot put them into words without seeming to use the terms of mockery. This I do affirm, that if the transmutation theory were proved true in the actual world, and we could hatch rats out of eggs of geese, it would still be difficult to account for the successive forms of organic life in the old world. They appear to me to give the lie to the theory of transmutation at every turn of the pages of Dame Nature's old book.

The limits of this letter compel me to omit any long discussion of the Tertiary mammals, of course including man at their head. On physical grounds, the transmutation theory is untrue, if we reason (as we ought to do) from the known to the unknown. To this rule, the Tertiary mammals offer us no exception. Nor is there any proof, either ethnographical or physical, of the bestial origin of man.

And now for a few words upon Darwin's long *interpolated periods* of geological ages. He has an eternity of past time to draw upon; and I am willing to give him ample measure; only let him use it logically, and in some probable accordance with facts and phenomena.

1st. I place the theory against facts viewed collectively. I see no proofs of enormous gaps of geological time (I say nothing of years or centuries), in those cases where there is a sudden change in the ancient fauna and flora. I am willing, out of the stock of past time, to lavish millions or billions upon each epoch, if thereby we can gain rational results from the operation of *true causes*. But time and "natural selection" can do nothing if there be not a *vera causa* working with them.[3] I must confine myself to a very small number of the collective instances.

2nd. Towards the end of the carboniferous period, there was a vast extinction of animal and vegetable life. We can, I think, account for this extinction mechanically. The old crust was broken up. The sea bottom underwent a great change. The old flora and fauna went out; and a new flora and fauna appeared, in the ground, now called permian, at the base of the new red sandstone, which overlies the carboniferous rocks. I take the fact as it is, and I have no difficulty. The time in which all this was brought about may have been very long, even upon a geological scale of time. But where do the intervening and connecting types exist, which are to mark the work of natural selection? We do not find them. Therefore, the step onwards gives no true resting-place to a baseless theory; and is, in fact, a stumbling-block in its way.

3rd. Before we rise through the new red sandstone, we find the muschel-kalk (wanting in England, though its place on the scale is well-known) with an *entirely new* fauna: where have we a proof of any enormous lapse of geological time to account for the change? We have no proof in the deposits themselves: the presumption they offer to our senses is of a contrary kind.

4th. If we rise from the muschel-kalk to the Lias, we find again a new fauna. All the anterior species are gone. Yet the passage through the upper members of the new red sandstone to the Lias is by insensible gradations, and it is no easy matter to fix the physical line of their demarcation. I think it would be a very rash assertion to affirm that a great geological interval took place between the formation of the upper part of the new red sandstone and the Lias. Physical evidence is against it. To support a baseless theory, Darwin would require a countless lapse of ages of which we have no commensurate physical monuments; and he is unable to supply any of the connecting organic links that ought to bind together the older fauna with that of the Lias.

I cannot go on any further with these objections. But I will not conclude without expressing my deep aversion to the theory; because of its unflinching materialism; because it has deserted the inductive track—the only track that leads to physical truth; because it utterly repudiates final causes, and thereby indicates a demoralized understanding on the part of its advocates. By the word, demoralized, I mean a want of capacity for comprehending the force of moral evidence, which is dependent on the highest faculties of our nature. What is it that gives us the sense of right and wrong, of law, of duty, of cause and effect? What is it that enables us to construct true theories on good inductive evidence? Theories which enable us, whether in the material or the moral world, to link together the past and the present. What is it that enables us to anticipate the future, to act wisely with reference to future good, to believe in a future state, to acknowledge the being of a God? These faculties, and many others of like kind, are a part of ourselves quite as much so as our organs of sense. All nature is subordinate to law. Every organ of every sentient being has its purpose bound up in the very law of its existence. Are the highest conceptions of man, to which he is led by the necessities of his moral nature, to have no counterpart or fruition? I say no, to all such questions; and fearlessly affirm that we cannot speculate on man's position in the actual world of nature, on his destinies, or on *his origin*, while we keep his highest faculties out of our sight. Strip him of these faculties, and he becomes entirely bestial; and he may well be (under such a false and narrow view) nothing better than the natural progeny of a beast, which has to live, to beget its likeness, and then die forever.

By gazing only on material nature, a man may easily have his very senses bewildered (like one under the cheatery of an electro-biologist); he may become so frozen up, by a too long continued and exclusively material study, as to lose his relish for moral truth, and his vivacity in apprehending it. I think I can see traces of this effect, both in the origin and in the details of certain portions of Darwin's theory; and, in confirmation of what I now write, I would appeal to all that he states about those marvelous structures—the comb of a common honey-bee and the eye of a mammal. His explanations make demands on our credulity, that are utterly beyond endurance, and do not give us one true natural step towards an explanation of the phenomena—viz. the perfection of the structures, and their adaptation to their office. There is a light by which a man may see and comprehend facts and truths such as these. But Darwin wilfully shuts it out from our senses; either because he does not apprehend its power, or because he disbelieves in its existence. This is the grand blemish of his work. Separated from his sterile and contracted theory, it contains very admirable details and beautiful views of nature—especially in those chapters which relate to the battle of life, the variations of species, and their diffusion through wide regions of the earth.

In some rare instances, Darwin shows a wonderful credulity. He seems to believe that a white bear, by being confined to the slops floating in the polar basin, might in time be turned into a whale; that a lemur might easily be turned into a bat; that a three-toed tapir might be the great-grandfather of a horse; or that the progeny of a horse may (in America) have gone back into the tapir... all interpreted hypothetically produces, in some minds, a kind of pleasing excitement, which predisposes them in its favor and if they are unused to careful reflection, and averse to the labor of accurate investigation, they will be likely to conclude that what is (apparently) *original* must be a production of original genius, and that anything very much opposed to prevailing notions must be a grand discovery—in short, that whatever comes from "the bottom of a well" must be the "truth" which has been long hidden there.

NOTES

1. It is worth remarking that though no species of the horse genus was found in America when discovered, two or three *fossil* species have been found there. Now, if these horses have (through some influence of climate) been transmuted into tapirs or buffalos, one might expect to see the *tendency* at least towards such a change in the numerous herds of wild horses—the descendants of those brought from Europe—which are now found in both South and North America.

2. I forebear to mention the stagonolepis, a very highly organized reptile, the remains of which were found by Sir R. I. Muchison, in a rock near Elgin, supposed to belong to the old red sandstone. Some doubts have been expressed about the age of the deposit. Should the first opinion prove true (and I think it will), we shalt then have one of the oldest reptiles of the world exhibiting, not a very low, but a very high organic type.

3. See reference on *Time*, in the *Annotations of Bacon's Essays.*

CHAPTER 5

THE ORIGIN OF SPECIES

THOMAS H. HUXLEY

The hypotheses respecting the origin of species which profess to stand upon a scientific basis, and, as such, alone demand serious attention, are of two kinds. The one, the "special creation" hypothesis, presumes every species to have originated from one or more stocks, these not being the result of modification of any other form of living matter—or rising by natural agencies —but being produced, as such, by a supernatural creative act.

The other, the so-called "transmutation" hypothesis, considers that all existing species are the result of the modification of preexisting species, and those of their predecessors, by agencies similar to those which at the present-day produce varieties and races, and therefore in an altogether natural way; and it is probable, but not a necessary consequence of this hypothesis, that all living beings have arisen from a single stock. With respect to the origin of this primitive stock, or stocks, the doctrine of the origin of species is obviously not necessarily concerned. The transmutation hypothesis, for example, is perfectly consistent either with the conception of a special creation of the primitive germ, or with the supposition of its having arisen, as a modification of inorganic matter, by natural causes.

From Thomas H. Huxley, "The Origin of Species," in *Darwiniana: Essays* (New York: D. Appleton & Company, 1896), pp. 53–59, 71–79.

The doctrine of special creation owes its existence very largely to the supposed necessity of making science accord with the Hebrew cosmogony; but it is curious to observe that, as the doctrine is at present maintained by men of science, it is as hopelessly inconsistent with the Hebrew view as any other hypothesis.

If there be any result which is, more clearly out of geological investigation and another, it is, that the vast series of extinct animals and plants is not divisible, as it was once supposed to be, into distinct groups, separated by sharply marked boundaries. There are no great gulfs between epochs and formations—no successive periods marked by the appearance of plants, of water animals, and of land animals, *en masse.* Every year adds to list of links between what the older geologists supposed to be widely separated epochs: witness the crags linking the drift with older tertiaries; the Maestricht beds linking the tertiaries with the chalk; the St. Cassian beds exhibiting an abundant fauna of mixed mesozoic and paleozoic types, in rocks of an epoch once supposed to be eminently poor in life; witness, lastly, the incessant disputes as to whether a given stratum shall be reckoned devonian or carboniferous, silurian or devonian, cambrian or silurian.

This truth is further illustrated in the most interesting manner by the impartial and highly competent testimony of M. Pictet, from whose calculations of what percentage of the genera of animals, existing in any formation, lived during the preceding formation, it results that in no case is the proportion less than *one-third*, or 33 percent. It is the triassic formation, or the commencement of the mesozoic epoch, which has received the smallest inheritance from preceding ages. The other formations not uncommonly exhibit 60, 80, or even 94 percent of genera in common with those whose remains are imbedded in their predecessor. Not only is this true, but the subdivisions of each formation exhibit new species characteristic of, and found only in, them; and, in many cases, as in the lias for example, the separate beds of these subdivisions are distinguished by well-marked and peculiar forms of life. A section, a hundred feet thick, will exhibit, at different heights, a dozen species of ammonite, none of which passes beyond its particular zone of limestone, or clay, into the zone below it or into that above it; so that those who adopt the doctrine of special creation must be prepared to admit, that at intervals of time, corresponding with the thickness of these beds, the Creator thought fit to interfere with the natural course of events for the purpose of making a new ammonite. It is not easy to transplant oneself into the frame of mind of those who can accept such a conclusion as this, on any evidence short of absolute demonstration; and it is difficult to see what is to be gained by so doing, since, as we have said, it is obvious that such a view of the origin of living beings is utterly opposed to the Hebrew cosmogony. Deserving no aid from the powerful arm of Bibliolatry, then, does the received form of the hypothesis of special creation derive any support from science or sound logic? Assuredly not much. The arguments brought forward in its favor all take one form: If species were not supernaturally created, we cannot understand the facts *x*, or *y*, or *z*; we cannot understand the structure of animals or plants, unless we suppose they were contrived for special ends; we cannot understand the structure of the eye, except by supposing it to have been made to see with; we cannot understand instincts, unless we suppose animals to have been miraculously endowed with them.

As a question of dialectics, it must be admitted that this sort of reasoning is not very formidable to those who are not to be frightened by consequences. It is an *argumentum ad ignorantiam*—take this explanation or be ignorant. But suppose we prefer to admit our ignorance rather than adopt a hypothesis at variance with all the teachings of Nature? Or, suppose for a moment we admit the explanation, and then seriously ask ourselves how much the wiser are we; what does the explanation explain? Is it any more than it were into a grandiloquent way of announcing the fact that you really know nothing about the matter? A phenomenon is explained when it is shown to be a case of some general law of Nature; but the supernatural interposition of the Creator can, by the nature of the case, exemplify no law, and if species have really arisen in this way, it is absurd to attempt to discuss their origin.

Or, lastly, let us ask ourselves whether any amount of evidence which the nature of our faculties permits us to attain, can justify us in asserting that any phenomenon is out of the region of natural causation. To this end it is obviously necessary that we should know all the consequences to which all possible combinations, continued to unlimited time, can give rise. If we knew these, and found none competent to originate species, we should have a good ground for denying their origin by natural causation. Till we know them, any hypothesis is better than which involves us in such miserable presumption.

But the hypothesis of special creation is not only a mere specious mask for our ignorance; its existence in Biology marks the youth and imperfection of the science. For what is the history of every science but the history of the elimination of the notion of creative, or other interferences, with the natural order of the phenomena which are the subject-matter of that science? When astronomy was young "the morning stars sang together for joy," and the planets were guided in their courses by celestial hands. Now, the harmony of the stars has resolved itself into gravitation according to the inverse squares of the distances, and the orbits of the planets are deducible from the laws of the forces which allow a schoolboy's stone to break a window. The lightning was the angel of the Lord; but it has pleased Providence, in these modern times, that science should make it the humble messenger of man, and we know that every flash that shimmers about the horizon on a summer's evening is determined by ascertainable conditions, and its direction and brightness might, if our knowledge of these were great enough, have been calculated.

The solvency of great mercantile companies rests on the validity of the laws which have been ascertained to govern the seeming irregularity of that human life which the moralist bewails as the most uncertain of things; plague, pestilence, and famine are admitted, by all but fools, to be the natural result of causes for the most part fully within human control, and not the unavoidable tortures inflicted by wrathful Omnipotence upon His helpless handiwork.

Harmonious order governing eternally continuous progress—the web and woof of matter and force interweaving by slow degrees, without a broken thread, that veil which lies between us in the Infinite—that universe which alone we know or can know; such is the picture which science draws of the world, and in proportion as any part of that picture is in unison with the rest, so may we feel sure that it is rightly painted. Shall Biology alone remain out of harmony with her sisters sciences?

***

The Darwinian hypothesis has the merit of being eminently simple and comprehensible in principle, and its essential positions may be stated in a very few words: all species have been produced by the development of varieties from common stocks; by the conversion of these, first into permanent races and then into new species, by the process of natural selection, which process is essentially identical with that artificial selection by which man has originated the races of domestic animals—the *struggle for existence* taking the place of man, and exerting, in the case of natural selection, that selective action which he performs in artificial selection.

The evidence brought forward by Mr. Darwin in support of this hypothesis is of three kinds. First, he endeavors to prove that species may be originated by selection; secondly, he attempts to show that natural causes are competent to exert selection; and thirdly, he tries to prove that the most remarkable and apparently anomalous phenomena exhibited by the distribution, development, and mutual relations of species, can be shown to be deducible from the general doctrine of their origin, which he propounds, combined with the known facts of geological change; and that, even if all these phenomena are not at present explicable by it, none are necessarily inconsistent with it.

There cannot be a doubt that the method of inquiry which Mr. Darwin has adopted is not only rigorously in accordance with the canons of scientific logic, but that it is the only adequate method. Critics exclusively trained in classics or in mathematics, who have never determined a scientific fact in their lives by induction from experiment or observation, prate learnedly about Mr. Darwin's method, which is not inductive enough, not Baconian enough, forsooth, for them. But even if practical acquaintance with the process of scientific investigation is denied them, they may learn, by the perusal of Mr. Mill's admirable chapter "On the Deductive Method," that there are multitudes of scientific inquiries in which the method of pure induction helps the investigator but a very little way:

> The mode of investigation which,...from the proved inapplicability of direct methods of observation and experiment, remains to us as the main source of the knowledge we possess, or can acquire, respecting the conditions and laws of recurrence of the more complex phænomena, is called, in its most general expression, the deductive method, and consists of three operations: the first, one of direct induction; the second, of ratiocination; and the third, of verification.

Now, the conditions which have determined the existence of species are not only exceedingly complex, but, so far as the great majority of them are concerned, are necessarily beyond our cognizance. But what Mr. Darwin has attempted to do is in exact accordance with the rule laid down by Mr. Mill; he has endeavored to determine certain great facts inductively, by observation and experiment; he has then reasoned from the data thus furnished; and lastly, he has tested the validity of his ratiocination by comparing his deductions with the observed facts of Nature. Inductively, Mr. Darwin endeavors to prove that species arise in a given way. Deductively, he desires to show that, if they arise in that way, the facts of distribution, development, classification, etc., may be accounted for, i.e., may be deduced from their mode of origin, combined with admitted changes in physical geography and climate, during an indefinite period. And this explanation, or coincidence of observed with deduced facts, is, so far as it extends, a verification of the Darwinian view.

There is no fault to be found with Mr. Darwin's method, then; but it is another question whether he has fulfilled all the conditions imposed by that method. Is it satisfactorily proved, in fact, that species may be originated by selection? That there is such a thing as natural selection? That none of the phenomena exhibited by species is inconsistent with the origin of species in this way? If these questions can be answered in the affirmative, Mr. Darwin's view steps out of the rank of hypotheses into those of proved theories; but, so long as the evidence at present adduced falls short of enforcing that affirmation, so long, to our minds, must the new doctrine be content to remain among the former—an extremely valuable, and in the highest degree probable, doctrine, indeed the only extant hypothesis which is worth anything in a scientific point of view; but still a hypothesis, and not yet the theory of species.

After much consideration, and with assuredly no bias against Mr. Darwin's views, it is our clear conviction that, as the evidence stands, it is not absolutely proven that a group of animals, having all the characters exhibited by species in Nature, has ever been originated by selection, whether artificial or natural. Groups having the morphological character of species—distinct and permanent races in fact—have been so produced over and over again; but there is no positive evidence, at present, that any group of animals has, by variation and selective breeding, given rise to another group which was, even in the least degree, infertile with the first. Mr. Darwin is perfectly aware of this weak point, and brings forth a multitude of ingenious and important arguments to diminish the force of the objection. We admit the value of these arguments to their fullest extent; nay, we will go so far as to express our belief that experiments, conducted by the skillful physiologist, would very probably obtain the desired production of mutually more or less infertile breeds from a common stock, in a comparatively few years; but still, as the case stands at present, this "little rift within the lute" is not to be disguised nor overlooked.

In the remainder of Mr. Darwin's argument our own private ingenuity has not hitherto enabled us to pick holes of any great importance; and judging by what we hear and read, other adventurers in the same field do not seem to have been much more fortunate. It is being urged, for instance, that in his chapters on the struggle for existence and on natural selection, Mr. Darwin does not so much prove that natural selection does occur, as that it must occur; but, in fact, no other sort of demonstration is attainable. A race does not attract our attention in Nature until it has, in all probability, existed for considerable time, and then it is too late to inquire into the conditions of its origin. Again, it is said that there is no real analogy between the selection which takes place under domestication, by human influence, and any operation which can be affected by Nature, for man interferes intelligently. Reduced to its elements, this argument implies that an effect produced with trouble by an intelligent agent must, *à fortiori*, be more troublesome, if not impossible, to an unintelligent agent. Even putting aside the question whether Nature, acting as she does according to definite and invariable laws, can be rightly called an intelligent agent, such a position as this is wholly untenable. Mix salt and sand, and it shall puzzle the wisest of men, with his mere natural appliances, to separate all the grains of sand from all the grains of salt; but a shower of rain will effect the same object in ten minutes. And so, while man may find it tax all his intelligence to separate any variety which arises, and to breed selectively from it, the destructive agencies incessantly at worked in Nature, if they find one variety to be more soluble in circumstances than the other, will inevitably, in the long run, eliminate it.

A frequent and a just objection to the Lamarckian hypothesis of the transmutation of species is based on the absence of transitional forms between many species. But against the Darwinian hypothesis this argument has no force. Indeed, when the most valuable and suggestive parts of Mr. Darwin's work is that in which he proves, that the frequent absence of transitions is a necessary consequence of his doctrine, and that the stock whence two or more species have sprung, need in no respect be intermediate between these species. If any two species have arisen from a common stock in the same way as the carrier and the pouter, say, have arisen from the rock-pigeon, then the common stock of these two species need to be no more intermediate between the two then the rock-pigeon is between the carrier and pouter. Clearly appreciate the force of this analogy, then all the arguments against the origin of species by selection, based in the absence of transitional forms, fall to the ground. And Mr. Darwin's position might, we think, have been even stronger than at this if he had not embarrassed himself with the aphorism, "*Natura non facit saltum*," which turns up so often in his pages. We believe, as we have said above, that Nature does make jumps now and then, and a recognition of the fact is of no small importance in disposing of many minor objections to the doctrine of transmutation.

But we must pause. The discussion of Mr. Darwin's arguments in detail would lead us far beyond the limits within which we proposed, at starting, to confine this article. Our object has been attained if we have given an intelligible, however brief, account of the established facts connected with species, and of the relation of the explanation of those facts offered by Mr. Darwin to the theoretical views held by his predecessors and his contemporaries, and, above all, to the requirements of scientific logic. We have ventured to point out that it does not, as yet, satisfy all those requirements; but we do not hesitate to assert that it is as superior to any preceding or contemporary hypothesis, in the extent of observational and experimental basis on which it rests, in its rigorously scientific method, and in its power of explaining biological phenomena, as was the hypothesis of Copernicus to the speculations of Ptolemy. But the planetary orbits turned out to be not quite circular after all, and, grand as was the service Copernicus rendered to science, Kepler and Newton had to come after him. What if the orbit of Darwinism should be a little too circular? What if species should offer residual phenomena, here and there, not explicable by natural selection? Twenty years hence naturalists may be in a position to say whether this is, or is not, the case; but in either event they will owe the author of "The Origin of Species" an immense debt of gratitude. We should leave a very wrong impression on the reader's mind if we permitted him to suppose that the value of that work depends wholly on the ultimate justification of the theoretical views which it contains. On the contrary, if they were disproved to-morrow, the book would still be the best of its kind—the most compendious statement of well-sifted facts bearing on the doctrine of species that has ever appeared. The chapters on Variation, on the Struggle for Existence, on Instinct, on Hybridism, in the Imperfection of the Geological Record, on Geographical Distribution, have not only no equals, but, so far as our knowledge goes, no competitors, within the range of biological literature. And viewed as a whole, we do not believe that, since the publication of Von Baer's "Researches on Development," thirty years ago, any work has appeared calculated to exert so large an influence, not only on the future of Biology, but in extending the domination of science over regions of thought into which she has, as yet, hardly penetrated.

CHAPTER 6
WHAT IS DARWINISM?
CHARLES HODGE

The great fact of experience is that the universe exists. The great problem which has ever pressed upon the human mind is to account for its existence. What was its origin? To what causes are the changes we witness around us to be referred? As we are a part of the universe, these questions concern ourselves. What are the origin, nature, and destiny of man? Professor Huxley is right in saying, "The question of questions for mankind—the problem which underlies all others, and is more interesting than any other—is the ascertainment of the place which Man occupies in nature and of his relation to the universe of things. Whence our race has come, what are the limits of our power over nature, and of nature's power over us, to what goal are we tending, are the problems which present themselves anew and with undiminished interest to every man born into the world."[1] Mr. Darwin undertakes to answer these questions. He proposes a solution of the problem which thus deeply concerns every living man. Darwinism is, therefore, a theory of the universe, at least so far as the living organisms on this earth are concerned. This being the case, it may be well to state, in few words, the other prevalent

---

From Charles Hodge, *What Is Darwinism?* (New York: Scribner, Armstrong, and Co., 1874), pp. 1–7, 22–30, 40–48, 125–42, 162–70.

theories on this great subject, that the points of agreement and of difference between them and the views of Mr. Darwin may be the more clearly seen.

THE SCRIPTURAL SOLUTION OF THE PROBLEM OF THE UNIVERSE.

That solution is stated in words equally simple and sublime: "In the beginning God created the heavens and the earth." We have here, first, the idea of God. The word God has in the Bible a definite meaning. It does not stand for an abstraction, for mere force, for law or ordered sequence. God is a spirit, and as we are spirits, we know from consciousness that God is, (1.) A Substance; (2.) That He is a person; and, therefore, a self-conscious, intelligent, voluntary agent. He can say I; we can address Him as Thou; we can speak of Him as He or Him. This idea of God pervades the Scriptures. It lies at the foundation of natural religion. It is involved in our religious consciousness. It enters essentially into our sense of moral obligation. It is inscribed ineffaceably, in letters more or less legible, on the heart of every human being. The man who is trying to be an atheist is trying to free himself from the laws of his being. He might as well try to free himself from liability to hunger or thirst.

The God of the Bible, then, is a Spirit, infinite, eternal, and unchangeable in his being, wisdom, power, holiness, goodness, and truth. As every theory must begin with some postulate, this is the grand postulate with which the Bible begins. This is the first point.

The second point concerns the origin of the universe. It is not eternal either as to matter or form. It is not independent of God. It is not an evolution of his being, or his existence form. He is extramundane as well as antemundane. The universe owes its existence to his will.

Thirdly, as to the nature of the universe; it is not a mere phenomenon. It is an entity, having real objective existence, or actuality. This implies that matter is a substance endowed with certain properties, in virtue of which it is capable of acting and of being acted upon. These properties being uniform and constant, are physical laws to which, as their proximate causes, all the phenomena of nature are to be referred.

Fourthly, although God is extramundane, He is nevertheless everywhere present. That presence is not only a presence of essence, but also of knowledge and power. He upholds all things. He controls all physical causes, working through them, with them, and without them, as He sees fit. As we, in our limited spheres, can use physical causes to accomplish our purposes, so God everywhere and always cooperates with them to accomplish his infinitely wise and merciful designs.

Fifthly, man a part of the universe, is, according to the Scriptures, as concerns his body, of the earth. So far, he belongs to the animal kingdom. As to his soul, he is a child of God, who is declared to be the Father of the spirits of all men. God is a spirit, and we are spirits. We are, therefore, of the same nature with God. We are God-like; so that in knowing ourselves we know God. No man conscious of his manhood can be ignorant of his relationship to God as his Father.

The truth of this theory of the universe rests, in the first place, so far as it has been correctly stated, on the infallible authority of the Word of God. In the second place, it is a satisfactory solution of the problem to be solved. (1) It accounts for the origin of the universe. (2) It accounts for all the universe contains, and gives a satisfactory explanation of the marvelous contrivances which abound in living organisms, of the adaptations of these organisms to conditions external to themselves, and for those provisions for the future, which on any other assumption are utterly inexplicable. (3) It is in conflict with no truth of reason and with no fact of experience.[2] (4) The Scriptural doctrine accounts for the spiritual nature of man, and meets all his spiritual necessities. It gives him an object of adoration, love, and confidence. It reveals the Being on whom his indestructible sense of responsibility terminates. The truth of this doctrine, therefore, rests not only on the authority of the Scriptures, but on the very constitution of our nature. The Bible has little charity for those who reject it. It pronounces them to be either derationalized or demoralized, or both.

. . .

THEISM IN UNSCRIPTURAL FORMS.

There are men who are constrained to admit the being of God, who depart from the Scriptural doctrine as to his relation to the world. According to some, God created matter and endowed it with certain properties, and then left it to itself to work out, without any interference or control on his part, all possible results. According to others, He created not only matter, but life, or living germs, one or more, from which without any divine intervention all living organisms have been developed. Others, again, refer not only matter and life, but mind also to the act of the Creator; but with creation his agency ceases. He has no more to do with the world, than a ship-builder has with the ship he has constructed, when it is launched and far off upon the ocean. According to all these views a creator is a mere *Deus ex machina*, an assumption to account for the origin of the universe.

Another general view of God's relation to the world goes to the opposite extreme. Instead of God doing nothing, He does everything. Second causes have no efficiency. The laws of nature are said to be the uniform modes of divine operation. Gravitation does not flow from the nature of matter, but is a mode of God's uniform efficiency. What are called chemical affinities are not due to anything in different kinds of matter, but God always acts in one way in connection with an acid, and in another way in connection with an alkali. If a man places a particle of salt or sugar on his tongue, the sensation which he experiences is not to be referred to the salt or sugar, but to God's agency. When this theory is extended, as it generally is by its advocates, from the external to the internal world, the universe of matter and mind, with all their phenomena, is a constant effect of the omnipresent activity of God. The minds of some men, as remarked above, are so constituted that they can pass from the theory that God does nothing, to the doctrine that He does everything, without seeing the difference. Mr. Russel Wallace, the companion and peer of Mr. Darwin, devotes a large part of his book on natural selection to prove that the organs of plants and animals are formed by blind physical causes. Toward the close of the volume he teaches that there are no such causes. He asks the question, What is matter? and answers, Nothing. We know, he says, nothing but force; and as the only force of which we have any immediate knowledge is mind-force, the inference is "that the whole universe is not merely dependent on, but actually is, the will of higher intelligences, or of one Supreme Intelligence."[3] This is a transition from virtual materialism to idealistic pantheism. The effect of this admission on the part of Mr. Wallace on the theory of natural selection, is what an explosion of its boiler would be to a steamer in mid-ocean, which should blow out its deck, sides, and bottom. Nothing would remain above water.

The Duke of Argyll seems at times inclined to lapse into the same doctrine. "Science," he says, "in the modern doctrine of conservation of energy and the convertibility of forces, is already getting a firm hold of the idea, that all kinds of force are but forms of manifestations of one central force issuing from some one fountain-head of power. Sir John Herschel has not hesitated to say, 'that it is but reasonable to regard the force of gravitation as the direct or indirect result of a consciousness or will existing somewhere.' And even if we cannot certainly identify force in all its forms with the direct energies of the one Omnipresent and All-pervading Will, it is at least in the highest degree unphilosophical to assert the contrary—to think or to speak, as if the forces of nature were either independent of, or even separate from the Creator's power."[4] The Duke, however, in the general tenor of his book, does not differ from the common doctrine, except in one point. He does not deny the efficiency of physical causes, or resolve them all into the efficiency of God; but he teaches that God, in this world at least, never acts except through those causes. He applies this doctrine even to miracles, which he regards as effects produced by second causes of which we are ignorant, that is, by some higher law of nature. The Scriptures, however, teach that God is not thus bound; that He operates through second causes, with them, or without them, as He sees fit. It is a purely arbitrary assumption, that when Christ raised the dead, healed the lepers, or gave sight to the blind, any second cause intervened between the effect and the efficiency of his will. What physical law, or uniformly acting force, operated to make the axe float at the command of the prophet? Or, in that greatest of all miracles, the original creation of the world.

MR. DARWIN'S THEORY.

We have not forgotten Mr. Darwin. It seemed desirable, in order to understand his theory, to see its relation to other theories of the universe and its phenomena, with which it is more or less connected. His work on the "Origin of

Species" does not purport to be philosophical. In this aspect it is very different from the cognate works of Mr. Spencer. Darwin does not speculate on the origin of the universe, on the nature of matter, or of force. He is simply a naturalist, a careful and laborious observer; skillful in his descriptions, and singularly candid in dealing with the difficulties in the way of his peculiar doctrine. He set before himself a single problem—namely, how are the fauna and flora of our earth to be accounted for? In the solution of this problem, he assumes:

1. The existence of matter, although he says little on the subject. Its existence however, as a real entity, is everywhere taken for granted.
2. He assumes the efficiency of physical causes, showing no disposition to resolve them into mind-force, or into the efficiency of the First Cause.
3. He assumes also the existence of life in the form of one or more primordial germs. He does not adopt the theory of spontaneous generation. What life is he does not attempt to explain, further than to quote (p. 326), with approbation, the definition of Herbert Spencer, who says, "Life depends on, or consists in, the incessant action and reaction of various forces," which conveys no very definite idea.
4. To account for the existence of matter and life, Mr. Darwin admits a Creator. This is done explicitly and repeatedly. Nothing, however, is said of the nature of the Creator and of his relation to the world, further than is implied in the meaning of the word.
5. From the primordial germ or germs (Mr. Darwin seems to have settled down to the assumption of only one primordial germ), all living organisms, vegetable and animal, including man, on our globe, through all the stages of its history, have descended.
6. As growth, organization, and reproduction are the functions of physical life, as soon as the primordial germ began to live, it began to grow, to fashion organs however simple, for its nourishment and increase, and for the reproduction, in some way, of living forms like itself. How all living things on earth, including the endless variety of plants, and all the diversity of animals—insects, fishes, birds, the ichthyosaurus, the mastodon, the mammoth, and man—have descended from the primordial animalcule, he thinks, may be accounted for by the operation of the following natural laws, viz:

First, the law of Heredity, or that by which like begets like. The offspring are like the parent.

Second, the law of Variation, that is, while the offspring are, in all essential characteristics, like their immediate progenitor, they nevertheless vary more or less within narrow limits, from their parent and from each other. Some of these variations are indifferent, some deteriorations, some improvements, that is, they are such as enable the plant or animal to exercise its functions to greater advantage.

Third, the law of Over Production. All plants and animals tend to increase in a geometrical ratio; and therefore tend to overrun enormously the means of support. If all the seeds of a plant, all the spawn of a fish, were to arrive at maturity, in a very short time the world could not contain them. Hence of necessity arises a struggle for life. Only a few of the myriads born can possibly live.

Fourth, here comes in the law of Natural Selection, or the Survival of the Fittest. That is, if any individual of a given species of plant or animal happens to have a slight deviation from the normal type, favorable to its success in the struggle for life, it will survive. This variation, by the law of heredity, will be transmitted to its offspring, and by them again to theirs. Soon these favored ones gain the ascendency, and the less favored perish; and the modification becomes established in the species. After a time another and another of such favorable variations occur, with like results. Thus very gradually, great changes of structure are introduced, and not only species, but genera, families, and orders in the vegetable and animal world, are produced. Mr. Darwin says he can set no limit to the changes of structure, habits, instincts, and intelligence, which these simple laws in the course of millions or milliards of centuries may bring into existence.

. . .

THE SENSE IN WHICH MR. DARWIN USES THE WORD "NATURAL."

We have not yet reached the heart of Mr. Darwin's theory. The main idea of his system lies in the word "natural." He uses that word in two senses: first, as antithetical to the word artificial. Men can produce very marked varieties as to structure and habits of animals. This is exemplified in the production of the different breeds of horses, cattle, sheep, and dogs; and specially, as Mr. Darwin seems to think, in the case of pigeons. Of these, he says, "The diversity of breeds is something astonishing." Some have long, and some very short bills; some have large feet, some small; some long necks, others long wings and tails, while others have singularly short tails; some have thirty, and even forty, tail-feathers, instead of the normal number of twelve or fourteen. They differ as much in instinct as they do in form. Some are carriers, some pouters, some tumblers, some trumpeters; and yet all are descendants of the Rock Pigeon, which is still extant. If, then, he argues, man, in a comparatively short time, has by artificial selection produced all these varieties, what might be accomplished on the boundless scale of nature, during the measureless ages of the geologic periods.

Secondly, he uses the word natural as antithetical to supernatural. Natural selection is a selection made by natural laws, working without intention and design. It is, therefore, opposed not only to artificial selection, which is made by the wisdom and skill of man to accomplish a given purpose, but also to supernatural selection, which means either a selection originally intended by a power higher than nature; or which is carried out by such power. In using the expression natural selection, Mr. Darwin intends to exclude design, or final causes. All the changes in structure, instinct, or intelligence, in the plants or animals, including man, descended from the primordial germ, or animalcule, have been brought about by unintelligent physical causes. On this point he leaves us in no doubt. He defines nature to be "the aggregate action and product of natural laws; and laws are the sequence of events as ascertained by us." It had been objected that he often uses teleological language, speaking of purpose, intention, contrivance, adaptation, etc. In answer to this objection, he says: "It has been said, that I speak of natural selection as a power or deity; but who objects to an author speaking of the attraction of gravity as ruling the movements of the planet?" He admits that in the literal sense of the words, natural selection is a false term; but "who ever objected to chemists, speaking of the elective affinities of various elements? And yet an acid cannot strictly be said to elect the base with which it in preference combines" (p. 93). We have here an affirmation and a negation. It is affirmed that natural selection is the operation of natural laws, analogous to the action of gravitation and of chemical affinities. It is denied that it is a process originally designed, or guided by intelligence, such as the activity which foresees an end and consciously selects and controls the means of its accomplishment. Artificial selection, then, is an intelligent process; natural selection is not.

There are in the animal and vegetable worlds innumerable instances of at least apparent contrivance, which have excited the admiration of men in all ages. There are three ways of accounting for them. The first is the Scriptural doctrine, namely, that God is a spirit, a personal, self-conscious, intelligent agent; that He is infinite, eternal, and unchangeable in his being and perfections; that He is ever present; that this presence is a presence of knowledge and power. In the external world there is always and everywhere indisputable evidence of the activity of two kinds of force: the one physical, the other mental. The physical belongs to matter, and is due to the properties with which it has been endowed; the other is the everywhere present and ever acting mind of God. To the latter are to be referred all the manifestations of design in nature, and the ordering of events in Providence. This doctrine does not ignore the efficiency of second causes; it simply asserts that God overrules and controls them. Thus the Psalmist says, "I am fearfully and wonderfully made.... My substance was not hid from thee, when I was made in secret, and curiously wrought (or embroidered) in the lower parts of the earth. Thine eyes did see my substance yet being imperfect; and in thy book all my members were written, which in continuance were fashioned, when as yet there were none of them." "He who fashioned the eye, shall not He see? He that formed the ear, shall not He hear? God makes the grass to grow, and herbs for the children of men." He sends rain, frost, and snow. He controls the winds and the waves. He determines the casting of the lot, the flight of an arrow, and the falling of a sparrow. This universal and constant control of God is not only one of the most patent and pervading doctrines of the Bible, but it is one of the fundamental principles of even natural religion.

The second method of accounting for contrivances in nature admits that they were foreseen and purposed by God, and that He endowed matter with forces which He foresaw and intended should produce such results. But here his agency stops. He never interferes to guide the operation of physical causes. He does nothing to control the course of nature, or the events of history. On this theory it may be said, (1) That it is utterly inconsistent with the Scriptures. (2) It does not meet the religious and moral necessities of our nature. It renders prayer irrational and inoperative. It makes it vain for a man in any emergency to look to God for help. (3) It is inconsistent with obvious facts. We see around us innumerable evidences of the constant activity of mind. This evidence of mind and of its operations, according to Lord Brougham and Dr. Whewell, is far more clear than that of the existence of matter and of its forces. If one or the other is to be denied, it is the latter rather than the former. Paley indeed says that if the construction of a watch be an undeniable evidence of design it would be a still more wonderful manifestation of skill, if a watch could be made to produce other watches; and, it may be added, not only other watches, but all kinds of time-pieces in endless variety. So it has been asked, if man can make a telescope, why cannot God make a telescope which produces others like itself? This is simply asking, whether matter can be made to do the work of mind? The idea involves a contradiction. For a telescope to make a telescope, supposes it to select copper and zinc in due proportions and fuse them into brass; to fashion that brass into inter-entering tubes; to collect and combine the requisite materials for the different kinds of glass needed; to melt them, grind, fashion, and polish them; adjust their densities and focal distances, etc., etc. A man who can believe that brass can do all this, might as well believe in God. The most credulous men in the world are unbelievers. The great Napoleon could not believe in Providence; but he believed in his star, and in lucky and unlucky days.

This banishing God from the world is simply intolerable, and, blessed be his name, impossible. An absent God who does nothing is, to us, no God. Christ brings God constantly near to us. He said to his disciples, "Consider the ravens, for they neither sow nor reap; which have neither store-house nor barn; and God feedeth them; how much better are ye than the fowls. And which of you by taking thought can add to his stature one cubit? Consider the lilies how they grow; they toil not, neither do they spin; and yet I say unto you that Solomon in all his glory was not arrayed like one of these. If then God so clothe the grass, which is to-day in the field, and to-morrow is cast into the oven; how much more will He clothe you, O ye of little faith." "And seek ye not what ye shall eat, or what ye shall drink, neither be ye of doubtful mind. For all these things do the nations of the world seek after; and your Father knoweth that ye have need of these things." It may be said that Christ did not teach science. True, but He taught truth; and science, so called, when it comes in conflict with truth, is what man is when he comes in conflict with God.

The advocates of these extreme opinions protest against being considered irreligious. Herbert Spencer says that his doctrine of an inscrutable, unintelligent, unknown force, as the cause of all things, is a much more religious doctrine than that of a personal, intelligent, and voluntary Being of infinite power and goodness. Matthew Arnold holds that an unconscious "power which makes for right," is a higher idea of God than the Jehovah of the Bible. Christ says, God is a Spirit. Holbach thought that he made a great advance on that definition, when he said, God is motion.

The third method of accounting for the contrivances manifested in the organs of plants and animals, is that which refers them to the blind operation of natural causes. They are not due to the continued cooperation and control of the divine mind, nor to the original purpose of God in the constitution of the universe. This is the doctrine of the Materialists, and to this doctrine, we are sorry to say, Mr. Darwin, although himself a theist, has given in his adhesion. It is on this account the Materialists almost deify him.

...

RELATION OF DARWINISM TO RELIGION.

The consideration of that subject would lead into the wide field of the relation between science and religion. Into that field we lack competency and time to enter; a few remarks, however, on the subject may not be out of place. Those remarks we would fain make in a humble way irenical. There is need of an Irenicum, for the fact is painfully notorious that there is an antagonism between scientific men as a class, and religious men as a class. Of course this opposition is neither felt nor expressed by all on either side. Nevertheless, whatever may be the cause of this antagonism, or whoever are to be blamed for it, there can be no doubt that it exists and that it is an evil.

The first cause of the alienation in question is, that the two parties, so to speak, adopt different rules of evidence, and thus can hardly avoid arriving at different conclusions. To understand this we must determine what is meant by science, and by scientific evidence. Science, according to its etymology, is simply knowledge. But usage has limited its meaning, in the first place, not to the knowledge of facts or phenomena, merely, but to their causes and relations. It was said of old, "[Greek: ὅτι] *scientiae fundamentum,* [Greek: διότι] *fastigium.*" No amount of materials would constitute a building. They must be duly arranged so as to make a symmetrical whole. No amount of disconnected data can constitute a science. Those data must be systematized in their relation to each other and to other things. In the second place, the word is becoming more and more restricted to the knowledge of a particular class of facts, and of their relations, namely, the facts of nature or of the external world. This usage is not universal, nor is it fixed. In Germany, especially, the word "*Wissenschaft*" is used of all kinds of ordered knowledge, whether transcendental or empirical. So we are accustomed to speak of mental, moral, social, as well as of natural science. Nevertheless, the more restricted use of the word is very common and very influential. It is important that this fact should be recognized. In common usage, a scientific man is distinguished specially from a metaphysician. The one investigates the phenomena of matter, the other studies the phenomena of mind, according to the old distinction between physics and metaphysics. Science, therefore, is the ordered knowledge of the phenomena which we recognize through the senses. A scientific fact is a fact perceived by the senses. Scientific evidence is evidence addressed to the senses. At one of the meetings of the Victoria Institute, a visitor avowed his disbelief in the existence of God. When asked what kind of evidence would satisfy him, he answered, "Just such evidence as I have of the existence of this tumbler which I now hold in my hand." The Rev. Mr. Henslow says, "By science is meant the investigation of facts and phenomena recognizable by the senses, and of the causes which have brought them into existence."[5] This is the main root of the trouble. If science be the knowledge of the facts perceived by the senses, and scientific evidence, evidence addressed to the senses, then the senses are the only sources of knowledge. Any conviction resting on any other ground than the testimony of the senses must be faith. Darwin admits that the contrivances in nature may be accounted for by assuming that they are due to design on the part of God. But, he says, that would not be science. Haeckel says that to science matter is eternal. If any man chooses to say, it was created, well and good; but that is a matter of faith, and faith is imagination. Ulrici quotes a distinguished German physiologist who believes in vital, as distinguished from physical forces; but he holds to spontaneous generation, not, as he admits, because it has been proved, but because the admission of any higher power than nature is unscientific.[6]

It is inevitable that minds addicted to scientific investigation should receive a strong bias to undervalue any other kind of evidence except that of the senses, i.e., scientific evidence. We have seen that those who give themselves up to this tendency come to deny God, to deny mind, to deny even self. It is true that the great majority of men, scientific as well as others, are so much under the control of the laws of their nature, that they cannot go to this extreme. The tendency, however, of a mind addicted to the consideration of one kind of evidence, to become more or less insensible to other kinds of proof, is undeniable. Thus even Agassiz, as a zoologist and simply on zoological grounds, assumed that there were several zones between the Ganges and the Atlantic Ocean, each having its own flora and fauna, and inhabited by races of men, the same in kind, but of different origins. When told by the comparative philologists that this was impossible, because the languages spoken through that wide region, demonstrated that its inhabitants must have had a common descent, he could only answer that as ducks quack everywhere, he could not see why men should not everywhere speak the same language.

A still more striking illustration is furnished by Dr. Lionel Beale, the distinguished English physiologist. He has written a book of three hundred and eighty-eight pages for the express purpose of proving that the phenomena of life, instinct, and intellect cannot be referred to any known natural forces. He avows his belief that in nature "mind governs matter," and "in the existence of a never-changing, all-seeing, power-directing and matter-guiding Omnipotence." He avows his faith in miracles, and "those miracles on which Christianity is founded." Nevertheless, his faith in all these points is provisional. He says that a truly scientific man, "if the maintenance, continuity, and nature of life on our planet should at some future time be fully explained without supposing the existence of any such supernatural omnipotent influence, would be bound to receive the new explanation, and might abandon the old conviction."[7] That is, all evidence of the truths of religion not founded on nature and perceived by the senses, amounts to nothing.

Now as religion does not rest on the testimony of the senses, that is on scientific evidence, the tendency of scientific men is to ignore its claims. We speak only of tendency. We rejoice to know or believe that in hundreds or thousands of scientific men, this tendency is counteracted by their consciousness of manhood—the conviction that the body is not the man—by the intuitions of the reason and the conscience, and by the grace of God. No class of men stands deservedly higher in public estimation than men of science, who, while remaining faithful to their higher nature, have enlarged our knowledge of the wonderful works of God.

A second cause of the alienation between science and religion, is the failure to make the due distinction between facts and the explanation of those facts, or the theories deduced from them. No sound minded man disputes any scientific fact. Religious men believe with Agassiz that facts are sacred. They are revelations from God. Christians sacrifice to them, when duly authenticated, their most cherished convictions. That the earth moves, no religious man doubts. When Galileo made that great discovery, the Church was right in not yielding at once to the evidence of an experiment which it did not understand. But when the fact was clearly established, no man sets up his interpretation of the Bible in opposition to it. Religious men admit all the facts connected with our solar system; all the facts of geology, and of comparative anatomy, and of biology. Ought not this to satisfy scientific men? Must we also admit their explanations and inferences? If we admit that the human embryo passes through various phases, must we admit that man was once a fish, then a bird, then a dog, then an ape, and finally what he now is? If we admit the similarity of structure in all vertebrates, must we admit the evolution of one from another, and all from a primordial germ? It is to be remembered that the facts are from God, the explanation from men; and the two are often as far apart as Heaven and its antipode.

These human explanations are not only without authority, but they are very mutable. They change not only from generation to generation, but almost as often as the phases of the moon. It is a fact that the planets move. Once it was said that they were moved by spirits, then by vortexes, now by self-evolved forces. It is hard that we should be called upon to change our faith with every new moon. The same man sometimes propounds theories almost as rapidly as the changes of the kaleidoscope. The amiable Sir Charles Lyell, England's most distinguished geologist, has published ten editions of his *Principles of Geology*, which so differ as to make it hard to believe that it is the work of the same mind. "In all the editions up to the tenth, he looked upon geological facts and geological phenomena as proving the fixity of species and their special creation in time. In the tenth edition, just published, he announces his change of opinion on this subject and his conversion to the doctrine of development by law."[8] "In the eighth edition of his work," says Dr. Bree, "Sir Charles Lyell, the Nestor of geologists, to whom the present generation is more indebted than to any other for all that is known of geology in its advanced stage, teaches that species have a real existence in nature, and that each was endowed at the time of its creation with the attributes and organization by which it is now distinguished." The change on the part of this eminent geologist, it is to be observed, is a mere change of opinion. There was no change of the facts of geology between the publication of the eighth and of the tenth edition of his work, neither was there any change in his knowledge of those facts. All the facts relied upon by evolutionists have long been familiar to scientific men. The whole change is a subjective one. One year the veteran geologist thinks the facts teach one thing, another year he thinks they teach another. It is now the fact, and it is feared it will continue to be a fact, that scientific men give the name of science to their explanations as well as to the facts. Nay, they are often, and naturally, more zealous for their explanations than they are for the facts. The facts are God's; the explanations are their own.

The third cause of the alienation between religion and science is the bearing of scientific men towards the men of culture who do not belong to their own class. When we, in such connections, speak of scientific men, we do not mean men of science as such, but those only who avow or manifest their hostility to religion. There is an assumption of superiority, and often a manifestation of contempt. Those who call their logic or their conjectures into question, are stigmatized as narrow-minded, bigots, old women, Bible worshippers, etc.

Professor Huxley's advice to metaphysicians and theologians is to let science alone. This is his Irenicum. But do he and his associates let metaphysics and religion alone? They tell the metaphysician that his vocation is gone; there is no such thing as mind, and of course no mental laws to be established. Metaphysics are merged into physics. Professor Huxley tells the religious world that there is over-whelming and crushing evidence (scientific evidence, of course) that no event has ever occurred on this earth which was not the effect of natural causes. Hence there have been no miracles, and Christ is not risen.[9] He says that the doctrine that belief in a personal God is necessary to any religion worthy of the name, is a mere matter of opinion. Tyndall, Carpenter, and Henry Thompson teach that prayer is a superstitious absurdity; Herbert Spencer, whom they call their "great philosopher," i.e., the man who does their thinking, labors to prove that there cannot be a personal God, or human soul or self; that moral laws are mere "generalizations of utility," or, as Carl Vogt says, that self-respect, and not the will of God, is the ground and rule of moral obligation. If any protest be made against such doctrines, we are told that scientific truth cannot be put down by denunciation (or as Vogt says, by barking). So doubtless the Pharisees, when our blessed Lord called them hypocrites and a generation of vipers, and said: "Ye compass sea and land to make one proselyte; and when he is made, ye make him twofold more the child of hell than yourselves," doubtless thought that that was a poor way to refute their theory, that holiness and salvation were to be secured by church-membership and church-rites. Nevertheless, as those words were the words of Christ, they were a thunderbolt which reverberates through all time and space, and still makes Pharisees of every name and nation tremble. Huxley's Irenicum will not do. Men who are assiduously poisoning the fountains of religion, morality, and social order, cannot be let alone.

Haeckel's Irenicum amounts to much the same as that of Professor Huxley. He forbids the right to speak on these vital subjects, to all who are not thoroughly versed in biology, and who are not entirely emancipated from the trammels of their long cherished traditional beliefs.[10] This, as the whole context shows, means that a man in order to be entitled to be heard on the evolution theory, must be willing to renounce his faith not only in the Bible, but in God, in the soul, in a future life, and become a monistic materialist.[11]

It is very reasonable that scientific men, in common with lawyers and physicians and other professional men, should feel themselves entitled to be heard with special deference on subjects belonging to their respective departments. This deference no one is disposed to deny to men of science. But it is to be remembered that no department of human knowledge is isolated. One runs into and overlaps another. We have abundant evidence that the devotees of natural science are not willing to confine themselves to the department of nature, in the common sense of that word. They not only speculate, but dogmatize, on the highest questions of philosophy, morality, and religion. And further, admitting the special claims to deference on the part of scientific men, other men have their rights. They have the right to judge of the consistency of the assertions of men of science and of the logic of their reasoning. They have the right to set off the testimony of one or more experts against the testimony of others; and especially, they have the right to reject all speculations, hypotheses, and theories, which come in conflict with well-established truths. It is ground of profound gratitude to God that He has given to the human mind intuitions which are infallible, laws of belief which men cannot disregard any more than the laws of nature, and also convictions produced by the Spirit of God which no sophistry of man can weaken. These are barriers which no man can pass without plunging into the abyss of outer darkness.

If there be any truth in the preceding remarks, then it is obvious that there can be no harmony between science and religion until the evils referred to be removed. Scientific men must come to recognize practically, and not merely in words, that there are other kinds of evidence of truth than the testimony of the senses. They must come to give due weight to the testimony of consciousness, and to the intuitions of the reason and conscience. They must cease to require the deference due to established facts to be paid to their speculations and explanations. And they must treat their fellow men with due respect. The Pharisees said to the man whose sight had been restored by Christ, "Thou wast altogether born in sin, and dost thou teach us!" Men of science must not speak thus. They must not say to every objector, Thou art not scientific, and therefore hast no right to speak. The true Irenicum is for all parties to give due heed to such words as these, "If any man would be wise, let him become a fool, that he may be wise"; or these, "Be converted, and become as little children"; or these, "The Spirit of Truth shall guide you in all truth." We are willing to hear this called cant. Nevertheless, these latter words fell from the lips of Him who spake as never man spake.

So much, and it is very little, on the general question of the relation of science to religion. But what is to be thought of the special relation of Mr. Darwin's theory to the truths of natural and revealed religion? We have already seen that Darwinism includes the three elements, evolution, natural selection, and the denial of design in nature. These points, however, cannot now be considered separately.

It is conceded that a man may be an evolutionist and yet not be an atheist and may admit of design in nature. But we cannot see how the theory of evolution can be reconciled with the declarations of the Scriptures. Others may see it, and be able to reconcile their allegiance to science with their allegiance to the Bible. Professor Huxley, as we have seen, pronounces the thing impossible. As all error is antagonistic to truth, if the evolution theory be false, it must be opposed to the truths of religion so far as the two come into contact. Mr. Henslow, indeed, says science and religion are not antagonistic because they are in different spheres of thought. This is often said by men who do not admit that there is any thought at all in religion; that it is merely a matter of feeling. The fact, however, is that religion is a system of knowledge, as well as a state of feeling. The truths on which all religion is founded are drawn within the domain of science, the nature of the first cause, its relation to the world, the nature of second causes, the origin of life, anthropology, including the origin, nature, and destiny of man. Religion has to fight for its life against a large class of scientific men. All attempts to prevent her exercising her right to be heard are unreasonable and vain.

...

...There is another consideration of decisive importance. Strauss says there are three things which have been stumbling-blocks in the way of science. First, the origin of life; second, the origin of consciousness; third, the origin of reason. These are equivalent to the gaps which, Principal Dawson says, exist in the theory of evolution. He states them thus: 1. That between dead and living matter. 2. That between vegetable and animal life. "These are necessarily the converse of each other: the one deoxidizes and accumulates, the other oxidizes and expends." 3. That "between any species of plant or animal, and any other species. It was this gap, and this only, which Darwin undertook to fill up by his great work on the origin of species, but, notwithstanding the immense amount of material thus expended, it yawns as wide as ever, since it must be admitted that no case has been ascertained in which an individual of one species has transgressed the limits between it and another species." 4.

"Another gap is between the nature of the animal and the self-conscious, reasoning, and moral nature of man" (pp. 325–28).

First, as to the gap between death and life; this is what Dr. Stirling calls the "gulf of all gulfs, which Mr. Huxley's protoplasm is as powerless to efface as any other material expedient that has ever been suggested."[12] This gulf Mr. Darwin does not attempt to bridge over. He admits that life owes its origin to the act of the Creator. This, however, the most prominent of the advocates of Darwinism say, is giving up the whole controversy. If you admit the intervention of creative power at one point, you may as well admit it in any other. If life owes its origin to creative power, why not species? If the stupendous miracle of creation be admitted, there is no show of reason for denying supernatural intervention in the operations of nature. Most Darwinians attempt to pass this gulf on the imaginary bridge of spontaneous generation. In other words, they say there is no gulf there. The molecules of matter, in one combination, may as well exhibit the phenomena of life, as in other combinations, any other kind of phenomena. The distinguished Sir William Thomson cannot trust himself to that bridge. "Dead matter," he says, "cannot become living matter without coming under the influence of matter previously alive. This seems to me as sure a teaching of science as the law of gravitation. . . . I am ready to adopt, as an article of scientific faith, true through all space and through all time, that life proceeds from life, and nothing but life."[13] He refers the origin of life on this earth to falling meteors, which bring with them from other planets the germs of living organisms; and from those germs all the plants and animals with which our world is now covered have been derived. Principal Dawson thinks that this was intended as irony. But the whole tone of the address, and specially of the closing portion of it, in which this idea is advanced, is far too serious to admit of such an explanation.

No one can read the address referred to without being impressed, and even awed, by the immensity and grandeur of the field of knowledge which falls legitimately within the domain of science. The perusal of that discourse produces a feeling of humility analogous to the sense of insignificance which every man experiences when he thinks of himself as a speck on the surface of the earth, which itself is but a speck in the immensity of the universe. And when a man of mere ordinary culture sees Sir William Thomson surveying that field with a mastery of its details and familiarity with all the recondite methods of its investigation, he feels as nothing in his presence. Yet this great man, whom we cannot help regarding with wonder, is so carried away by the spirit of his class as to say, "Science is bound, by the everlasting law of honor, to face fearlessly every problem which can fairly be brought before it. If a probable solution, consistent with the ordinary course of nature, can be found, we must not invoke an abnormal act of Creative Power." And, therefore, instead of invoking Creative Power, he accounts for the origin of life on earth by falling meteors. How he accounts for its origin in the places whence the meteors came, he does not say. Yet Sir William Thomson believes in Creative Power; and in a subsequent page, we shall quote his explicit repudiation of the atheistic element in the Darwinian theory.

Strauss quotes Dubois-Reymond, a distinguished naturalist, as teaching that the first of these great problems, viz. the origin of life, admits of explanation on scientific (i.e., in his sense, materialistic) principles; and even the third, viz. the origin of reason; but the second, or the origin of consciousness, he says, "is perfectly inscrutable." Dubois-Reymond holds that "the most accurate knowledge of the essential organism reveals to us only matter in motion; but between this material movement and my feeling pain or pleasure, experiencing a sweet taste, seeing red, with the conclusion 'therefore I exist,' there is a profound gulf; and it 'remains utterly and forever inconceivable why to a number of atoms of carbon, hydrogen, etc., it should not be a matter of indifference how they lie or how they move; nor, can we in any wise tell how consciousness should result from their concurrent action.' Whether," adds Strauss, "these *Verba Magistri* are indeed the last word on the subject, time only can tell."[14] But if it is inconceivable, not to say absurd, that sense-consciousness should consist in the motion of molecules of matter, or be a function of such molecules, it can hardly be less absurd to account for thought, conscience, and religious feeling and belief on any such hypothesis. It may be said that Mr. Darwin is not responsible for these extreme opinions. That is very true. Mr. Darwin is not a Monist, for in admitting creation, he admits a dualism as between God and the world. Neither is he a Materialist, inasmuch as he assumes a supernatural origin for the infinitesimal modicum of life and intelligence in the primordial animalcule, from which without divine purpose or agency, all living things in the whole history of our earth have descended. All the innumerable varieties of plants, all the countless forms of animals, with all their instincts and faculties, all the varieties of men with their intellectual endowments, and their moral and religious nature, have, according to Darwin, been evolved by the agency of the blind, unconscious laws of nature. This infinitesimal spark of supernaturalism in Mr. Darwin's theory would inevitably have gone out of itself, had it not been rudely and contemptuously trodden out by his bolder, and more logical successors.

The grand and fatal objection to Darwinism is this exclusion of design in the origin of species, or the production of living organisms. By design is meant the intelligent and voluntary selection of an end, and the intelligent and voluntary choice, application, and control of means appropriate to the accomplishment of that end. That design, therefore, implies intelligence, is involved in its very nature. No man can perceive this adaptation of means to the accomplishment of a preconceived end, without experiencing an irresistible conviction that it is the work of mind. No man does doubt it, and no man can doubt it. Darwin does not deny it. Haeckel does not deny it. No Darwinian denies it. What they do is to deny that there is any design in nature. It is merely apparent, as when the wind of the Bay of Biscay, as Huxley says, "selects the right kind of sand and spreads it in heaps upon the plains." But in thus denying design in nature, these writers array against themselves the intuitive perceptions and irresistible convictions of all mankind—a barrier which no man has ever been able to surmount. Sir William Thomson, in the address already referred to, says:

> I feel profoundly convinced that the argument of design has been greatly too much lost sight of in recent zoological speculations. Reaction against the frivolities of teleology, such as are to be found, not rarely, in the notes of the learned commentators on Paley's *Natural Theology*, has, I believe, had a temporary effect of turning attention from the solid irrefragable argument so well put forward in that excellent old book. But overpowering proof of intelligence and benevolent design lie all around us, and if ever perplexities, whether metaphysical or scientific, turn us away from them for a time, they come back upon us with irresistible force, showing to us through nature the influence of a free will, and teaching us that all living beings depend upon one ever-acting Creator and Ruler.

NOTES

1. Thomas H. Huxley, *Evidences of Man's Place in Nature* (London: 1864), p. 57.

2. The two facts which are commonly urged as inconsistent with theism are the existence of misery in the world and the occurrence of undeveloped or useless organs, as teeth in the jaws of the whale and mammae on the breast of a man. As to the former objection, sin, which is the only real evil, is accounted for by the voluntary apostasy of man; and as to undeveloped organs they are regarded as evidences of the great plan of structure which can be traced in the different orders of animals. These unused organs were—says Professor Joseph Le Conte, in his interesting volume *Religion and Science* (New York: 1874), p. 54—regarded as blunders in nature, until it was discovered that use is not the only end of design. "By further patient study of nature," he says, "came the recognition of another law beside use—a law of order underlying and conditioning the law of use. Organisms are, indeed, contrived for use, but according to a preordained plan of structure, which must not be violated." It is of little moment whether this explanation be considered satisfactory or not. It would certainly be irrational to refuse to believe that the eye was made for the purpose of vision, because we cannot tell why a man has mammae. A man might as well refuse to admit that there is any meaning in all the writings of Plato, because there is a sentence in them which he cannot understand.

3. Alfred Russel Wallace, *The Theory of Natural Selection* (London: 1870), p. 368.

4. The Duke of Argyle, *Reign of Law*, fifth edition (London: 1867), p. 123.

5. Rev. George Henslow, "Science and Scripture not Antagonistic, because Distinct in their Spheres of Thought" (lecture, London, 1873).

6. Gott und Natur, p. 200.

7. Lionel S. Beale, *Protoplasm; or, Matter and Life*, third edition (London and Philadelphia: 1874), p. 345, and the whole chapter on Design.

8. C. R. Bree, *Fallacies in the Hypothesis of Mr. Darwin* (London: 1872), p. 290.

9. When Professor Huxley says, as quoted above, that he does not deny the possibility of miracles, he must use the word miracle in a sense peculiar to himself.

10. Jenaer Literaturzeitung, January 3, 1874. In this number there is a notice by Doctor Haeckel of two books: Oscar Schmidt, *Descendenzlehre und Darwinismus* (Leipzig: 1873); and J. W. Spengel, *Die Fortschritte des Darwinismus* (Cöln and Leipzig: 1874), in which he says: "Erstens, um in Sachen der Descendenz-Theorie mitreden zu können, ein gewisser Grad von tieferer biologischer (sowohl morphologischer als physiologischer) Bildung unentbehrlich, den die meistzen von jenen Auctoren (the opposers of the theory) nicht besitzen. Zweitens ist für ein klares und zutreffendes

Urtheil in diesem Sachen eine rücksichtslose Hingabe an vernunftgemässe Erkenntniss und eine dadurch bedingte Resignation auf uralte, liebgewordene und tief vererbte Vorurtheile erforderlich, zu welcher sich die wenigsten entschliesen können."

11. In his *Natürlische Schöpfungsgeschichte*, Haeckel is still more exclusive. When he comes to answer the objections to the evolution, or, as he commonly calls it, the descendence theory, he dismisses the objections derived from religion, as unworthy of notice, with the remark that all Glaube ist Aberglaube; all faith is superstition. The objections from a priori, or intuitive truths, are disposed of in an equally summary manner, by denying that there are any such truths, and asserting that all our knowledge is from the senses. The objection that so many distinguished naturalists reject the theory, he considers more at length. First, many have grown old in another way of thinking and cannot be expected to change. Second, many are collectors of facts, without studying their relations, or are destitute of the genius for generalization. No amount of material makes a building. Others, again, are specialists. It is not enough that a man should be versed in one department; he must be at home in all: in Botany, Zoology, Comparative Anatomy, Biology, Geology, and Paleontology. He must be able to survey the whole field. Fourthly, and mainly, naturalists are generally lamentably deficient in philosophical culture and in a philosophical spirit. "The immovable edifice of the true, monistic science, or what is the same thing, natural science, can only arise through the most intimate interaction and mutual interpenetration of philosophy and observation (Philosophie und Empirie)," pp. 638–41. It is only a select few, therefore, of learned and philosophical monistic materialists, who are entitled to be heard on questions of the highest moment to every individual man, and to human society.

12. James H. Stirling, *As Regards Protoplasm in Relation to Professor Huxley's Essay on the Physical Basis of Life* [Scholarly Publishing Office, University of Michigan Library, 2005]. See, also, L. S. Beale, *Physiological Anatomy and Physiology of Man*; also, L. S. Beale, *The Mystery of Life in Reply to Dr. Gull's Attack on the Theory of Vitality* (1871).

13. The address delivered by Sir William Thomson, as President of the British Association at its meeting in Edinburgh, 1871.

14. The Old Faith and the New. Prefatory Postscript, p. xxi.

CHAPTER 7

DARWINISM AS A METAPHYSICAL RESEARCH PROGRAM

KARL POPPER

I have always been extremely interested in the theory of evolution, and very ready to accept evolution as a fact. I have also been fascinated by Darwin as well as by Darwinism—though somewhat unimpressed by most of the evolutionary philosophers; with the one great exception, that is, of Samuel Butler.[1]

My *Logik der Forschung* contained a theory of the growth of knowledge by trial and error-elimination, that is, by Darwinian selection rather than Lamarckian instruction; this point (at which I hinted in that book) increased, of course, my interest in the theory of evolution. Some of the things I shall have to say spring from an attempt to utilize my methodology and its resemblance to Darwinism to throw light on Darwin's theory of evolution.

The Poverty of Historicism[2] contains my first brief attempt to deal with some epistemological questions connected with the theory of evolution. I continued to work on such problems, and I was greatly encouraged when I later found that I had come to results very similar to some of Schrödinger's.[3]

In 1961 I gave the Herbert Spencer Memorial Lecture in Oxford, under

---

From Karl Popper, "Darwinism as a Metaphysical Research Program," in *Unended Quest* (LaSalle, IL: Open Court, 1976), pp. 167–79, 234–35. Reprinted by permission of the author's estate.

the title "Evolution and the Tree of Knowledge."[4] In this lecture I went, I believe, a little beyond Schrödinger's ideas; and I have since developed further what I regard as a slight improvement on Darwinian theory,[5] while keeping strictly within the bounds of Darwinism as opposed to Lamarckism—within natural selection, as opposed to instruction.

I tried also in my Compton lecture (1966)[6] to clarify several connected questions; for example, the question of the *scientific status* of Darwinism. It seems to me that Darwinism stands in just the same relation to Lamarckism as does:

Deductivism	to	*Inductivism*
Selection	to	*Instruction by Reception*
Critical Error Elimination	to	*Justification*

The logical untenability of the ideas on the right-hand side of this table establishes a kind of logical explanation of Darwinism (i.e., of the left-hand side). Thus it could be described as "almost tautological"; or it could be described as applied logic—at any rate, as applied *situational logic* (as we shall see).

From this point of view the question of the scientific status of Darwinian theory—in the widest sense, the theory of trial and error-elimination—becomes an interesting one. I have come to the conclusion that Darwinism is not a testable scientific theory, but a *metaphysical research program*—a possible framework for testable scientific theories.[7]

Yet there is more to it: I also regard Darwinism as an application of what I call "situational logic." Darwinism as situational logic can be understood as follows.

Let there be a world, a framework of limited constancy, in which there are entities of limited variability. Then some of the entities produced by variation (those which "fit" into the conditions of the framework) may "survive," while others (those which clash with the conditions) may be eliminated.

Add to this the assumption of the existence of a special framework—a set of perhaps rare and highly individual conditions—in which there can be life or, more especially, self-reproducing but nevertheless variable bodies. Then a situation is given in which the idea of trial and error-elimination, or of Darwinism, becomes not merely applicable, but almost logically necessary. This does not mean that either the framework or the origin of life is necessary. There may be a framework in which life would be possible, but in which the trial which leads to life has not occurred, or in which all those trials which led to life were eliminated. (The latter is not a mere possibility but may happen at any moment: there is more than one way in which all life on earth might be destroyed.) What is meant is that if a life-permitting situation occurs, and if life originates, then this total situation makes the Darwinian idea one of situational logic.

To avoid any misunderstanding: it is not in every possible situation that Darwinian theory would be successful; rather, it is a very special, perhaps even a unique situation. But even in a situation without life Darwinian selection can apply to some extent: atomic nuclei which are relatively stable (in the situation in question) will tend to be more abundant than unstable ones; and the same may hold for chemical compounds.

I do not think that Darwinism can explain the origin of life. I think it quite possible that life is so extremely improbable that nothing can "explain" why it originated; for statistical explanation must operate, *in the last instance*, with very high probabilities. But if our high probabilities are merely low probabilities which have become high because of the immensity of the available time (as in Boltzmann's "explanation"), then we must not forget that in this way it is possible to "explain" almost everything.[8] Even so, we have little enough reason to conjecture that any explanation of this sort is applicable to the origin of life. But this does not affect the view of Darwinism as situational logic, once life and its framework are assumed to constitute our "situation."

I think that there is more to say for Darwinism than that it is just one metaphysical research program among others. Indeed, its close resemblance to situational logic may account for its great success, in spite of the almost tautological character inherent in the Darwinian formulation of it, and for the fact that so far no serious competitor has come forward.

Should the view of Darwinian theory as situational logic be acceptable, then we could explain the strange similarity between my theory of the growth of knowledge and Darwinism: both would be cases of situational logic. The new and special element in the *conscious scientific approach to knowledge*—conscious criticism of tentative conjectures, and a conscious building up of selection pressure on these conjectures (by criticizing them)—would be a consequence of the emergence of a descriptive and argumentative language; that is, of a descriptive language whose descriptions can be criticized.

The emergence of such a language would face us here again with a highly improbable and possibly unique situation, perhaps as improbable as life itself. But given this situation, the theory of the growth of exosomatic knowledge through a conscious procedure of conjecture and refutation follows "almost" logically: it becomes part of the situation as well as part of Darwinism.

As for Darwinian theory itself, I must now explain that I am using the term "Darwinism" for the modern forms of this theory, called by various names, such as "neo-Darwinism" or (by Julian Huxley) "The New Synthesis." It consists essentially of the following assumptions or conjectures, to which I will refer later.

(1) The great variety of the forms of life on earth originate from very few forms, perhaps even from a single organism: there is an evolutionary tree, an evolutionary history.

(2) There is an evolutionary theory which explains this. It consists in the main of the following hypotheses:

(a) Heredity: The offspring reproduce the parent organisms fairly faithfully.

(b) Variation: There are (perhaps among others) "small" variations. The most important of these are the "accidental" and hereditary mutations.

(c) Natural selection: There are various mechanisms by which not only the variations but the whole hereditary material is controlled by elimination. Among them are mechanisms which allow only "small" mutations to spread; "big" mutations (hopeful monsters) are as a rule lethal, and thus eliminated.

(d) Variability: Although *variations* in some sense—the presence of different competitors—are for obvious reasons prior to selection, it may well be the case that variability—the scope of variation—is controlled by natural selection; for example, with respect to the frequency as well as the size of variations. A gene theory of heredity and variation may even admit special genes controlling the variability of other genes. Thus we may arrive at a hierarchy, or perhaps at even more complicated interaction structures. (We must not be afraid of complications; for they are known to be there. For example, from a selectionist point of view we are bound to assume that something like the genetic code method of controlling heredity is itself an early product of selection, and that it is a highly sophisticated product.)

Assumptions (1) and (2) are, I think, essential to Darwinism (together with some assumptions about a changing environment endowed with some regularities). The following point (3) is a reflection of mine on point (2).

(3) It will be seen that there is a close analogy between the "conservative" principles (a) and (d) and what I have called dogmatic thinking; and likewise between (b) and (c), and what I have called critical thinking.

I now wish to give some reasons why I regard Darwinism as metaphysical, and as a research program.

It is metaphysical because it is not testable. One might think that it is. It seems to assert that, if ever on some planet we find life which satisfies conditions (a) and (b), then (c) will come into play and bring about in time a rich variety of distinct forms. Darwinism, however, does not assert as much as this. For assume that we find life on Mars consisting of exactly three species of bacteria with a genetic outfit similar to that of three terrestrial species. Is Darwinism refuted? By no means. We shall say that these three species were the only forms among the many mutants which were sufficiently well adjusted to survive. And we shall say the same if there is only one species (or none). Thus Darwinism does not really predict the evolution of variety. It therefore cannot really explain it. At best, it can *predict* the evolution of variety under "favorable conditions." But it is hardly possible to describe in general terms what favorable conditions are—except that, in their presence, a variety of forms will emerge.

And yet I believe I have taken the theory almost at its best—almost in its most testable form. One might say that it "almost predicts" a great variety of forms of life.[9] At first sight natural selection appears to explain it, and in a way it does; but hardly in a scientific way. To say that a species now living is adapted to its environment is, in fact, almost tautological. Indeed we use the terms "adaptation" and "selection" in such a way that we can say that, if the species were not adapted, it would have been eliminated by natural selection.

Similarly, if a species has been eliminated it must have been ill adapted to the conditions. Adaptation or fitness is defined by modern evolutionists as survival value, and can be measured by actual success in survival: there is hardly any possibility of testing a theory as feeble as this.[10]

And yet, the theory is invaluable. I do not see how, without it, our knowledge could have grown as it has done since Darwin. In trying to explain experiments with bacteria which become adapted to, say, penicillin, it is quite clear that we are greatly helped by the theory of natural selection. Although it is metaphysical, it sheds much light upon very concrete and very practical researches. It allows us to study adaptation to a new environment (such as a penicillin-infested environment) in a rational way: it suggests the existence of a mechanism of adaptation, and it allows us even to study in detail the mechanism at work. And it is the only theory so far which does all that.

This is, of course, the reason why Darwinism has been almost universally accepted. Its theory of adaptation was the first nontheistic one that was convincing; and theism was worse than an open admission of failure, for it created the impression that an ultimate explanation had been reached.

Now to the degree that Darwinism creates the same impression, it is not so very much better than the theistic view of adaptation; it is therefore important to show that Darwinism is not a scientific theory, but metaphysical. But its value for science as a metaphysical research program is very great, especially if it is admitted that it may be criticized, and improved upon.

Let us now look a little more deeply into the research program of Darwinism, as formulated above under points (1) and (2).

First, though (2), that is, Darwin's theory of evolution, does not have sufficient explanatory power to *explain* the terrestrial evolution of a great variety of forms of life, it certainly suggests it, and thereby draws attention to it. And it certainly does predict that if such an evolution takes place, it will be *gradual.*

The nontrivial *prediction of gradualness* is important, and it follows immediately from (2)(a)–(2)(c); and (a) and (b) and at least the smallness of the mutations predicted by (c) are not only experimentally well supported, but known to us in great detail.

Gradualness is thus, from a logical point of view, the central prediction of the theory. (It seems to me that it is its only prediction.) Moreover, as long as changes in the genetic base of the living forms are gradual, they are—at least "in principle"—explained by the theory; for the theory does predict the occurrence of small changes, each due to mutation. However, "explanation in principle"[11] is something very different from the type of explanation which we demand in physics. While we can explain a particular eclipse by predicting it, we cannot predict or explain any particular evolutionary change (except perhaps certain changes in the gene population *within* one species); all we can say is that if it is not a small change, there must have been some intermediate steps—an important suggestion for research: a research program.

Moreover, the theory predicts *accidental* mutations, and thus *accidental* changes. If any "direction" is indicated by the theory, it is that throwback mutations will be comparatively frequent. Thus we should expect evolutionary sequences of the random-walk type. (A random walk is, for example, the track described by a man who at every step consults a roulette wheel to determine the direction of his next step.)...

...

Another suggestion concerning evolutionary theory which may be worth mentioning is connected with the idea of "survival value," and also with teleology. I think that these ideas may be made a lot clearer in terms of problem solving.

Every organism and every species is faced constantly by the threat of extinction; but this threat takes the form of concrete problems which it has to solve. Many of these concrete problems are not as such survival problems. The problem of finding a good nesting place may be a concrete problem for a pair of birds without being a survival problem for these birds, although it may turn into one for their offspring; and the species may be very little affected by the success of these particular birds in solving the problem here and now. Thus I conjecture that most problems are posed not so much by survival, but by *preferences*, especially *instinctive preferences*, and even if the instincts in question (p-genes) should have evolved under external selection pressure, the problems posed by them are not as a rule survival problems.

It is for reasons such as these that I think it is better to look upon organisms as problem-solving rather than as end-pursuing: as I have tried to show in "Of Clouds and Clocks,"[12] we may in this way give a rational account—"in principle," of course—of *emergent evolution.*

I conjecture that the origin of *life* and the origin of *problems* coincide. This is not irrelevant to the question whether we can expect biology to turn out to be reducible to chemistry and further to physics. I think it not only possible but likely that we shall one day be able to re-create living things from non-living ones. Although this would, of course, be extremely exciting in itself[13] (as well as from the reductionist point of view), it would not establish that biology can be "reduced" to physics or chemistry. For it would not establish a physical explanation of the emergence of problems—any more than our ability to produce chemical compounds by physical means establishes a physical theory of the chemical bond or even the existence of such a theory.

My position may thus be described as one that upholds a theory of *irreducibility and emergence*, and it can perhaps best be summarized in this way:

(1) I conjecture that there is no biological process which cannot be regarded as correlated in detail with a physical process or cannot be progressively analyzed in physicochemical terms. But no physicochemical theory can explain the emergence of a new problem, and no physicochemical process can as such solve a *problem.* (Variational principles in physics, like the principle of least action or Fermat's principle, are perhaps similar but they are not solutions to problems. Einstein's theistic method tries to use God for similar purposes.)

(2) If this conjecture is tenable it leads to a number of distinctions. We must distinguish from each other:

a physical problem = a physicist's problem;
a biological problem = a biologist's problem;
an organism's problem = a problem like: How am I to survive? How am I to propagate? How am I to change? How am I to adapt?
a man-made problem = a problem like: How do we control waste?

From these distinctions we are led to the following thesis: the problems of organisms are not physical; they are neither physical things, nor physical laws, nor physical facts. They are specific biological realities; they are "real" in the sense that their existence may be the cause of biological effects.

(3) Assume that certain physical bodies have "solved" their problem of reproduction: that they can reproduce themselves; either exactly, or, like crystals, with minor faults which may be chemically (or even functionally) inessential. Still, they might not be "living" (in the full sense) if they cannot adjust themselves: they need reproduction plus genuine variability to achieve this.

(4) The "essence" of the matter is, I propose, *problem solving*. (But we should not talk about "essence"; and the term is not used here seriously.) Life as we know it consists of physical "bodies" (more precisely, structures) which are problem solving. This the various species have "learned" by natural selection, that is to say by the method of reproduction plus variation, which itself has been learned by the same method. This regress is not necessarily infinite—indeed, it may go back to some fairly definite moment of emergence.

Thus men like Butler and Bergson, though I suppose utterly wrong in their theories, were right in their intuition. Vital force ("cunning") does, of course, exist—but it is in its turn a product of life, of *selection*, rather than anything like the "essence" of life. It is indeed the preferences *which lead the way*. Yet the way is not Lamarckian but Darwinian.

This emphasis on *preferences* (which, being dispositions, are not so very far removed from propensities) in my theory is, clearly, a purely "objective" affair: we *need not* assume that these preferences are conscious. But they *may* become conscious; at first, I conjecture, in the form of states of well-being and of suffering (pleasure and pain).

My approach, therefore, leads almost necessarily to a research program that asks for an explanation, in objective biological terms, of the emergence of states of consciousness.

NOTES

1. Samuel Butler has suffered many wrongs from the evolutionists, including a serious wrong from Charles Darwin himself who, though greatly upset by it, never put things right. They were put right, as far as possible, by Charles's son Francis, after

Butler's death. The story, which is a bit involved, deserves to be retold. See pp. 167–219 of Nora Barlow, ed., *The Autobiography of Charles Darwin* (London: Collins, 1958), especially p. 219, where references to most of the other relevant material will be found.

2. See [Karl Popper, "The Poverty of Historicism, III," *Economica* 12 (1945)]: section 27; [Karl Popper, *The Poverty of Historicism* (London: Routledge & Kegan Paul, and Boston, MA: Beacon Press, 1957)] and later editions, especially pp. 106–108.

3. I am alluding to Schrödinger's remarks on evolutionary theory in *Mind and Matter*, especially those indicated by his phrase "Feigned Lamarckism"; see E. Schrödinger, *Mind and Matter*, p. 26.

4. The lecture [Karl Popper, "Evolution and the Tree of Knowledge," (Herbert Spencer Lecture, Oxford, October 30, 1961)] was delivered on October 31, 1961, and the manuscript was deposited on the same day in the Bodleian Library. It now appears in a revised version with an addendum, as chapter 7 of Karl Popper, *Objective Knowledge: An Evolutionary Approach* (Oxford: Clarendon Press, 1972).

5. See [Karl Popper, *Of Clouds and Clocks: An Approach to the Problem of Rationality and the Freedom of Man* (St. Louis, MO: Washington University Press, 1966)]; now chapter 6 of [Popper, *Objective Knowledge: An Evolutionary Approach*].

6. See [Popper, *Of Clouds and Clocks: An Approach to the Problem of Rationality and the Freedom of Man*].

7. The term "metaphysical research programme" was used in my lectures from about 1949 on, if not earlier but it did not get into print until 1958, though it is the main topic of the last chapter of the postscript (in galley proofs since 1957). I made the postscript available to my colleagues, and Professor Lakatos acknowledges that what he calls "scientific research programmes" are in the tradition of what I described as "metaphysical research programmes" ("metaphysical" because nonfalsifiable). See p. 183 of his paper "Falsification and the Methodology of Scientific Research Programmes," in *Criticism and the Growth of Knowledge*, ed. Imrê Lakatos and Alan Musgrave (Cambridge: Cambridge University Press, 1970).

8. See L.Sc.D., section 67.

9. For the problem of "degrees of prediction," see F. A. Hayek, "Degrees of Explanation," first published in 1955 and now chap. 1 of his F. A. Hayek, *Studies in Philosophy. Politics and Economics* (London: Routledge & Kegan Paul, 1967), chap. 1; see especially ii. 4 on p. 9. For Darwinism and the production of "a great variety of structures," and for its irrefutability, see especially ibid., p. 32.

10. Darwin's theory of sexual selection is partly an attempt to explain falsifying instances of this theory; such things, for example, as the peacock's tail, or the stag's antlers.

11. For the problem of "explanation in principle" (or "of the principle") in con-

trast to "explanation in detail," see Hayek, *Philosophy. Politics and Economics*, chap. 1, especially section V1, pp. 11–14.

12. See [Popper, *Of Clouds and Clocks: An Approach to the Problem of Rationality and the Freedom of Man*], pp. 20–26, especially pp. 24f., point (II). Now [Popper, *Objective Knowledge: An Evolutionary Approach*], p. 244.

13. See [Karl Popper, "A Realist View of Logic, Physics, and History," in *Physics, Logic and History*, ed. Wolfgang Yourgrau and Allen D. Breck (New York and London: Plenum Press, 1970)], esp. pp. 5–10; [and Popper, *Objective Knowledge: An Evolutionary Approach*], pp. 289–95.

CHAPTER 8

KARL POPPER'S PHILOSOPHY OF BIOLOGY

MICHAEL RUSE

Although Sir Karl Popper has not yet given us a full-length philosophical treatment of evolutionary biology, enough of his general position has been sketched to make possible a preliminary evaluation. There are at least two reasons why such an evaluation seems worthwhile. First, a number of biologists are taking seriously Popper's views on science generally and biology in particular. (See, for example, Ayala et al. 1974; Monod 1975.) Secondly, Popper has labeled his overall epistemology "evolutionary," and has drawn a very strong analogy between what he takes to be biological evolution and the evolution of scientific knowledge (if indeed they are not for him part and parcel of the same thing). "The theory of knowledge which I wish to propose is a largely Darwinian theory of the growth of knowledge" (Popper 1975, 261). Hence, such an evaluation of Popper's work, one which concentrates primarily on his views on biology rather than on his wider position about the growth of knowledge, is the aim of this [chapter]. In the concluding section, however, I shall make some brief remarks about the implications of the biological discussion for Popper's more general position.

From Michael Ruse, "Karl Popper's Philosophy of Biology," *Philosophy of Science* 44, no. 4 (1977): 638–61. Reprinted by permission of the Philosophy of Science Association.

I

Popper's first detailed and significant comments about evolutionary biology are to be found in *The Poverty of Historicism* (1963). A primary aim of that work is to deny that there are any human laws of progress or destiny—"there can be no prediction of the course of human history by scientific or any other rational methods" (iv.)—and in the course of his argument Popper poses the question: "Is there a law of evolution?" Although, obviously, Popper's major concern is with evolutionary laws as applied to humans, he keeps his discussion sufficiently general that some of his views about biological evolution become apparent.

Broadly speaking, as might be expected from one concerned to deny laws of human destiny, Popper's position is that there are no such laws of evolution. "There are neither laws of succession, nor laws of evolution" (117). Moreover, Popper's reason for taking such a position is simple and clear. Laws require repeatability. However: "The evolution of life on earth, or of human society, is a unique historical process" (108). Hence, there can be no evolutionary laws.

At this point the scientist or philosopher concerned with the welfare of evolutionary biology may perhaps be undergoing simultaneous emotions of déjà vu and depression. If indeed there are no evolutionary laws, then, as the many critics of evolutionary biology who have made reference to the uniqueness argument have been happy to point out, this surely says little for and much against evolutionary biology as a science. (See, for example, Manser 1965, 31.) However, as equally many commentators have noted, such an approach to evolutionary biology is even more surely misguided. (See, for example, Ruse 1973, 89–91.) Although indeed the evolution of (say) elephants is unique, this is no more of a bar to evolutionary laws than is the uniqueness of Earth a bar to astronomical laws. The evolutionary biology sympathizer will point out that one must draw a distinction between the unique history of life on earth (involving "phylogenies") and the biologists' theory of evolution through natural selection: a theory which speaking generally argues in a law-like way that, given groups of organisms, one gets a differential reproduction which, combined with the constant injection of new variation (through random mutation), leads to a "selecting" and eventually to an evolution of forms.

Nevertheless, careful study of what Popper has to say shows that, were one to fault him for confusing unique phylogenies with general evolutionary claims, one would be doing him an injustice. His position is more subtle than a casual reading implies, and he is indeed making important distinctions and claims that, regretfully, have not always been made by biological commentators writing since he did. Indeed, even more regretfully, at least one such commentator has mistakenly cited Popper's authority in support of the above given rather crude argument to the nonlawlike nature of evolutionary claims from the uniqueness of evolutionary phenomena. (See Goudge 1961, 124.)

For a start, Popper is fully and explicitly aware of the distinction to be drawn between the unique history of the evolution of organisms and a theory, perhaps giving a mechanism, of evolutionary change. Moreover, although he replies in the negative to his question about a law of evolution, he makes it clear that he does not deny the possibility of laws being involved in a theory of evolutionary change. Far from it: "Such a process [as the evolution of life on earth], we may assume, proceeds in accordance with all kinds of causal laws, for example, the laws of mechanics, of chemistry, of heredity and segregation, of natural selection, etc." (Popper 1963, 108). What Popper is concerned to deny is overall extrapolations from the course of evolution—extrapolations pointing to a general progression in the course of evolutionary history, and the like. And this is a denial in which the great majority of evolutionists would no doubt join with Popper. Certainly, modern evolutionists seem to find little room in their theorizing for progressive speculations. Apart from anything else, the theory of evolution through natural selection implies that change is opportunistic. Even though a change from a human viewpoint might be "degenerative," say the loss of some complex organ, it is quite possible that such might happen if the conditions favor a degenerative change. One thinks here, for example, of the loss of sight of cave dwelling mammals, or, a case to be mentioned later, the loss of flying ability of oceanic island dwelling insects.

But even so, one might feel that Popper is still being unnecessarily firm in his stand against evolutionary laws (excluding from consideration now laws to do with selection, and the like). After all, one might argue, although overall laws of progress and such things may not hold, may indeed be just not the thing that a phylogenetic description is about, it is well known that evolutionary history shows many *trends* (Simpson 1953). For instance, one frequently sees (in the fossil record) a trend towards larger bodily size—so frequently, in fact, that it is referred to (following its discoverer) as "Cope's Rule." Could one not argue that such as Cope's Rule or "Dollo's Law" (that evolution never reverses itself) give the lie to Popper's denial of evolutionary laws—that here we go beyond the unique?

There are two points to be made in reply, both of which are in Popper's favor. First, such "rules" or "laws" of evolution tend to have an awful lot of exceptions. A recent discussant of Cope's rule, S. M. Stanley, admits candidly that: "Because numerous exceptions are known, recognition of the concept as a law has been rejected by most workers" (1973, 1). Similarly, in a spirited attack which regretfully is not always as careful in its treatment of uniqueness as is Popper, the paleontologist G. G. Simpson points out that some evolutionary "laws" have up to 36 percent exceptions! (1963, 29). Even Boyle's law has a better batting average than this, and one might well feel that such frail reeds need raise no qualms for a Popperian denial of evolutionary laws.

But this brings me to the second point. Suppose one insisted in regarding some of these claims about evolutionary trends as being at least quasilaws or near laws. There is some justification for doing this. For a start, some of the claims do not have that many exceptions. Having denied that Cope's rule is a law, Stanley nevertheless continues: "Still, it [i.e., Cope's rule] has been widely upheld as a valid empirical generalization, and of the definitions for 'rule' listed by Webster, 'a generally prevailing condition,' describes it accurately" (1973, 1). Second, by a process of excluding exceptions specifically, so common a practice with laws, one can make such claims even more accurate. Cope's rule seems to break down primarily when a new type of environment is being invaded, as when the first amphibians and the first birds evolved. Hence, by restating Cope's rule as excluding such instances, one has even more accurate general claims. Third, and most important, many evolutionary "laws" have the kind of theoretical justificatory backing which tends to distinguish laws from mere accidental generalizations. Stanley, for example, argues that there are good adaptive reasons why major evolutionary breakthroughs usually involve comparatively small body sizes, which sizes after the breakthrough can then be increased. Moreover, he argues also that there are reasons why this might not hold in the exceptional cases. (See Ruse 1973, for more argumentation in this vein.)

However, despite these arguments, which seem to have some strength,

Popper's position is unscathed—because he allows for and points to the position they are supporting! First, Popper points out that merely showing that there are evolutionary trends does not prove the existence of evolutionary laws. Laws do not assert existence. "*But trends are not laws.* A statement asserting the existence of a trend is existential, not universal" (1963, 115, his italics). However, secondly, in reply to the obvious counter, Popper then goes on to say: "If we succeed in determining the complete or sufficient singular conditions *c* of a singular trend *t*, then we can formulate the universal law: 'Whenever there are conditions of the kind *c* there will be a trend of the kind *t*'" (129n). In other words, if evolutionists really can set up the conditions for a trend, then Popper will allow them a law—although he does have some doubts about testing such a law. Hence, the way still seems open for the possibility of something like Cope's rule being made into a law, and Popper points out this way.

In short, although one might feel—given what he does allow—that Popper is a little paradoxical in his firm denial of evolutionary laws,[1] there is much of real value in his first excursion into the philosophy of evolutionary biology.

II

We come now to Popper's more recent comments about evolutionary biology. These are to be found in his *Objective Knowledge: An Evolutionary Approach* (Popper 1975), and in his autobiographical contribution to the *Philosophy of Karl Popper* (Popper 1974). Since, not surprisingly, many of the points made in the two works coincide, I shall here base my discussion on the second, fuller contribution, subtitled "Darwinism as a Metaphysical Research Program," feeling free to refer where necessary to the first contribution for points of clarification and expansion. (There are also some comments in Popper's contribution to the 1973 Herbert Spencer Lecture Series [Popper 1975]. However these seem not to add anything else.)

Put simply, it is Popper's claim that in an important sense neo-Darwinian evolutionary theory, the modern theory of evolution, is not a genuine scientific theory. He argues that the theory is not properly testable, and then, true to his most fundamental philosophical tenets, he concludes that the theory is metaphysical. "I have come to the conclusion that Darwinisin is not a testable scientific theory but a *metaphysical research program*—a possible framework for testable scientific theories" (Popper 1974, 134, his italics). One should add that in calling the theory "metaphysical," unlike a logical positivist. Popper is making a philosophical point and not thereby implying condemnation. Indeed, he puts Darwinisin in the same column as Deductivism (as opposed to the column with Lamarckism and inductivism), and there can surely be no higher Popperian praise than that—although perhaps revealingly Popper does later refer to evolutionary theory as a "feeble" theory (1974, 137), and perhaps even more revealingly he makes suggestions for "an enrichment of Darwinism" (1974, 138). In what follows I shall therefore first consider Popper's reasons for calling neo-Darwinism "metaphysical," and then I shall look at his suggested improvements.

III

Popper begins with an argument about possible life on Mars. Popper argues (correctly I think) that the Darwinian evolutionist would make at least the following three claims. First, organisms reproduce in kind fairly faithfully; second, there are small, accidental, hereditary mutations (causing change); third, there is a process of natural selection.[2] Now, argues Popper, the evolutionist would seem to be committed to the view that if ever on some planet we find life satisfying the first two claims, selection will come into play and cause a wide variety of organic forms. Hence, evolutionary theory would seem to make predictions that are testable. Therefore, evolutionary theory would seem to have genuine scientific content and would seem to offer the possibility of genuine scientific explanations. However, argues Popper, Darwinian evolutionary theory does not really make such a claim about a variety of forms.

> For assume that we find life on Mars consisting of exactly three species of bacteria with a genetic outfit similar to that of terrestrial species. Is Darwinism refuted? By no means. We shall say that these three species were the only forms among the many mutants which were sufficiently well adjusted to survive. And we shall say the same if there is only one species (or none).

> Thus Darwinism does not really *predict* the evolution of variety. It therefore cannot really *explain* it. (Popper 1974, 136, his italics)

I make two comments about this argument. First, although evolutionists do believe that natural selection working on random variation will normally lead to variety, they do not think that this is something which must follow necessarily. Selection can act to keep a population or species absolutely stable by eliminating all new mutations. (See Lewontin 1974, especially chapter 4.) But this leads on to the second point. Variety will come about when and only when there is, as it were, some advantage to or cause for such variety—when different ecological niches, for example, can be used. Now this claim, it seems, does lead to predictions testable at least in principle. Popper's Mars example is perhaps a little unfair, because we know already that Mars is not going to be very hospitable to life,[3] so one is hardly going to expect so much organic variety as here on earth (remember, here on earth one gets a lot more variety in the jungle than in the desert). But let us consider for a moment some of the reasons or conditions for variety, in particular (following Popper) let us consider some of the reasons why one might get different species created. (A species is a group of organisms, breeding between themselves, but potentially or actually isolated reproductively from all others. The differences between species are a reflection of and dependent upon differences in genetic constitution, "gene pools." Different species therefore have different gene pools, although how different may vary and is indeed a matter of some debate. We can sidestep this issue here. See Mayr 1963, for more details.)

First, there is the question of isolation or separation. There is, in fact, some controversy between evolutionary biologists about exactly how or why speciation occurs, and the role played by isolation—in particular, some feel that speciation (between two groups originally of the same species) always requires a period of geographical separation. Others believe that although most speciation may be of this kind, "allopatric" speciation, it can occur between groups not separated ("sympatric" speciation). But even those who allow sympatric speciation often demand some kind of ecological separation—say, speciating groups being on different parts of the same host. So, separation or isolation seems most important for speciation. (See Lewontin 1974 and Mayr 1963 for details.)

Obviously however, whether speciation is allopatric or sympatric, more is needed. What is needed is some reason or reasons to push apart the genes of the two speciating groups. Selection is clearly going to be the main thing operative here—for instance, the ecological and geographical conditions of the two groups may be very different, and these in turn might well lead to different selective pressures. And finally, let us mention something that may well be of great importance in speciation, namely, the so-called founder principle (Dobzhansky 1951; Lewontin 1974; Mayr 1963). It is generally believed today that species of organisms contain a great deal of genetic variability—there is no such thing as a standard typical member, for nearly all members will have some genes that are not common and perhaps not have some genes that are fairly common. The reasons why such variability is supposed are several—one reason, for instance, is the probable widespread occurrence of superior heterozygote fitness (i.e., the heterozygote for two alleles is fitter than either homozygote), which in turn leads to different genes being "balanced" and maintained in a population. But the consequence of this variability is that if a small group (of "founders") is isolated from the main population (say, on an island) they will not be typical, because nothing is typical. Hence, not only will they a fortiori be on the way to being a different gene group, but as they breed between themselves and "shake down" into a cohesive groups they may alter drastically the various selective fitnesses of the genes they do possess, thus leading to rapid change. For instance, a gene A_1 may have selective value when it is rare (as in the total population) but little at all when it is common (as in the founder population).

Now, let us start putting some of these points together in possible models.[4] Suppose for a start one came across a planet where the chances of allopatric speciation seemed difficult rather than otherwise—suppose, for instance, the planet were fairly small and uniformly covered with water (without freakish currents, and so on). One suspects that were but a few aquatic species discovered on such a planet, no evolutionist would be desperately perturbed. On the other hand, suppose one came across a planet with conditions that seemed tailor-made for speciation. Suppose, for instance, one had an area with fairly large populations that investigation showed to be variable genetically (which in turn manifested itself as phenotypic variation). Suppose also on such a planet one had other isolated areas, with differing conditions—cold, warm, dry, wet, and the like. And suppose finally there seemed possible rare (but only rare) ways in which organisms might go from the main area to the isolated areas (which isolated areas were now inhabited). Had one reason to believe that life on the planet was fairly old (e.g., through the fossil record or general complexity of structure), yet were one to find that absolutely no speciation at all had occurred, then I suggest that, contra Popper, modern evolutionists would be worried. Their theory, parts of it at least, would have been falsified. The claims that they make about speciation would seem not to hold.

Of course, talk of hypothetical planets at best makes evolutionary theory testable in principle, but this seems all that is necessary to counter Popper here, since his argument is at the hypothetical level. One would add, however, that there is empirical evidence from this world which seems both to support and test evolutionists' claims about speciation. Repeatedly through the world have been found and are found cases where populations, isolated from the main group under the kinds of conditions described above, have evolved into new species. The classic case is probably that of the finches on the Galapagos Archipelago off the coast of South America, those very finches which made Darwin an evolutionist (Lack 1947). Moreover, there is experimental evidence based on populations of captive fruit flies that supports the founder principle hypothesis in particular, apart from more general evidence supporting claims about the genetic variability always in populations, which claims are so crucial to modern thinking about speciation. (See Dobzhansky 1970 and Lewontin 1974 for more details.)

IV

Next, in his campaign to show that evolutionary theory is metaphysical, Popper brings up that popular suggestion that adaptation and selection are just about vacuous.

> Take "Adaptation." At first sight natural selection appears to explain it, and in a way it does, but it is hardly a scientific way. To say that a species now living is adapted to its environment is, in fact, almost tautological. Indeed we

> use the terms "adaptation" and "selection" in such a way that we can say that, if the species were not adapted, it would have been eliminated by natural selection. Similarly, if a species has been eliminated it must have been ill adapted to the conditions. Adaptation or fitness is *defined* by modern evolutionists as survival value, and can be measured by actual success in survival: there is hardly any possibility of testing a theory as feeble as this. (Popper 1974, 137, his italics)

Before attempting a criticism of this passage, one must in fairness to Popper point out that in making a claim of this kind he is in very distinguished biological company. Indeed, one of today's leading evolutionists, R. C. Lewontin, at one point used virtually the same language as Popper, although in later writings he has backtracked somewhat. (See Lewontin 1962, 309, and Lewontin 1969, 41–42.) Because of such ambiguity, what one starts to suspect is that Popper (and Lewontin) are right in thinking that there is something tautological or analytic surrounding selection—a definition perhaps—but that Popper is wrong in thinking that this is all there is to the matter. There are probably some very solid empirical claims being made. And this, I would suggest, is in fact the case.

First, take natural selection itself. This is a systematic differential reproduction between organisms, ultimately brought about on the one hand by organisms' tendency to increase in number in geometric fashion, and on the other hand by the inevitable limitations of space and food supply. Now, in pointing to the fact that there is a differential reproduction—that not all organisms which are born survive and reproduce (offspring which are in turn viable)—we hardly have something that is tautological. The differential reproduction may be as "obvious" as the roundness of the earth, but neither is empirically empty—certainly the differential reproduction makes evolutionary theory testable. If we all just budded off one and only one offspring asexually, evolutionary theory would be false. At least, it would be false inasmuch as one tried to apply it to this world of ours, which is precisely what evolutionists do want and try to do. (In the short term, it must be allowed that one could have a differential reproduction, even though all organisms reproduce. It would just be a question of one organism or kind having more [viable] offspring than another organism or kind.)

Secondly, we have the point about selection that seems to cause so much trouble. Now, it certainly seems to be the case that evolutionists do link up adaptive value and fitness in terms of survival value (or, more precisely, in terms of reproductive value), and that what we have here are definitions—analytic or tautological statements. But even here we have the evolutionists doing more than just making straight analytic definitions. Evolutionists always emphasize that natural selection is *systematic*—the differential reproduction is not a random matter. Overall success is believed to be on average a function of organisms' peculiar characteristics, and so not only do we have the very nonanalytic matter of which characteristics aid survival and reproduction—if we were all identical there could be no selection—but also we have the claim that things of adaptive value in one situation, will also be of value in similar situations. This may be difficult to test, but it is an empirical claim and could be false. Suppose one found on an island a group of men who had lost their sight, and who were not troglodytes or in any other way peculiar. Once again evolutionists would have to rethink their theory. That no such men have been found does not prove the theory analytic—it may just be that the theory is true. Falsifiability should not be confused with being falsified.

Of course, once again in practice it may be difficult actually to test this aspect of evolutionary theory. No doubt it would not be easy to be able to rule out possible adaptive reasons for men losing their sight. But evolutionists do have positive evidence, from experiments and nature, that there is a kind of uniformity about adaptive value in similar situations. For example, winglessness seems to be of value to insects and other small animals on oceanic islands (because they stand less chance of being blown away), and experiments bear this out (Dobzhansky 1951)—and after all, animals do not lose their sight (in a systematic way) except for good reason, as with moles.

Moreover, even granting something analytic about the way in which adaptation is defined, we still have empirical questions surrounding the whole problem of evolutionary change. Why, for example, a characteristic is adaptive at one point in time or space, but not at a later time or different place? Suppose an animal color in a species changes from light to dark, as in the famous case of moths in industrial Britain (Sheppard 1975). In this particular instance, it has been shown that years ago light color was an adaptation, whereas today dark color is adaptive. The reasons supposed however, that because of pollution trees have become darker and that avian moth predators more easily see moths that do not match their backgrounds, clearly take one beyond the analytic.[5]

Thirdly, Popper ignores entirely the fact that evolutionists allow that it can be the less well-adapted that can survive and the more well-adapted that fails to survive. For a start, it is percentages that count not individuals—does one group on average have a better record than another, not does one individual survive rather than another. Secondly, Popper ignores entirely the hypothesis of genetic drift, which supposes that in certain special situations fortuitously the less well-adapted (or neutral) can succeed where the more well-adapted gets eliminated. Genetic drift is still a highly contentious issue (Lewontin 1974), but the way Popper argues it would be ruled out as contradictory, whereas even its strongest critics seem to feel the need to mount empirical counterarguments. It should be added, lest it be thought that the possible existence of drift makes evolutionary theory unfalsifiable in the sense that all characteristics necessarily have an explanation—selection or drift—that no one today denies that selection fashioned major characteristics like the hand, the eye, and so on. Conversely, the above given hypothetical islanders could not have lost through drift so important a characteristic as their sight. Hence, drift is not a ubiquitous escape clause against falsification. (See Ruse 1973, 115.)

Finally, let us mention that in evolutionary theory, linked with selection although perhaps not really part of it, we have the claim that the selected characteristics will be passed on from one generation to the next. This is obviously necessary for evolution, for were there no such transmission selection would have no effect. And the claim is clearly empirical—logically it is quite possible that the strong, sexy, or otherwise advantaged individuals always have puny, ugly, or otherwise disadvantaged offspring. Although an evolutionary theory based on selection does not necessarily have to use a theory of inheritance stemming from Mendelian or neo-Mendelian claims and findings—Darwin's did not, for example—it seems today that most evolutionists do work with an amalgam of Darwinian selection and Mendelian genetics. (See Ruse 1973 for details.)

All things considered therefore, it seems ridiculous to keep claiming that evolutionary theory has at its heart a devastating tautology. The time has come to lay this misconception quietly to rest.[6]

V

Popper continues his analysis by taking up the gradualness of evolution as forecast by Darwinian evolutionary theory. He allows that the theory "certainly does *predict* that if such an evolution takes place, it will be *gradual*" (Popper 1974). However, Popper goes on to say:

> Gradualness is thus, from a logical point of view, the central prediction of the theory. (It seems to me that it is its only prediction.) Moreover, as long as changes in the genetic base of the living forms are gradual, they are—at least "in principle"—explained by the theory; for the theory does predict the occurrence of small changes, each due to mutation. However, "explanation in principle" is something very different from the type of explanation which we demand in physics. While we can explain a particular eclipse by predicting it, we cannot predict or explain any particular evolutionary change (except perhaps certain changes in the gene population *within* one species); all we can say is that if it is not a small change, there must have been some intermediate steps—an important suggestion for research: a research program (Popper 1974, 137–38, his italics)

I am not quite sure what to make of this argument, because it seems to me to be so unfair. Either evolutionary theory predicts that change will be gradual, or it does not. In fact, for most cases it does, therefore we have a prediction, therefore it is testable, therefore it is not metaphysical. One may argue that the explanation is not of very much, but it is of something—although in fact if one looks at the matter historically one finds that the gradualists had a terrific battle to win over the nongradualists, the saltationists. Of course, the explanation is "in principle" in the sense that until one turns to an actual case, specific details are lacking, but this is the same as any theory until one turns to actual cases. The gradualness at least is no more in principle than are eclipses. One may agree with Popper that one has little more than a "research program," a start to explanations not an end, but by Popper's own philosophy the program is not metaphysical. (Incidentally, Popper seems unaware that evolutionists believe that in the plant world, evolution can occur nongradually in steps due to combining of complete sets of parental chromosomes in offspring (Stebbins 1950). He seems also unaware of the great amount of explanatory information there is about some actual cases of gradual evolution, filling out the details—for example, in the case of the horse (Simpson 1953).[7]

VI

Finally in considering Popper's arguments about the nature of evolutionary theory, before turning to his suggestion for improving the theory, one might add that Popper's own position seems paradoxical—almost contradictory. Popper believes that neo-Darwinian evolutionary theory is metaphysical because it is unfalsifiable. Yet, Popper seems to believe not merely that it leads to falsifiable predictions, but that these are false! He writes:

> It is clear that a sufficient increase in fecundity depending fundamentally on genetical factors, or a shortening of the period of immaturity, may have the same survival value as, or even a greater survival value than, say, an increase in skill or in intelligence. From this point of view it may be a little hard to understand why natural selection should have produced anything beyond a general increase in rates of reproduction and the elimination of all but the most fertile breeds. (Popper 1975, 271)

Then, in a footnote, Popper comments: "This is only one of the countless difficulties of Darwin's theory to which some Neo-Darwinists seem to be almost blind" (1975, 271n).

If we pursue this line of argument then Popper's point seems to be that selection theory is not merely not tautological but false, because there are in fact many breeds that are far less fertile with respect to numbers than others—elephants as opposed to herring, for instance. Of course, the truth of the matter is that selection theory does not necessarily have the implications that Popper seems to think it has, and that neo-Darwinists are far from blind to this whole question of reproductive rates. Indeed, there has been considerable attention paid to and controversy surrounding the problem of reproductive rates—why it is that organisms have the particular rates that they do.

All biologists seem to agree that controlled reproductive rates are a function of organisms being able to survive only with limited numbers in the face of limited resources, but that with such control these organisms can exploit those resources more successfully perhaps than faster breeders. However, some biologists have argued for a kind of "group selection," claiming that when one has (say) limited resources, there is selection on the group to keep reproductive rates down, or the whole group may perish. (See particularly Wynne-Edwards 1962.) Other biologists argue that such group selection is chimerical and that individual selection is the cause (Williams 1966). Plovers, for instance, have a very stable clutch size, and they do not reproduce to their possible limit (if eggs are removed from the nest, then the female brings the number back up to the norm). It has been argued by David Lack simply that there is selection on the individual plover for such clutch size—more eggs and the individuals collectively are less fit, and a fewer egg number does not make for fitness, which compensates the fewness in number (Lack 1954, 1966). Incidentally, there is nothing unfalsifiable about the evolutionists' position here—that whatever the reproduction rates may be evolutionary theory will necessarily have an answer. If experiments showed (what they do not) that plovers given additional eggs have more fit offspring, then Lack would be wrong.

In short, at this point we can save Popper from the unwelcome implications of his own arguments. Evolutionists have paid attention to the problem of reproductive rates, and they do have arguments showing why there might be selective advantages to keeping such rates at a (comparatively) low level. Curiously however, despite his criticisms of evolutionists at this point, Popper seems to know the truth all along. Immediately after the quotation given above (from the main text), he writes "[There may be many different factors involved in the processes which determine the rates of reproduction and of mortality, for instance the ecological conditions of the species, its interplay with other species, and the balance of the two (or more) populations]" (Popper 1975, 271, his square brackets). Even if Popper thinks these claims ill-founded, and he gives no arguments to such an effect, it is a little odd that he should accuse neo-Darwinians of blindness on this point. (Wilson 1975 contains a discussion of recent work on the group versus individual selection controversy.)

. . .

VIII

It is odd that one who, as we have seen, at one point dealt so sensitively with evolutionary biology, should now seem so determined to undervalue its nature and achievements. My suspicion is that, in part, Popper's attitude comes simply from the fact that he has not yet accepted the Darwinian revolution. Although at one point Popper says (truly) that Darwin's great move was to show how design and purpose in the organic world can be explained in purely physical terms, almost at once Popper then goes on to say that the difficulty with Darwinism is explaining something like the eye in terms of accidental mutations, and it is clear (by his own admission) that Popper is trying to get away from the accidental in evolution, and to give some direction to change (Popper 1975, 270). But this is the whole point—unless one can see that it is selection acting on "accidents" bringing about designlike effects, one misses entirely the force of Darwinism. Somehow one feels that Popper is in a tradition that started as soon as Darwin's *Origin* appeared—a tradition that includes such men as Charles Lyell and St. George Jackson Mivart, who were evolutionists but who felt that in order to account for the designlike effects of the organic world one must supplement selection with other mechanisms. Unlike them Popper may have no theological axe to grind; but, he seems a direct intellectual descendent. (Details of this tradition can be found in Ruse 1975, 1977.)

Possibly also Popper's desire to see links between organic evolution and scientific theory growth leads him astray. This could be in at least two ways. First, fairly obviously, Popper's theory of theory growth is not itself scientific. It is, in a nonpejorative sense, metaphysical. Therefore, inasmuch as biological evolutionary theory can be shown metaphysical, links can be established between Popper's philosophical views and the beliefs of biologists. Second, the actual way in which Popper analyzes theory growth may be influencing his approach to biology. Using "*P*" for problem, "*TS*" for tentative solutions, "*EE*" for error-elimination, Popper sees all evolutionary sequences as following this pattern:

$$P_1 \rightarrow TS \rightarrow EE \rightarrow P_2 \text{ (Popper 1975, 243)}$$

But the fact of the matter is that tentative knowledge solutions are frequently fairly large (saltationary) and often designed—think, for example, of Darwin's solution to the organic origins' problem. Could it be that (mistakenly) Popper is reading features of the evolution of knowledge into the evolution of organisms, and that it is for this reason that he wants to supplement biological evolutionary theory in the ways he suggests?

But even if my surmises are correct and my criticisms of Popper's views about biological evolutionary theory are well-taken, what then does this all imply? In particular, what does it imply about Popper's philosophical theory about scientific theory growth? In one sense, not a great deal. Darwinians do not have a monopoly on the word "evolution"; hence nothing I have argued can properly stop Popper characterizing his views as evolutionary. Nor has anything here proven his general philosophical theory mistaken, although this is not necessarily to say that it is true. However, I suggest my arguments do show one most important thing. No longer ought Popper claim close ties between his philosophical evolutionary theory and biological evolutionary theory, or feel that somehow some of the legitimacy of the latter rubs off on the former. The relationship between the two theories is at best one of weak analogy. In important respects, Popperian scientific theory evolution and neo-Darwinian biological evolution are different.

NOTES

1. In Popper (1975), Popper again somewhat paradoxically asserts: "There are no Darwinian laws of evolution" (267). In Ruse (1975) I consider the place in Darwin's theory of laws, which I think Popper would allow; and I do the same in Ruse (1970, 1972, 1973) for modern evolutionary theory. A valuable discussion of these problems, with a slight different emphasis, is Hull (1974).

2. Popper sees the evolutionist making another claim, about variability, but this is irrelevant to his present argument.

3. However, recent space probes show that it is apparently not as inhospitable as was earlier thought.

4. Discussions of speciation usually cover the presumed events if and when isolated groups again come into contact. For brevity these will be ignored here.

5. Curiously, given his belief that selection is tautological, Popper does allow that finding out what are adaptations and why is an empirical matter.

6. In a previously mentioned passage that influenced my analysis (and that is I trust captured by my analysis), Lewontin writes: "Evolution is the necessary consequence of three observations about the world.... They are: (1) There is phenotypic variation, the members of a species do not all look and act alike. (2) There is a correlation between parents and offspring.... (3) Different phenotypes leave different numbers of offspring in *remote* generations.... These are three contingent statements, all of which are true about at least some part of the biological world.... There is nothing tautological here" (Lewontin 1969, 42–43, his italics). Incidentally, in my reply to Popper I ignore the way Popper has blurred "adaptation" with being "adapted"—an organism can have an adaptation like the eye and still be ill-adapted to its environment. I ignore also the way Popper speaks only of inter specific selection and ignores intra specific selection.

7. In Ruse (1969) and (1973), I discuss in detail the problems of testing evolutionary theory. See also Williams (1973).

REFERENCE LIST

Ayala, F. J., M. L. Tracey, L. O. Barr, J. F. McDonald, and S. Pevez-Salas. 1974. "Genetic Variation in Natural Populations of Five Drosophila Species and the Hypothesis of the Selective Neutrality of Protein Polymorphism." *Genetics* 77: 343–84.

Dobzhansky, T. 1951. *Genetics and the Origin of Species* (3rd ed.). New York: Columbia University Press.

———. 1970. *Genetics of the Evolutionary Process.* New York: Columbia University Press.

Goldschimdt, R. 1940. *The Material Basis of Evolution.* New Haven: Yale University Press.

———. 1952. "Evolution as Viewed by One Geneticist." *American Scientist* 40: 84–135.

Goudge, T. A. 1961. The Ascent of Life. Toronto: University of Toronto Press.

Hull, D. 1974. *The Philosophy of Biological Science.* Englewood Cliffs, NJ: Prentice-Hall.

Lack, D. 1947. *Darwin's Finches.* Cambridge: Cambridge University Press.

———. 1954. *The Natural Regulation of Animal Numbers.* Oxford: Oxford University Press.

———. 1966. *Population Studies of Birds.* Oxford: Oxford University Press.

Lewontin, R. C. 1962. "Evolution and the Theory of Games." *Journal of Theoretical Biology* 1: 382–403.

———. 1969. "The Bases of Conflict in Biological Explanation." *Journal of the History of Biology* 2: 35–45.

———. 1974. *The Genetic Basis of Evolutionary Change.* New York: Columbia University Press. Notes: XXV in the Columbia Biology Series

Manser, A. 1965. "The Concept of Evolution." *Philosophy* 40: 18–34.

Mayr, E. 1960. "The Emergence of Evolutionary Novelties." In *Evolution After Darwin,* edited by S. Tax, 349–80. Chicago: University of Chicago Press.

———. 1963. *Animal Species and Evolution.* Cambridge, MA: Harvard University Press.

Monod, J. 1975. "On the Molecular Theory of Evolution." In *Problems of Scientific Revolution: Progress and Obstacles to Progress in the Sciences,* edited by Ron Harré. Oxford: Oxford University Press.

Popper, K. R. 1957. *The Poverty of Historicism.* London: Routledge and Kegan Paul. Corrected version, 1963.

———. 1972. *Objective Knowledge.* Oxford: Oxford University Press.

———. 1974. "Darwinism as a Metaphysical Research Programme." In *The Philosophy of Karl Popper,* edited by P. A. Schilpp, 133–43. Vol. 1. LaSalle, IL: Open Court.

———. 1975. "The Rationality of Scientific Revolutions." In *Problems of Scientific Revolution: Progress and Obstacles to Progress in the Sciences,* edited by Ron Harré. Oxford: Oxford University Press.

Ruse, M. 1969. "Confirmation and Falsification of Theories of Evolution." *Scientia* 104: 329–57.

———. 1970. "Are There Laws in Biology?" *Australasian Journal of Philosophy* 48: 234–346.

———. 1972. "Is the Theory of Evolution Different?" *Scientia* 106: 765-83; 1069–93.

———. 1975a. "Charles Darwin's Theory of Evolution: An Analysis." *Journal of the History of Biology* 8: 219–41.

———. 1975b. "The Relationship between Science and Religion in Britain, 1830–1870." *Church History* 44: 505–22.

———. 1977. "William Whewell and the Argument from Design." *Monist* 60: 244–68.

Sheppard, P. M. 1958. *Natural Selection and Heredity.* London: Hutchinson.

Simpson, G. G. 1953. *The Major Features of Evolution.* New York: Columbia University Press.

———. 1963. "Historical Science." In *Historical Science,* edited by C. C. Albritton, 24–48. Stanford: Freeman, Cooper.

Stanley, S. M. 1973. "An Wxplanation for Cope's Rule." *Evolution* 27: 1–26.

Stebbins, G. L. 1950. *Variation and Evolution in Plants.* New York: Columbia University Press.

Waddington, C. H. 1957. *The Strategy of the Genes.* London: Allen and Unwin.

Williams, G. C. 1966. *Adaptation and Natural Selection.* Princeton, NJ: Princeton University Press.

Williams, M. B. 1973. "Falsifiable Predictions of Evolutionary Theory." *Philosophy of Science* 40: 518–37.

Wilson, E. O. 1975. *Sociobiology: The New Synthesis.* Cambridge, MA: Harvard University Press.

Wynne-Edwards, V. C. 1962. *Animal Dispersion in Relation to Social Behaviour.* Edinburgh: Oliver and Boyd.

CHAPTER 9

HUMAN NATURE: ONE EVOLUTIONIST'S VIEW

FRANCISCO J. AYALA

"It is a disgraceful and dangerous thing for an infidel to hear a Christian, while presumably giving the meaning of Holy Scripture, talking nonsense. We should take all means to prevent such an embarrassing situation, in which people show up vast ignorance in a Christian and laugh it to scorn [...] If they find a Christian mistaken in a field which they themselves know well, and hear him maintaining his foolish opinions about the Scriptures, how then are they going to believe those Scriptures in matters concerning the resurrection of the dead, the hope of eternal life, and the kingdom of heaven?"

—St. Augustine, *The Literal Meaning of Genesis*, book 1, chapter 19.

"IN THE IMAGE OF GOD"

The book of Genesis dramatically sets forth humans' lofty uniqueness within the natural world: "So God created man in his own image, in the image of God created he him; male and female created he them" (Genesis 1: 27).

It does not take a great deal of biological expertise to realize that humans have organs and limbs similar to those of other animals; that we bear our young like other mammals; that, bone by bone, there is a precise correspondence

Original essay. Reprinted by permission of the author.

between the skeletons of a chimpanzee and a human. But it does not take much reflection to notice the distinct uniqueness of our species. There is the bipedal gait and the enlarged brain. Much more conspicuous than the anatomical differences are the distinct behaviors and their outcomes. Humans have elaborate social and political institutions, codes of law, literature and art, ethics and religion; humans build roads and cities, travel by motorcars, ships, and airplanes, and communicate by means of telephones, computers, and televisions.

I will first, in the pages that follow, set forth the biological continuity between humans and animals. I will outline what we currently know about the evolutionary history of humans for the last four million years, from bipedal but small-brained *Australopithecus* to modern *Homo sapiens*, our species, through the prolific tool-maker *Homo habilis* and the continent-wanderer *Homo erectus*. The genes of living humans manifest that our ancestors were no fewer than several thousand individuals at any one time in the history of these hominid species.

I shall, then, identify anatomical traits that distinguish us from other animals, and point out our two kinds of heredity, the biological and the cultural. Biological inheritance is based on the transmission of genetic information, in humans very much the same as in other sexually reproducing organisms. But cultural inheritance is distinctively human, based on transmission of information by a teaching and learning process, which is in principle independent of biological parentage. Cultural inheritance makes possible the cumulative transmission of experience from generation to generation. Cultural heredity is a swifter and more effective (because it can be designed) mode of adaptation to the environment than the biological mode. The advent of cultural heredity ushered in cultural evolution, which transcends biological evolution.

In the latter part of this essay, I will explore ethical behavior as a model case of a distinctive human trait, and seek to ascertain the causal connections between human ethics and human biology. My conclusions are that (1) the proclivity to make ethical judgments, that is, to evaluate actions as either good or evil, is rooted in our (biological) nature, a necessary outcome of our exalted intelligence; but (2) the moral codes which guide our decisions as to which actions are good and which ones are evil, are products of culture, including social and religious traditions. This second conclusion contradicts evolutionists and sociobiologists who claim that the morally good is simply that which is promoted by the process of biological evolution.

HUMANKIND'S BIOLOGICAL ORIGINS

Mankind is a biological species that has evolved from other species that were not human. In order to understand human nature, we must know our biological makeup and whence we come, the story of our humbler beginnings. For a century after the publication of Darwin's *On the Origin of Species* in 1859, the story of evolution was reconstructed with evidence from paleontology (the study of fossils), biogeography (the study of the geographical distribution of organisms), and from the comparative study of living organisms: their morphology, development, physiology, and the like. Since the mid-twentieth century we have, in addition, molecular biology, the most informative and precise discipline for reconstructing the ancestral relationships of living species.

Our closest biological relatives are the great apes and, among them, the chimpanzees, who are more related to us than they are to the gorillas, and much more than to the orangutans. The hominid lineage diverged from the chimpanzee lineage 5–7 million years ago (Mya) and it evolved exclusively in the African continent until the emergence of *Homo erectus*, somewhat before 1.8 Mya. The first known hominid, *Ardipithecus ramidus*, lived 4.4 Mya, but it is not certain that it was bipedal or in the direct line of descent to modern humans, *Homo sapiens*. The recently described *Australopithecus anamensis*, dated 3.9–4.2 Mya, was bipedal and has been placed in the line of descent to *Australopithecus afarensis*, *Homo habilis*, *H. erectus*, and *H. sapiens*. Other hominids, not in the direct line of descent to modern humans, are *Australopithecus africanus*, *Paranthropus aethiopicus*, *P. boisei*, and *P. robustus*, who lived in Africa at various times between 3 and 1 Mya, a period when three or four hominid species lived contemporaneously in the African continent.

Shortly after its emergence in tropical or subtropical eastern Africa, *H. erectus* spread to other continents. Fossil remains of H. erectus are known from Africa, Indonesia (Java), China, the Middle East, and Europe. *H. erectus* fossils from Java have been dated 1.81+0.04 and 1.66+0.04 Mya, and from Georgia between 1.6 and 1.8 Mya. Anatomically distinctive *H. erectus* fossils have been found in Spain, deposited before 780,000 years ago, the oldest in southern Europe.

The transition from *H. erectus* to *H. sapiens* occurred around 400,000 years ago, although this date is not well determined owing to uncertainty as to whether some fossils are erectus or "archaic" forms of *sapiens. H. erectus* persisted for some time in Asia, until 250,000 years ago in China and perhaps until 100,000 ago in Java, and thus was coetaneous with early members of its descendant species, *H. sapiens.* Fossil remains of Neanderthal hominids (*Homo neanderthalensis*), with brains as large as those of *H. sapiens*, appeared in Europe around 200,000 years ago and persisted until thirty or forty thousand years ago. The Neanderthals were thought to be ancestral to anatomically modern humans, but now we know that modern humans appeared at least 100,000 years ago, much before the disappearance of the Neanderthals. Moreover, in caves in the Middle East, fossils of modern humans have been found dated 120,000–100,000 years ago, as well as Neanderthals dated at 60,000 and 70,000 years ago, followed again by modern humans dated at 40,000 years ago. It is unclear whether the two forms repeatedly replaced one another by migration from other regions, or whether they coexisted in the same areas. Recent genetic evidence indicates that interbreeding between *sapiens* and *neanderthalensis* never occurred.

There is considerable controversy about the origin of modern humans. Some anthropologists argue that the transition from *H. erectus* to archaic *H. sapiens* and later to anatomically modern humans occurred consonantly in various parts of the Old World. Proponents of this "multiregional model" emphasize fossil evidence showing regional continuity in the transition from *H. erectus* to archaic and then modern *H. sapiens.* In order to account for the transition from one to another species (something which cannot happen independently in several places), they postulate that genetic exchange occurred from time to time between populations, so that the species evolved as a single gene pool, even though geographic differentiation occurred and persisted, just as geographically differentiated populations exist in other animal species, as well as in living humans. This explanation depends on the occurrence of persistent migrations and interbreeding between populations from different continents, of which no direct evidence exists. Moreover, it is difficult to reconcile the multiregional model with the contemporary existence of different species or forms in different regions, such as the persistence of *H. erectus* in China and Java for more than 100,000 years after the emergence of *H. sapiens.*

Other scientists argue instead that modern humans first arose in Africa or in the Middle East somewhat prior to 100,000 years ago and from there spread throughout the world, replacing elsewhere the preexisting populations of *H. erectus* or archaic *H. sapiens*. Some proponents of this "African replacement" model claim further that the transition from archaic to modern *H. sapiens* was associated with a very narrow bottleneck, consisting of only two or very few individuals who are the ancestors of all modern mankind. This particular claim of a narrow bottleneck is supported, erroneously as I will soon show, by the investigation of a peculiar small fraction of our genetic inheritance, the mitochondrial DNA (mtDNA). The African (or Middle East) origin of modern humans is, however, supported by a wealth of recent genetic evidence and is, therefore, favored by many evolutionists.

THE MYTH OF THE MITOCHONDRIAL EVE

The genetic information we inherit from our parents is encoded in the linear sequence of the DNA's four nucleotide components (represented by A, C, G, T) in the same fashion as semantic information is encoded in the sequence of letters of a written text. Most of the DNA is contained in the chromosomes inside the cell nucleus. The total amount of DNA in a human cell nucleus consists of six thousand million nucleotides, half in each set of 23 chromosomes inherited from each parent. A relatively small amount of DNA, about 16,000 nucleotides, exists in the mitochondria, cell organelles outside the nucleus. The mtDNA is inherited in a peculiar manner, that is, exclusively along the maternal line. The inheritance of the mtDNA is a gender mirror image of the inheritance of the family name. Sons and daughters inherit their mtDNA from their mother but only the daughters transmit it to their progeny, just as sons and daughters receive the family name of the father, but only the sons transmit it to their children.

Analysis of the mtDNA from ethnically diverse individuals has shown that the mtDNA sequences of modern humans coalesce to one ancestral sequence, the "mitochondrial Eve" that existed in Africa about 200,000 years ago.[1] This Eve, however, is not the one mother from whom all humans descend, but an mtDNA molecule (or the woman carrier of that molecule) from whom all modern mtDNA *molecules* descend.

Some science writers have drawn the inference that all humans descend from only one, or very few women, but this is based on a confusion between gene genealogies and individual genealogies. Gene genealogies gradually coalesce toward a unique DNA ancestral sequence (in a similar fashion as living species, such as humans, chimpanzees, and gorillas, coalesce into one ancestral species). Individual genealogies, on the contrary, increase by a factor of two in each ancestral generation: an individual has two parents, four grandparents, and so on.[2] Coalescence of a gene genealogy into one ancestral gene, originally present in one individual, does not disallow the contemporary existence of many other individuals, who are also our ancestors, and from whom we have inherited the other genes.

This conclusion can be illustrated with an analogy. My family name is shared by many people, who live in Spain, Mexico, the Philippines, and other countries. A historian of our family name has concluded that all Ayalas descend from Don Lope Sánchez de Ayala, grandson of Don Vela, vasal of King Alfonso VI, who established the domain ("señorío") de Ayala in the year 1085, in the now Spanish Basque province of Alava. Don Lope is the Adam from whom we all descend on the paternal line, but we also descend from many other men and women who lived in the eleventh century, as well as earlier and later.

The inference warranted by the mtDNA analysis is that the mitochondrial Eve is the ancestor of modern humans in the *maternal line.* Any person has a single ancestor in the maternal line in any given generation. Thus a person inherits the mtDNA from the mother, from the maternal grandmother, from the great grandmother on the maternal line, and so on. But the person also inherits other genes from other ancestors. The mtDNA that we have inherited from the mitochondrial Eve represents one-four-hundred-thousandth of the DNA present in any modern human (sixteen thousand out of six billion nucleotides). The rest of the DNA, 400,000 times more than the mtDNA, we have inherited from other contemporaries of the mitochondrial Eve.

From how many contemporaries? The issue of how many human ancestors we had in the past has been elucidated by investigating the genes of the human immune system.[3] The genes of the human leucocyte antigen (HLA) complex exist in multiple versions, which provide people with the diversity necessary to confront bacteria and other pathogens that invade the body. The evolutionary history of some of these genes shows that they coalesce into one ancestral gene

30–60 Mya, that is, much before the divergence of humans and apes. (Indeed, humans and apes share many of these genes.) The mathematical theory of gene coalescence makes it possible to estimate the number of ancestors that must have lived in any one generation in order to account for the preservation of so many diverse genes through hundreds of thousands of generations. The estimated *effective* number is about 100,000 individuals per generation. This "effective" number of individuals is an average rather than a constant number, but it is a peculiar kind of average (a "harmonic mean"), compatible with much larger but not much smaller numbers of individuals in different generations. Thus, through millions of years our ancestors existed in populations that were 100,000 individuals strong, or larger. Population bottlenecks may have occurred in rare occasions. But the genetic evidence indicates that human populations never consisted of fewer than several thousand individuals.

HUMAN UNIQUENESS

The most distinctive human anatomical traits are erect posture and large brain. We are the only vertebrate species with a bipedal gait and erect posture; birds are bipedal, but their backbone stands horizontal rather than vertical. Brain size is generally proportional to body size; relative to body mass, humans have the largest (and most complex) brain. The chimpanzee's brain weighs less than a pound; a gorilla's slightly more. The human male adult brain is 1400 cubic centimeters (cc), about three pounds in weight.

Evolutionists used to raise the question whether bipedal gait or large brain came first, or whether they evolved consonantly. The issue is now resolved. Our *Australopithecus* ancestors had, since four million years ago, a bipedal gait, but a small brain, about 450 cc, a pound in weight. Brain size starts to increase notably with our *Homo habilis* ancestors, about 2.5 Mya, who had a brain about 650 cc and also were prolific tool-makers (hence the name *habilis*). Between one and two million years afterwards, there lived *Homo erectus*, with adult brains about 1200 cc. Our species, *Homo sapiens*, has a brain about three times as large as that of *Australopithecus*, 1300–1400 cc, or some three pounds of gray matter. Our brain is not only much larger than that of chimpanzees or gorillas, but also much more complex. The cerebral cortex, where the higher cognitive functions are processed, is in humans disproportionally much greater than the rest of the brain when compared to apes.

Erect posture and large brain are not the only anatomical traits that distinguish us from nonhuman primates, even if they may be the most obvious. A list of our most distinctive anatomical features includes the following (of which the last five items are not detectable in fossils):

- Erect posture and bipedal gait (entail changes of the backbone, hipbone, and feet)
- Opposing thumbs and arm and hand changes (make possible precise manipulation)
- Large brain
- Reduction of jaws and remodeling of face
- Changes in skin and skin glands
- Reduction in body hair
- Cryptic ovulation (and continuous female sexual receptivity)
- Slow development
- Modification of vocal tract and larynx
- Reorganization of the brain

Humans are notably different from other animals not only in anatomy but also and no less importantly in their behavior, both as individuals and socially. A list of distinctive human behavioral traits includes the following:

- Subtle expression of emotions
- Intelligence: abstract thinking, categorizing, and reasoning
- Symbolic (creative) language
- Self-awareness and death-awareness
- Tool-making and technology
- Science, literature, and art
- Ethics and religion
- Social organization and cooperation (division of labor)
- Legal codes and political institutions

BIOLOGICAL EVOLUTION AND CULTURAL EVOLUTION

Humans live in groups that are socially organized, and so do other primates. But primate societies do not approach the complexity of human social organization. A distinctive human social trait is culture, which may be understood as the set of nonstrictly biological human activities and creations. Culture includes social and political institutions, ways of doing things, religious and ethical traditions, language, common sense and scientific knowledge, art and literature, technology, and in general all the creations of the human mind. The advent of culture has brought with it cultural evolution, a superorganic mode of evolution superimposed on the organic mode, and that has in the last few millennia become the dominant mode of human evolution. Cultural evolution has come about because of cultural change and inheritance, a distinctively human mode of achieving adaptation to the environment and transmitting it through the generations.

There are in mankind two kinds of heredity—the biological and the cultural, which may also be called organic and superorganic, or *endosomatic* and *exosomatic* systems of heredity. Biological inheritance in humans is very much like that in any other sexually reproducing organism; it is based on the transmission of genetic information encoded in DNA from one generation to the next by means of the sex cells. Cultural inheritance, on the other hand, is based on transmission of information by a teaching-learning process, which is in principle independent of biological parentage. Culture is transmitted by instruction and learning, by example and imitation, through books, newspapers and radio, television and motion pictures, through works of art, and by any other means of communication. Culture is acquired by every person from parents, relatives, and neighbors, and from the whole human environment.

Cultural inheritance makes possible for people what no other organism can accomplish—the cumulative transmission of experience from generation to generation. Animals can learn from experience, but they do not transmit their experiences, their "discoveries" (at least not to any large extent) to the following generations. Animals have individual memory, but they do not have a "social memory." Humans, on the other hand, have developed a culture because they can transmit cumulatively their experiences from generation to generation.

Cultural inheritance makes possible cultural evolution, that is, the evolution of knowledge, social structures, ethics, and all other components that make up human culture. Cultural inheritance makes possible a new mode of adaptation to the environment that is not available to nonhuman organisms—adaptation by means of culture. Organisms in general adapt to the environment by means of natural selection, by changing over generations their genetic constitution to suit the demands of the environment. But humans, and humans alone, can also adapt by changing the environment to suit the needs of their genes. (Animals build nests and modify their environment also in other ways, but the manipulation of the environment by any nonhuman species is trivial compared to mankind's.) For the last few millennia humans have been adapting the environments to their genes more often than their genes to the environments.

In order to extend its geographical habitat, or to survive in a changing environment, a population of organisms must become adapted, through slow accumulation of genetic variants sorted out by natural selection, to the new climatic conditions, different sources of food, different competitors, and so on. The discovery of fire and the use of shelter and clothing allowed humans to spread from the warm tropical and subtropical regions of the Old World to the whole earth, except for the frozen wastes of Antarctica, without the anatomical development of fur or hair. Humans did not wait for genetic mutants promoting wing development; they have conquered the air in a somewhat more efficient and versatile way by building flying machines. People travel the rivers and the seas without gills or fins. The exploration of outer space has started without waiting for mutations providing humans with the ability to breathe with low oxygen pressures or to function in the absence of gravity; astronauts carry their own oxygen and specially equipped pressure suits. From their obscure beginnings in Africa, humans have become the most widespread and abundant species of mammal on earth. It was the appearance of culture as a superorganic form of adaptation that made mankind the most successful animal species.

Cultural adaptation has prevailed in mankind over biological adaptation because it is a more rapid mode of adaptation and because it can be directed. A favorable genetic mutation newly arisen in an individual can be transmitted to a sizable part of the human species only through innumerable generations.

However, a new scientific discovery or technical achievement can be transmitted to the whole of mankind, potentially at least, in less than one generation. Moreover, whenever a need arises, culture can directly pursue the appropriate changes to meet the challenge. On the contrary, biological adaptation depends on the accidental availability of a favorable mutation, or of a combination of several mutations, at the time and place where the need arises.

AN EVOLUTIONARY ACCOUNT OF HUMAN NATURE

Erect posture and large brain are distinctive anatomical features of modern humans. High intelligence, symbolic language, religion, and ethics are some of the behavioral traits that distinguish us from other animals. The account of human origins that I have sketched above implies a continuity in the evolutionary process that goes from our nonhuman ancestors of eight million years ago through primitive hominids to modern humans. A scientific explanation of that evolutionary sequence must account for the emergence of human anatomical and behavioral traits in terms of natural selection together with other distinctive biological causes and processes. One explanatory strategy is to focus on a particular human feature and seek to identify the conditions under which this feature may have been favored by natural selection. Such a strategy may lead to erroneous conclusions as a consequence of the fallacy of selective attention: some traits may have come about not because they are themselves adaptive, but rather because they are associated with traits that are favored by natural selection.

Geneticists have long recognized the phenomenon of "pleiotropy," the expression of a gene in different organs or anatomical traits. It follows that a gene that becomes changed owing to its effects on a certain trait will result in the modification of other traits as well. The changes of these other traits are epigenetic consequences of the changes directly promoted by natural selection. The cascade of consequences may be, particularly in the case of humans, very long and far from obvious in some cases. Literature, art, science, and technology are among the behavioral features that may have come about not because they were adaptively favored in human evolution, but because they are expressions of the high intellectual abilities present in modern humans:

what may have been favored by natural selection (its "target") was an increase in intellectual ability rather than each one of those particular activities.

I now will briefly explore ethics and ethical behavior as a model case of how we may seek the evolutionary explanation of a distinctively human trait. I select ethical behavior because morality is a human trait that seems remote from biological processes. My goal is to ascertain whether an account can be advanced of ethical behavior as an outcome of biological evolution and, if such is the case, whether ethical behavior was directly promoted by natural selection, or has rather come about as an epigenetic manifestation of some other trait that was the target of natural selection.

I will argue that ethical behavior (the proclivity to judge human actions as either good or evil) has evolved as a consequence of natural selection, not because it was adaptive in itself, but rather as a pleiotropic consequence of the high intelligence characteristic of humans. I will first point out that the question whether ethical behavior is biologically determined may refer either to (1) the capacity for ethics (i.e., the proclivity to judge human actions as either right or wrong) and which I will refer to as "ethical behavior," or (2) the moral *norms* or moral codes accepted by human beings for guiding their actions. My theses are that: (1) the capacity for ethics is a necessary attribute of human nature, and thus a product of biological evolution; but (2) moral norms are products of cultural evolution, not of biological evolution.

My first thesis is grounded on the argument that humans exhibit ethical behavior because their biological makeup determines the presence of the three necessary, and jointly sufficient, conditions for ethical behavior; namely, the ability to anticipate the consequences of one's own actions, to make value judgments, and to choose between alternative courses of action. I thus maintain that ethical behavior came about in evolution not because it is adaptive in itself, but as a necessary consequence of man's eminent intellectual abilities, which are an attribute directly promoted by natural selection.

My second thesis contradicts the proposal of many distinguished evolutionists who, since Darwin's time, have argued that the norms of morality are derived from biological evolution. It also contradicts the sociobiologists, who have recently developed a subtle version of that proposal. The sociobiologists' argument is that human ethical norms are sociocultural correlates of behaviors fostered by biological evolution. I argue that such proposals are misguided and do not escape the naturalistic fallacy. It is true that both natural selection and moral norms sometimes coincide on the same behavior; that is, the two are consistent. But this isomorphism between the behaviors promoted by natural selection and those sanctioned by moral norms exists only with respect to the consequences of the behaviors; the underlying causations are completely disparate.

I shall now develop these ideas.

ETHICS AND LANGUAGE: A PARALLEL DISTINCTION

I have just noted that the question of whether ethical behavior is biologically determined may refer to either one of the following issues: (1) Is the capacity for ethics—the proclivity to judge human actions as either right or wrong—determined by the biological nature of human beings? (2) Are the systems or codes of ethical norms accepted by human beings biologically determined? A similar distinction can be made with respect to language. The issue whether the capacity for symbolic language is determined by our biological nature is different from the question of whether the particular language we speak (English, Spanish, or Japanese) is biologically necessary.

The first question posed is more fundamental; it asks whether or not the biological nature of *Homo sapiens* is such that humans are necessarily inclined to make moral judgments and to accept ethical values, to identify certain actions as either right or wrong. Affirmative answers to this first question do not necessarily determine what the answer to the second question should be. Independently of whether or not humans are necessarily ethical, it remains to be determined whether particular moral prescriptions are in fact determined by our biological nature, or whether they are chosen by society, or by individuals. Even if we were to conclude that people cannot avoid having moral standards of conduct, it might be that the choice of the particular standards used for judgment would be arbitrary. Or that it depended on some other, nonbiological criteria. The need for having moral values does not necessarily tell us what these moral values should be, just as the capacity for language does not determine which language we shall speak.

The thesis that I propose is that humans are ethical beings by their biological nature. Humans evaluate their behavior as either right or wrong, moral or immoral, as a consequence of their eminent intellectual capacities which include self-awareness and abstract thinking. These intellectual capacities, are products of the evolutionary process, but they are distinctively human. Thus, I maintain that ethical behavior is not causally related to the social behavior of animals, including kin and reciprocal "altruism."

A second thesis that I put forward is that the moral norms according to which we evaluate particular actions as morally either good or bad (as well as the grounds that may be used to justify the moral norms) are products of cultural evolution, not of biological evolution. The norms of morality belong, in this respect, to the same category of phenomena as the languages spoken by different peoples, their political and religious institutions, and the arts, sciences, and technology. The moral codes, like these other products of human culture, are often consistent with the biological predispositions of the human species, dispositions we may to some extent share with other animals. But this consistency between ethical norms and biological tendencies is not necessary or universal: it does not apply to all ethical norms in a given society, much less in all human societies.

Moral codes, like any other dimensions of cultural systems, depend on the existence of human biological nature and must be consistent with it in the sense that they could not counteract it without promoting their own demise. Moreover, the acceptance and persistence of moral norms is facilitated whenever they are consistent with biologically conditioned human behaviors. But the moral norms are independent of such behaviors in the sense that some norms may not favor, and may hinder, the survival and reproduction of the individual and its genes, which are the targets of biological evolution. Discrepancies between accepted moral rules and biological survival are, however, necessarily limited in scope or would otherwise lead to the extinction of the groups accepting such discrepant rules.

THE NECESSARY CONDITIONS FOR ETHICAL BEHAVIOR

The question whether ethical behavior is determined by our biological nature must be answered in the affirmative. By "ethical behavior" I mean here to refer to the urge of *judging* human actions as either good or bad, which is not the same as "good behavior" (i.e., *doing* what is perceived as good instead of what is perceived as evil). Humans exhibit ethical behavior by nature because their biological constitution determines the presence in them of the three necessary, and jointly sufficient, conditions for ethical behavior. These conditions are: (a) the ability to anticipate the consequences of one's own actions; (b) the ability to make value judgments; and (c) the ability to choose between alternative courses of action. I shall briefly examine each of these abilities and show that they exist as a consequence of the eminent intellectual capacity of human beings.

The ability to anticipate the consequences of one's own actions is the most fundamental of the three conditions required for ethical behavior. Only if I can anticipate that pulling the trigger will shoot the bullet, which in turn will strike and kill my enemy, can the action of pulling the trigger be evaluated as nefarious. Pulling a trigger is not in itself a moral action; it becomes so by virtue of its relevant consequences. My action has an ethical dimension only if I do anticipate these consequences.

The ability to anticipate the consequences of one's actions is closely related to the ability to establish the connection between means and ends; that is, of seeing a mean precisely as mean, as something that serves a particular end or purpose. This ability to establish the connection between means and their ends requires the ability to anticipate the future and to form mental images of realities not present or not yet in existence.

The ability to establish the connection between means and ends happens to be the fundamental intellectual capacity that has made possible the development of human culture and technology. The evolutionary roots of this capacity may be found in the evolution of bipedal gait, which transformed the anterior limbs of our ancestors from organs of locomotion into organs of manipulation. The hands thereby gradually became organs adept for the construction and use of objects for hunting and other activities that improved survival and reproduction, that is, that increased the reproductive fitness of their carriers.

The construction of tools, however, depends not only on manual dexterity, but in perceiving them precisely as tools, as objects that help to perform certain actions, that is, as means that serve certain ends or purposes: a knife for cutting, an arrow for hunting, an animal skin for protecting the body from the cold. The hypothesis I am propounding is that natural selection promoted the intellectual capacity of our bipedal ancestors, because increased intelligence facilitated the perception of tools as tools, and therefore their construction and use, with the ensuing amelioration of biological survival and reproduction.

The development of the intellectual abilities of our ancestors took place over two million years or longer, gradually increasing the ability to connect means with their ends and, hence, the possibility of making ever more complex tools serving remote purposes. The ability to anticipate the future, essential for ethical behavior, is therefore closely associated with the development of the ability to construct tools, an ability that has produced the advanced technologies of modern societies and that is largely responsible for the success of mankind as a biological species.

The second condition for the existence of ethical behavior is the ability to make value judgments, to perceive certain objects or deeds as more desirable than others. Only if I can see the death of my enemy as preferable to his or her survival (or vice versa) can the action leading to his or her demise be thought of as moral. If the alternative consequences of an action are neutral with respect to value, the action cannot be characterized as ethical. The ability to make value judgments depends on the capacity for abstraction, that is, on the capacity to perceive actions or objects as members of general classes. This makes it possible to compare objects or actions with one another and to perceive some as more desirable than others. The capacity for abstraction, necessary to perceive individual objects or actions as members of general classes, requires an advanced intelligence such as it exists in humans and apparently in them alone. Thus, I see the ability to make value judgments primarily as an implicit consequence of the enhanced intelligence favored by natural selection in human evolution. Nevertheless, valuing certain objects or actions and choosing them over their alternatives can be of biological consequence; doing this in terms of general categories can be beneficial in practice.

Moral judgments are a particular class of value judgments; namely, those where preference is not dictated by one's own interest or profit, but by regard for others, which may cause benefits to particular individuals (altruism), or take into consideration the interests of a social group to which one belongs. Value judgments indicate preference for what is perceived as good and rejection of what is perceived as bad; good and bad may refer to monetary, aesthetic, or all sorts of other kinds of values. Moral judgments concern the values of right and wrong in human conduct.

The third condition necessary for ethical behavior is the ability to choose between alternative courses of action. Pulling the trigger can be a moral action only if I have the option not to pull it. A necessary action beyond our control is not a moral action: the circulation of the blood or the digestion of food are not moral actions.

Whether there is free will has been much discussed by philosophers... and this is not the appropriate place to review the arguments. I will only advance two considerations based on our commonsense experience. One is our profound personal conviction that the possibility of choosing between alternatives is genuine rather than only apparent.[4] The second consideration is that when we confront a given situation that requires action on our part, we are able mentally to explore alternative courses of action, thereby extending the field within which we can exercise our free will. In any case, if there were no free will, there would be no ethical behavior; morality would only be an illusion. The point that I wish to make here is, however, that free will is dependent on the existence of a well-developed intelligence, which makes it possible to explore alternative courses of action and to choose one or another in view of the anticipated consequences.

In summary, my proposal is that ethical behavior is an attribute of the biological makeup of humans and is, in that sense, a product of biological evolution. But I see no evidence that ethical behavior developed because it was adaptive in itself. I find it hard to see how evaluating certain actions as either good or evil (not just choosing some actions rather than others, or evaluating them with respect to their practical consequences) would promote the reproductive fitness of the evaluators. Nor do I see how there might be some form of "incipient" ethical behavior that would then be further promoted by natural selection. The three necessary conditions for there being ethical behavior are manifestations of advanced intellectual abilities.

It rather seems that the likely target of natural selection may have been the development of these advanced intellectual capacities. This development was favored by natural selection because the construction and use of tools improved the strategic position of our bipedal ancestors. Once bipedalism evolved and tool-using and tool-making became possible, those individuals more effective in these functions had a greater probability of biological success. The biological advantage provided by the design and use of tools persisted long enough so that intellectual abilities continued to increase, eventually yielding the eminent development of intelligence that is characteristic of *Homo sapiens.*

EVOLUTIONARY THEORIES OF MORALITY

There are many theories concerned with the rational grounds for morality, such as deductive theories that seek to discover the axioms or fundamental principles that determine what is morally correct on the basis of direct moral intuition. There also are theories, like logical positivism or existentialism, which negate rational foundations for morality, reducing moral principles to emotional decisions or to other irrational grounds. Since the publication of Darwin's theory of evolution by natural selection, philosophers as well as biologists have attempted to find in the evolutionary process the justification for moral norms. The common ground to all such proposals is that evolution is a natural process that achieves goals that are desirable and thereby morally good; indeed it has produced humans. Proponents of these ideas claim that only the evolutionary goals can give moral value to human action: whether a human deed is morally right depends on whether it directly or indirectly promotes the evolutionary process and its natural objectives.

Herbert Spencer[5] was perhaps the first philosopher seeking to find the grounds of morality in biological evolution. More recent attempts include those of the distinguished evolutionists J. S. Huxley[6] and C. H. Waddington[7] and of Edward O. Wilson,[8] founder of sociobiology as an independent discipline engaged in discovering the biological foundations of social behavior. I have argued elsewhere[9] that the moral theories proposed by Spencer, Huxley, and Waddington are mistaken and fail to avoid the naturalistic fallacy.[10] These authors argue, in one or other fashion, that the standard by which human actions are judged good or evil derives from the contribution the actions make to evolutionary advancement or progress. A blunder of this argumentation is that it is based on value judgments about what is or is not progressive in (particularly human) evolution.[11] There is nothing objective in the evolutionary process itself that makes the success of bacteria, which have persisted for more than three billion years and in enormous diversity and numbers, less "progressive" than that of the vertebrates, even though the latter are more complex.[12] Nor are the insects, of which more than one million species exist, less successful or less progressive from a purely biological perspective than humans or any other mammal species. Moreover, the proponents of evolution-grounded moral codes fail to demonstrate why the promotion of biological evolution by itself should be the standard to measure what is morally good.

The most recent and most subtle attempt to ground the moral codes on the evolutionary process emanates from the sociobiologists, particularly from E. O. Wilson,[13] who starts by proposing that "scientists and humanists should consider together the possibility that the time has come for ethics to be removed temporarily from the hands of the philosophers and biologicized."[14] The sociobiologists' argument is that the perception that morality exists is an epigenetic manifestation of our genes, which so manipulate humans as to make them believe that some behaviors are morally "good" so that people behave in ways that are good for their genes. Humans might not otherwise pursue these behaviors (altruism, for example) because their genetic benefit is not apparent (except to sociobiologists after the development of their discipline).[15]

As I have argued elsewhere, the sociobiologists' account of the evolution of the moral sense is misguided.[16] As I have argued above, we make moral judgments as a consequence of our eminent intellectual abilities, not as an innate way for achieving biological gain. Moreover, the sociobiologists' position may be interpreted as calling for the supposition that those norms of morality should be considered supreme that achieve the most biological (genetic) gain (because that is, in their view, why the moral sense evolved at all). This, in turn, would justify social preferences, including racism and even genocide, that many of us (sociobiologists included) judge morally obtuse and even heinous.

The evaluation of moral codes or human actions must take into account biological knowledge, but biology is insufficient for determining which moral codes are, or should be, accepted. This may be reiterated by returning to the analogy with human languages. Our biological nature determines the sounds that we can or cannot utter and also constrains human language in other ways. But a language's syntax and vocabulary are not determined by our biological nature (otherwise, there could not be a multitude of tongues), but are products of human culture. Likewise, moral norms are not determined by biological processes, but by cultural traditions and principles that are products of human history.

THE ROAD TRAVERSED

The portrait of human nature that I have presented here abides by Augustine's injunction against "talking nonsense" and "maintaining foolish opinions about the Scriptures." It is cognizant of science achievements, particularly the discoveries of evolutionary biology and the continuity of descent between humans and other animals. Further, it is an image that acknowledges that human nature transcends biology.

I started this essay summoning evidence for the evolutionary continuity between humans and our primate ancestors. Human nature, I argued, is biological nature, but is also much more. In mankind, biological evolution has transcended itself and has ushered in a new mode, cultural evolution, more rapid and effective than the biological mode. A complete portrait of human nature must integrate mankind's two dimensions, the biological and the cultural, its biological continuity with the living world and its radical distinctness. Mankind's uniqueness is embodied in a suite of features that include ethical behavior and religious beliefs. These distinctly human characteristics (and surely others, like art, science, technology, and sociopolitical institutions) emanate as epigenetic outcome of mankind's enhanced intelligence, which may very well be the fundamental and most distinctive trait targeted by natural selection in the process that brought about *Homo sapiens.*

NOTES

1. A. C. Wilson and R. L. Cann, "The Recent African Genesis of Humans," *Scientific American*, April 1992, pp. 68–73.

2. The theoretical number of ancestors for any one individual becomes enormous after some tens of generations, but "inbreeding" occurs: after some generations, ancestors appear more than once in the genealogy.

3. F.J. Ayala, "The Myth of Eve: Molecular Biology and Human Origins," *Science* 270 (1995): 1930–36.

4. That free will is a universal and inalienable human attribute was thus conveyed by Confucius: "One may rob an army of 30 its commander-in-chief; one cannot deprive the humblest man of his free will." *The Analects of Confucius*, translation and notes by Simon Leys (New York: Norton, 1996).

5. H. Spencer, *The Principles of Ethics* (London, 1893).

6. T. H. Huxley and J. S. Huxley, *Touchstone for Ethics* (New York: Harper, 1947); J. S. Huxley, *Evolution in Action* (New York: Harper, 1953).

7. C. H. Waddington, *The Ethical Animal* (London: Allen & Unwin, 1960).

8. E. O. Wilson, *Sociobiology: The New Synthesis* (Cambridge, MA: Harvard University Press, 1975); and E. O. Wilson, *On Human Nature* (Cambridge, MA: Harvard University Press, 1978).

9. F.J. Ayala, "The Biological Roots of Morality," *Biology and Philosophy* 2 (1987): 235–52.

10. The "naturalistic fallacy" consists in identifying what "is" with what "ought" to be (G. E. Moore, *Principia Ethica* [Cambridge: Cambridge University Press, 1903]). This error was already pointed out by Hume: "In every system of morality which I have hitherto met with I have always remarked that the author proceeds for some time in the ordinary way of reasoning... when of a sudden I am surprised to find, that instead of the usual copulations of propositions, is and is not, I meet with no proposition that is not connected with an ought or ought not. This change is imperceptible; but is, however, of the last consequence. For as this ought or ought not expresses some new relation or affirmation, it is necessary that it should be observed and explained; and at the same time a reason should be given, for what seems altogether inconceivable, how this new relation can be a deduction from others, which are entirely different from it" (D. Hume, *Treatise of Human Nature* [Oxford: Oxford University Press, 1740], p. 1978).

11. Ibid.

12. See S.J. Gould, *Full House. The Spread of Excellence from Plato to Darwin* (New York: Harmony Books, 1996).

13. Wilson, *Sociobiology: The New Synthesis*; and Wilson, *On Human Nature.*

14. Wilson, *Sociobiology: The New Synthesis*, p. 562.

15. M. Ruse, *Taking Darwin Seriously: A Naturalistic Approach to Philosophy* (Oxford: Basil Blackwell, 1986); M. Ruse, "Evolutionary Ethics: A Phoenix Arisen," *Zygon* 21 (1986): 95–112; M. Ruse and E. O. Wilson, "Moral Philosophy as Applied Science," *Philosophy: Journal of the Royal Institute of Philosophy* 61 (1986): 173–92.

16. Ayala, "The Biological Roots of Morality"; F.J. Ayala, "The Difference of Being Human: Ethical Behavior as an Evolutionary Byproduct," in *Biology, Ethics and the Origin of Life*, ed. H. Rolston III (Boston and London: Jones and Bartlett, 1995), pp. 113–35.

CHAPTER 10

UNIVERSAL DARWINISM

RICHARD DAWKINS

It is widely believed on statistical grounds that life has arisen many times all around the universe (Asimov 1979; Billingham 1981). However varied in detail alien forms of life may be, there will probably be certain principles that are fundamental to all life, everywhere. I suggest that prominent among these will be the principles of Darwinism. Darwin's theory of evolution by natural selection is more than a local theory to account for the existence and form of life on Earth. It is probably the only theory that can adequately account for the phenomena that we associate with life.

My concern is not with the details of other planets. I shall not speculate about alien biochemistries based on silicon chains, or alien neurophysiologies based on silicon chips. The universal perspective is my way of dramatizing the importance of Darwinism for our own biology here on Earth, and my examples will be mostly taken from Earthly biology. I do, however, also think that "exobiologists" speculating about extraterrestrial life should make more use of evolutionary reasoning. Their writings have been rich in speculation about how extraterrestrial life might work, but poor in discussion about how it might

---

From Richard Dawkins, "Universal Darwinism," in *Evolution from Molecules to Men*, ed. S. Bendall (Cambridge: Cambridge University Press, 1983), pp. 403–25. Reprinted with permission of Cambridge University Press.

evolve. This essay should, therefore, be seen firstly as an argument for the general importance of Darwin's theory of natural selection; secondly as a preliminary contribution to a new discipline of "evolutionary exobiology."

The "growth of biological thought" (Mayr 1982) is largely the story of Darwinism's triumph over alternative explanations of existence. The chief weapon of this triumph is usually portrayed as *evidence.* The thing that is said to be wrong with Lamarck's theory is that its assumptions are factually wrong. In Mayr's words: "Accepting his premises. Lamarck's theory was as legitimate a theory of adaptation as that Darwin. Unfortunately, these premises turned out to be invalid." But I think we can say something stronger: *even accepting his premises*, Lamarck's theory is *not* as legitimate a theory of adaptation as that of Darwin because, unlike Darwin's, it is *in principle* incapable of doing the job we ask of it—explaining the evolution of organized, adaptive complexity. I believe this is so for all theories that have ever been suggested for the mechanism of evolution except Darwinian natural selection, in which case Darwinism rests on a securer pedestal than that provided by facts alone.

Now, I have made reference to theories of evolution "doing the job we ask of them." Everything turns on the question of what that job is. The answer may be different for different people. Some biologists, for instance, get excited about "the species problem," while I have never mustered much enthusiasm for it as a "mystery of mysteries." For some, the main thing that any theory of evolution has to explain is the diversity of life—cladogenesis. Others may require of their theory an explanation of the observed changes in the molecular constitution of the genome. I would not presume to try to convert any of these people to my point of view. All I can do is to make my point of view clear, so that the rest of my argument is clear. I agree with Maynard Smith (1969) that "The main task of any theory of evolution is to explain adaptive complexity, i.e., to explain the same set of facts which Paley used as evidence of a Creator." I suppose people like me might be labeled neo-Paleyists, or perhaps "transformed Paleyists." We concur with Paley that adaptive complexity demands a very special kind of explanation: either a Designer as Paley taught, or something such as natural selection that does the job of a designer. Indeed, adaptive complexity is probably the best diagnostic of the presence of life itself.

ADAPTIVE COMPLEXITY AS A DIAGNOSTIC CHARACTER OF LIFE

If you find something, anywhere in the universe, whose structure is complex and gives the strong appearance of having been designed for a purpose, then that something either is alive, or was once alive, or is an artifact created by something alive. It is fair to include fossils and artifacts since their discovery on any planet would certainly be taken as evidence for life there.

Complexity is a statistical concept (Pringle 1951). A complex thing is a statistically improbable thing, something with a very low a priori likelihood of coming into being. The number of possible ways of arranging the 10^{27} atoms of a human body is obviously inconceivably large. Of these possible ways, only very few would be recognized as a human body. But this is not, by itself, the point. Any existing configuration of atoms is, a posteriori, unique, as "improbable," with hindsight, as any other. The point is that, of all possible ways of arranging those 10^{27} atoms, only a tiny minority would constitute anything remotely resembling a machine that worked to keep itself in being, and to reproduce its kind. Living things are not just statistically improbable in the trivial sense of hindsight; their statistical improbability is limited by the a priori constraints of design. They are *adaptively* complex.

The term "adaptationist" has been coined as a pejorative name for one who assumes "without further proof that all aspects of the morphology, physiology and behavior of organisms are adaptive optimal solutions to problems" (Lewontin 1979). I have responded to this elsewhere (Dawkins 1982a, chapter 3). Here, I shall be an adaptationist in the much weaker sense that I shall only be concerned with those aspects of the morphology, physiology, and behavior of organisms that are undisputedly adaptive solutions to problems. In the same way a zoologist may specialize on vertebrates without denying the existence of invertebrates. I shall be preoccupied with undisputed adaptations because I have defined them as my working diagnostic characteristic of all life, anywhere in the universe, in the same way as the vertebrate zoologist might be preoccupied with backbones because backbones are the diagnostic character of all vertebrates. From time to time I shall need an example of an undisputed adaptation, and the time-honored eye will serve the purpose as well as ever (Paley 1828; Darwin 1859; any fundamentalist tract). "As far as the examination of the instrument goes, there is precisely the same proof that the eye was made for vision, as there is that the telescope was made for assisting it. They are made upon the same principles; both being adjusted to the laws by which the transmission and refraction of rays of light are regulated" (Paley 1828, vol. 1, p. 17).

If a similiar instrument were found upon another planet, some special explanation would be called for. Either there is a God, or, if we are going to explain the universe in terms of blind physical forces, those blind physical forces are going to have to be deployed in a very peculiar way. The same is not true of nonliving objects, such as the moon or the solar system (see below). Paley's instincts here were right.

> My opinion of Astronomy has always been, that it is *not* the best medium through which to prove the agency of an intelligent Creator.... The very simplicity of [the heavenly bodies'] appearance is against them.... Now we deduce design from relation, aptitude, and correspondence of *parts*. Some degree therefore of *complexity* is necessary to render a subject fit for this species of argument. But the heavenly bodies do not, except perhaps in the instance of Saturn's ring, present themselves to our observation as compounded of parts at all. (Paley 1828, vol. 2, 146–47)

A transparent pebble, polished by the sea, might act as a lens, focusing a real image. The fact that it is an efficient optical device is not particularly interesting because, unlike an eye or a telescope, it is too simple. We do not feel the need to invoke anything remotely resembling the concept of design. The eye and the telescope have many parts, all coadapted and working together to achieve the same functional end. The polished pebble has far fewer coadapted features: the coincidence of transparency, high refractive index and mechanical forces that polish the surface in a curved shape. The odds against such a threefold coincidence are not particularly great. No special explanation is called for.

Compare how a statistician decides what *P* value to accept as evidence for an effect in an experiment. It is a matter of judgment and dispute, almost of taste, exactly when a coincidence becomes too great to stomach. But, no matter whether you are a cautious statistician or a daring statistician, there are some complex adaptations whose "*P* value," whose coincidence rating, is so impressive that nobody would hesitate to diagnose life (or an artifact designed by a living thing). My definition of living complexity is, in effect, "that complexity which is too great to have come about through a single coincidence." For the purposes of this [chapter], the problem that any theory of evolution has to solve is how living adaptive complexity comes about.

In the book referred to previously, Mayr (1982) helpfully lists what he sees as the six clearly distinct theories of evolution that have ever been proposed in the history of biology. I shall use this list to provide me with my main headings in this [chapter]. For each of the six, instead of asking what the evidence is, for or against, I shall ask whether the theory is in principle capable of doing the job of explaining the existence of adaptive complexity. I shall take the six theories in order, and will conclude that only theory 6, Darwinian selection, matches up to the task.

Theory 1. Built-in capacity for, or drive toward, increasing perfection

To the modern mind this is not really a theory at all, and I shall not bother to discuss it. It is obviously mystical, and does not explain anything that it does not assume to start with.

Theory 2. Use and disuse plus inheritance of acquired characters

It is convenient to discuss this in two parts.

Use and Disuse

It is an observed fact that on this planet living bodies sometimes become better adapted as a result of use. Muscles that are exercised tend to grow bigger. Necks that reach eagerly toward the treetops may lengthen in all their parts. Conceivably, if on some planet such acquired improvements could be incorporated into the hereditary information, adaptive evolution could result. This is the theory often associated with Lamarck, although there was more to what Lamarck said. Crick (1982, 59) says of the idea: "As far as I know, no one has given general theoretical reasons why such a mechanism must be less efficient than natural selection. . . ." In this section and the next I shall give two general theoretical objections to Lamarckism of the sort which, I suspect,

Crick was calling for. I have discussed both before (Dawkins 1982b), so will be brief here. First the shortcomings of the principle of use and disuse.

The problem is the crudity and imprecision of the adaptation that the principle of use and disuse is capable of providing. Consider the evolutionary improvements that must have occurred during the evolution of an organ such as an eye, and ask which of them could conceivably have come about through use and disuse. Does "use" increase the transparency of a lens? No, photons do not wash it clean as they pour through it. The lens and other optical parts must have reduced, over evolutionary time, their spherical and chromatic aberration; could this come about through increased use? Surely not. Exercise might have strengthened the muscles of the iris, but it could not have built up the fine feedback control system that controls those muscles. The mere bombardment of a retina with colored light cannot call color-sensitive cones into existence, nor connect up their outputs so as to provide color vision.

Darwinian types of theory, of course, have no trouble in explaining all these improvements. Any improvement in visual accuracy could significantly affect survival. Any tiny reduction in spherical aberration may save a fast flying bird from fatally misjudging the position of an obstacle. Any minute improvement in an eye's resolution of acute colored detail may crucially improve its detection of camouflaged prey. The genetic basis of any improvement, however slight, will come to predominate in the gene pool. The relationship between selection and adaptation is a direct and close-coupled one. The Lamarckian theory, on the other hand, relies on a much cruder coupling: the rule that the more an animal uses a certain bit of itself, the bigger that bit ought to be. The rule occasionally might have some validity but not generally, and, as a sculptor of adaptation it is a blunt hatchet in comparison to the fine chisels of natural selection. This point is universal. It does not depend on detailed facts about life on this particular planet. The same goes for my misgivings about the inheritance of acquired characters.

Inheritance of Acquired Characters

The problem here is that acquired characters are not always improvements. There is no reason why they should be, and indeed the vast majority of them are injuries. This is not just a fact about life on earth. It has a universal rationale. If you have a complex and reasonably well-adapted system, the number of things you can do to it that will make it perform less well is vastly greater than the number of things you can do to it that will improve it (Fisher 1958). Lamarckian evolution will move in adaptive directions only if some mechanism—selection—exists for distinguishing those acquired characters that are improvements from those that are not. Only the improvements should be imprinted into the germ line.

Although he was not talking about Lamarckism, Lorenz (1966) emphasized a related point for the case of learned behavior, which is perhaps the most important kind of acquired adaptation. An animal learns to be a better animal during its own lifetime. It learns to eat sweet foods, say, thereby increasing its survival chances. But there is nothing inherently nutritious about a sweet taste. Something, presumably natural selection, has to have built into the nervous system the arbitrary rule: "treat sweet taste as reward," and this works because saccharine does not occur in nature whereas sugar does.

Similarly, most animals learn to avoid situations that have, in the past, led to pain. The stimuli that animals treat as painful tend, in nature, to be associated with injury and increased chance of death. But again the connection must ultimately be built into the nervous system by natural selection, for it is not an obvious, necessary connection (M. Dawkins 1980). It is easy to imagine artificially selecting a breed of animals that enjoyed being injured, and felt pain whenever their physiological welfare was being improved. If learning is adaptive *improvement*, there has to be, in Lorenz's phrase, an innate teaching mechanism, or "innate schoolmarm." The principle holds even where the reinforcers are "secondary," learned by association with primary reinforcers (Bateson 1983).

It holds, too, for morphological characters. Feet that are subjected to wear and tear grow tougher and more thick-skinned. The thickening of the skin is an acquired adaptation, but it is not obvious why the change went in this direction. In man-made machines, parts that are subjected to wear get thinner not thicker, for obvious reasons. Why does the skin on the feet do the opposite? Because, fundamentally, natural selection has worked in the past to ensure an adaptive rather than a maladaptive response to wear and tear.

The relevance of this for would-be Lamarckian evolution is that there has to be a deep Darwinian underpinning even if there is a Lamarckian surface structure: a Darwinian choice of which potentially acquirable characters shall in fact be acquired and inherited. As I have argued before (Dawkins 1982a, 164–77), this is true of a recent, highly publicized immunological theory of Lamarckian adaptation (Steele 1979). Lamarckian mechanisms cannot be fundamentally responsible for adaptive evolution. Even if acquired characters are inherited on some planet, evolution there will still rely on a Darwinian guide for its adaptive direction.

Theory 3. Direct induction by the environment

Adaptation, as we have seen, is a fit between organism and environment. The set of conceivable organisms is wider than the actual set. And there is a set of conceivable environments wider than the actual set. These two subsets match each other to some extent, and the matching is adaptation. We can reexpress the point by saying that information from the environment is present in the organism. In a few cases this is vividly literal—a frog carries a picture of its environment around on its back. Such information is usually carried by an animal in the less literal sense that a trained observer, dissecting a new animal, can reconstruct many details of its natural environment.

Now, how could the information get from the environment into the animal? Lorenz (1966) argues that there are two ways, natural selection and reinforcement learning, but that these are both *selective* processes in the broad sense (Pringle 1951). There is, in theory, an alternative method for the environment to imprint its information on the organism, and that is by direct "instruction" (Danchin 1979). Some theories of how the immune system works are "instinctive": antibody molecules are thought to be shaped directly by molding themselves around antigen molecules. The currently favored theory is, by contrast, selective (Burnet 1969). I take "instruction" to be synonymous with the "direct induction by the environment" of Mayr's theory 3. It is not always clearly distinct from theory 2.

Instruction is the process whereby information flows directly from its environment into an animal. A case could be made for treating imitation learning, latent learning, and imprinting (Thorpe 1963) as instructive, but for clarity it is safer to use a hypothetical example. Think of an animal on some planet, deriving camouflage from its tigerlike stripes. It lives in long dry "grass," and its stripes closely match the typical thickness and spacing of local grass blades. On our own planet such adaptation would come about through the selection of random genetic variation, but on the imaginary planet it comes about through direct instruction. The animals go brown except where their skin is shaded from the "sun" by blades of grass. Their stripes are therefore adapted with great precision, not just to any old habitat, but to the precise habitat in which they have sunbathed, and it is this same habitat in which they are going to have to survive. Local populations are automatically camouflaged against local grasses. Information about the habitat, in this case about the spacing patterns of the grass blades, has flowed into the animals, and is embodied in the spacing pattern of their skin pigment.

Instructive adaptation demands the inheritance of acquired characters if it is to give rise to permanent or progressive evolutionary change. "Instruction" received in one generation must be "remembered" in the genetic (or equivalent) information. This process is in principle cumulative and progressive. However, if the genetic store is not to become overloaded by the accumulations of generations, some mechanism must exist for discarding unwanted "instructions," and retaining desirable ones. I suspect that this must lead us, once again, to the need for some kind of selective process.

Imagine, for instance, a form of mammal-like life in which a stout "umbilical nerve" enabled a mother to "dump" the entire contents of her memory in the brain of her fetus. The technology is available even to our nervous systems: the corpus callosum can shunt large quantities of information from right hemisphere to left. An umbilical nerve could make the experience and wisdom of each generation automatically available to the next, and this might seem very desirable. But without a selective filter, it would take few generations for the load of information to become unmanageably large. Once again we come up against the need for a selective underpinning. I will leave this now, and make one more point about instructive adaptation (which applies equally to all Lamarckian types of theory).

The point is that there is a logical linkup between the two major theories of adaptive evolution—selection and instruction—and the two major theories of embryonic development—epigenesis and preformationism. Instructive evolution can work only if embryology is preformationistic. If embryology is epigenetic, as it is on our planet, instructive evolution cannot work. I have expounded the argument before (Dawkins 1982a, 174–76), so I will abbreviate it here.

If acquired characters are to be inherited, embryonic processes must be reversible: phenotypic change has to be read back into the genes (or equivalent). If embryology is preformationistic—the genes are a true blueprint—then it may indeed be reversible. You can translate a house back into its blueprint. But if embryonic development is epigenetic: if, as on this planet, the genetic information is more like a recipe for a cake (Bateson 1976) than a blueprint for a house, it is irreversible. There is no one-to-one mapping between bits of genome and bits of phenotype, any more than there is mapping between crumbs of cake and words of recipe. The recipe is not a blueprint that can be reconstructed from the cake. The transformation of recipe into cake cannot be put into reverse, and nor can the process of making a body. Therefore acquired adaptations cannot be read back into the "genes," on any planet where embryology is epigenetic.

This is not to say that there could not, on some planet, be a form of life whose embryology was preformationistic. That is a separate question. How likely is it? The form of life would have to be very different from ours, so much so that it is hard to visualize how it might work. As for reversible embryology itself, it is even harder to visualize. Some mechanism would have to scan the detailed form of the adult body, carefully noting down, for instance, the exact location of brown pigment in a sun-striped skin, perhaps turning it into a linear stream of code numbers, as in a television camera. Embryonic development would read the scan out again, like a television receiver. I have an intuitive hunch that there is an objection in principle to this kind of embryology, but I cannot at present formulate it clearly. All I am saying here is that, if planets are divided into those where embryology is preformationistic and those, like Earth, where embryology is epigenetic, Darwinian evolution could be supported on both kinds of planet, but Lamarckian evolution, even if there were no other reasons for doubting its existence, could be supported only on the preformationistic planets—if there are any.

The close theoretical link that I have demonstrated between Lamarckian evolution and preformationistic embryology gives rise to a mildly entertaining irony. Those with ideological reasons for hankering after a neo-Lamarckian view of evolution are often especially militant partisans of epigenetic, "interactionist," ideas of development, possibly—and here is the irony—for the very same ideological reasons (Koestler 1967; Ho and Saunders 1982).

Theory 4. Saltationism

The great virtue of the idea of evolution is that it explains, in terms of blind physical forces, the existence of undisputed adaptations whose statistical improbability is enormous, without recourse to the supernatural or the mystical. Since we define an undisputed adaptation as an adaptation that is too complex to have come about by chance, how is it possible for a theory to invoke only blind physical forces in explanation? The answer—Darwin's answer—is astonishingly simple when we consider how self-evident Paley's Divine Watchmaker must have seemed to his contemporaries. The key is that the coadapted parts do not have to be assembled *all at once.* They can be put together in small stages. But they really do have to be *small* stages. Otherwise we are back again with the problem we started with: the creation by chance of complexity that is too great to have been created by chance!

Take the eye again, as an example of an organ that contains a large number of independent coadapted parts, say, *N*. The a priori probability of any one of these *N* features coming into existence by chance is low, but not incredibly low. It is comparable to the chance of a crystal pebble being washed by the sea so that it acts as a lens. Any one adaptation on its own could, plausibly, have come into existence through blind physical forces. If each of the *N* coadapted features confers some slight advantage on its own, then the whole many-parted organ can be put together over a long period of time. This is particularly plausible for the eye—ironically in view of that organ's niche of honor in the creationist pantheon. The eye is, par excellence, a case where a fraction of an organ is better than no organ at all; an eye without a lens or even a pupil, for instance, could still detect the looming shadow of a predator.

To repeat, the key to the Darwinian explanation of adaptive complexity is the replacement of instantaneous, coincidental, multidimensional luck, by gradual, inch by inch, smeared-out luck. Luck is involved, to be sure. But a theory that bunches the luck up into major steps is more incredible than a theory that spreads the luck out in small stages. This leads to the following general principle of universal biology. Wherever in the universe adaptive complexity shall be found, it will have come into being gradually through a series of small alterations, never through large and sudden increments in adaptive complexity. We must reject Mayr's fourth theory, saltationism, as a candidate for explanation of the evolution of complexity.

It is almost impossible to dispute this rejection. It is implicit in the definition of adaptive complexity that the only alternative to gradualistic evolution is supernatural magic. This is not to say that the argument in favor of gradualism is a worthless tautology, an unfalsifiable dogma of the sort that creationists and philosophers are so fond of jumping about on. It is not *logically* impossible for a full-fashioned eye to spring de novo from virgin bare skin. It is just that the possibility is statistically negligible.

Now it has recently been widely and repeatedly publicized that some modern evolutionists reject "gradualism," and espouse what Turner (1982) has called theories of evolution by jerks. Since these are reasonable people without mystical leanings, they must be gradualists in the sense in which I am here using the term: the "gradualism" that they oppose must be defined differently. There are actually two confusions of language here, and I intend to clear them up in turn. The first is the common confusion between "punctuated equilibrium" (Eldredge and Gould 1972) and true saltationism. The second is a confusion between two theoretically distinct kinds of saltation.

Punctuated equilibrium is not macromutation, not saltation at all in the traditional sense of the term. It is, however, necessary to discuss it here, because it is popularly regarded as a theory of saltation, and its partisans quote, with approval, Huxley's criticism of Darwin for upholding the principle of *Natura non facit saltum* (Gould 1980). The punctuationist theory is portrayed as radical and revolutionary and at variance with the "gradualistic" assumptions of both Darwin and the neo-Darwinian synthesis (e.g., Lewin 1980). Punctuated equilibrium, however, was originally conceived as what the orthodox neo-Darwinian synthetic theory should truly predict, on a palaeontological timescale, if we take its embedded ideas of allopatric speciation seriously (Eldredge and Gould 1972). It derives its "jerks" by taking the "stately unfolding" of the neo-Darwinian synthesis, and inserting long periods of stasis separating brief bursts of gradual, albeit rapid, evolution.

The plausibility of such "rapid gradualism" is dramatized by a thought experiment of Stebbins (1982). He imagines a species of mouse, evolving larger body size at such an imperceptibly slow rate that the differences between the means of successive generations would be utterly swamped by sampling error. Yet even at this slow rate Stebbin's mouse lineage would attain the body size of a large elephant in about sixty thousand years, a time-span so short that it would be regarded as instantaneous by palaeontologists. Evolutionary change too *slow* to be detected by microevolutionists can nevertheless be too *fast* to be detected by microevolutionists. What a palacontologist sees as a "saltation" can in fact be a smooth and gradual change so slow as to be undetectable to the microevolutionist. This kind of palaeontological "saltation" has nothing to do with the one-generation macromutations that, I suspect, Huxley and Darwin had in mind when they debated *Natura non facit saltum.* Confusion has arisen here, possibly because some individual champions of punctuated equilibrium have also, incidentally, championed macromutation (Gould 1982). Other "punctuationists" have either confused their theory with macromutationism, or have explicitly invoked macromutation as one of the mechanisms of punctuation (e.g., Stanley 1981).

Turning to macromutation, or true saltation itself, the second confusion that I want to clear up is between the two kinds of macromutation that we might conceive of. I could name them, unmemorably, saltation (1) and saltation (2), but instead I shall pursue an earlier fancy for airliners as metaphors, and label them "Boeing 747" and "Stretched DC-8" saltation. 747 saltation is the inconceivable kind. It gets its name from Sir Fred Hoyle's much quoted metaphor for his own cosmic misunderstanding of Darwinism (Hoyle and Wickramasinghe 1981). Hoyle compared Darwinian selection to a tornado, blowing through a junkyard and assembling a Boeing 747 (what he overlooked, of course, was the point about luck being "smeared-out" in small steps—see above). Stretched DC-8 saltation is quite different. It is not in principle hard to believe in at all. It refers to large and sudden changes in magnitude of some biological measure, without an accompanying large increase in adaptive information. It is named after an airliner that was made by elongating the fuselage of an existing design, not adding significant new complexity. The change from DC-8 is a big change in magnitude—a saltation not a gradualistic series of tiny changes. But, unlike the change from junk-heap to 747, it is not a big increase in information content or complexity, and that is the point I am emphasizing by the analogy.

An example of DC-S saltation would be the following. Suppose the giraffe's neck shot out in one spectacular mutational step. Two parents had necks of standard antelope length. They had a freak child with a neck of modern giraffe length, and all giraffes are descended from this freak. This is unlikely to be true on Earth, but something like it may happen elsewhere in the universe. There is no objection to it in principle, in the sense that there is a profound objection to the (747) idea that a complex organ like an eye could arise from bare skin by a single mutation. The crucial difference is one of complexity.

I am assuming that the change from short antelope's neck to long giraffe's neck is not an increase in complexity. To be sure, both necks are exceedingly complex structures. You couldn't go from no-neck to either kind of neck in one step: that would be 747 saltation. But once the complex organization of the antelope's neck already exists, the step to giraffe's neck is just an elongation: various things have to grow faster at some stage in embryonic development; existing complexity is preserved. In practice, of course, such a drastic change in magnitude would be highly likely to have deleterious repercussions that would render the macromutant unlikely to survive. The existing antelope heart probably could not pump the blood up to the newly elevated giraffe head. Such practical objections to evolution by "DC-8 saltation" can only help my case in favor of gradualism, but I still want to make a separate, and more universal, case against 747 saltation.

It may be argued that the distinction between 747 and DC-8 saltation is impossible to draw in practice. After all, DC-8 saltations, such as the proposed macromutational elongation of the giraffe's neck, may appear very complex: myotomes, vertebrae, nerves, blood vessels, all have to elongate together. Why does this not make it a 747 saltation, and therefore rule it out? But although this type of "coadaptation" has indeed often been thought of as a problem for any evolutionary theory, not just macromutational ones (see Ridley 1982 for a history), it is so only if we take an impoverished view of developmental mechanisms. We know that single mutations can orchestrate changes in growth rates of many diverse parts of organs, and, when we think about developmental processes, it is not in the least surprising that this should be so. When a single mutation causes a *Drosophila* to grow a leg where an antenna ought to be, the leg grows in all its formidable complexity. But this is not mysterious or surprising, not a 747 saltation, because the organization of a leg is already present in the body before the mutation. Wherever, as in embryogenesis, we have a hierarchically branching tree of causal relationships, a small alteration at a senior node of the tree can have large and complex ramified effects on the tips of the twigs. But although the change may be large in magnitude, there can be no large and sudden increments in adaptive information. If you think you have found a particular example of a large and sudden increment in adaptively complex information in practice, you can be certain the adaptive information was already there, even if it is an atavistic "throwback" to an earlier ancestor.

There is not, then, any objection in principle to theories of evolution by jerks, even the theory of hopeful monsters (Goldschmidt 1940), provided that it is DC-8 saltation, not 747 saltation that is meant. Gould (1982) would clearly agree: "I regard forms of macromutation which include the sudden origin of new species with all their multifarious adaptations intact *ab initio*, as illegitimate." No educational biologist actually believes in 747 saltation, but not all have been sufficiently explicit about the distinction between DC-8 and 747 saltation. An unfortunate consequence is that creationists and their journalistic fellow-travelers have been able to exploit saltationist-sounding statements of respected biologists. The biologist's intended meaning may have been what I am calling DC-8 saltation, or even nonsaltatory punctuation; but the creationist assumes saltation in the sense that I have dubbed 747, and 747 saltation would, indeed, be a blessed miracle.

I also wonder whether an injustice is not being done to Darwin, owing to this same failure to come to grips with the distinction between DC-8 and 747 saltation. It is frequently alleged that Darwin was wedded to gradualism, and therefore that, if some form of evolution by jerks is proved, Darwin will have been shown wrong. This is undoubtedly the reason for the ballyhoo and publicity that has attended the theory of punctuated equilibrium. But was Darwin really opposed to all jerks? Or was he, as I suspect, strongly opposed only to 747 saltation?

As we have already seen, punctuated equilibrium has nothing to do with saltation, but anyway I think it is not at all clear that, as is often alleged, Darwin would have been discomfited by punctuationist interpretations of the fossil record. The following passage, from later editions of the *Origin*, sounds like something from a current issue of *Paleobiology*: "the periods during which species have been undergoing modification, though very long as measured by years, have probably been short in comparison with the periods during which these same species remained without undergoing any change."

Gould (1982) shrugs this off as somehow anomalous and away from the mainstream of Darwin's thought. As he correctly says: "You cannot do history by selective quotation and search for qualifying footnotes. General tenor and historical impact are the proper criteria. Did his contemporaries or descendants ever read Darwin as a saltationist?" Certainly nobody ever accused Darwin of being a saltationist. But to most people saltation means macromutation, and, as Gould himself stresses, "Punctuated equilibrium is not a theory of macromutation." More importantly, I believe we can reach a better understanding of Darwin's general gradualistic bias if we invoke the distinction between 747 and DC-8 saltation.

Perhaps part of the problem is that Darwin himself did not have the distinction. In some antisaltation passages it seems to be DC-8 saltation that he has in mind. But on those occasions he does not seem to feel very strongly about it: "About sudden jumps," he wrote in a letter in 1860, "I have no objection to them—they would aid me in some cases. All I can say is, that I went into the subject and found no evidence to make me believe in jumps [as a source of new species] and a good deal pointing in the other direction" (quoted in Gillespie 1979). This does not sound like a man fervently opposed, in principle, to sudden jumps. And of course there is no reason why he should have been fervently opposed, if he only had DC-8 saltations in mind.

But at other times he really is pretty fervent, and on those occasions, I suggest, he is thinking of 747 saltation: "... it is impossible to imagine so many co-adaptations being formed all by a chance blow" (quoted in Ridley 1982). As the historian Neal Gillespie puts it: "For Darwin, monstrous births, a doctrine favored by Chambers, Owen, Argyll, Mivart, and others, from clear theological as well as scientific motives, as an explanation of how new species or even higher taxa, had developed, was no better than a miracle: 'it leaves the case of the co-adaptation of organic beings to each other and to their physical conditions of life, untouched and unexplained.' It was 'no explanation' at all, of no more scientific value than creation 'from the dust of the earth'" (Gillespie 1979, 118).

As Ridley (1982) says of the "religious tradition of idealist thinkers [who] were committed to the explanation of complex adaptive contrivances by intelligent design," "The greatest concession they could make to Darwin was that the Designer operated by tinkering with the generation of diversity, designing the variation." Darwin's response was: "If I were convinced that I required such additions to the theory of natural selection, I would reject it as rubbish. ...I would give nothing for the theory of Natural selection, if it requires miraculous additions at any one stage of descent."

Darwin's hostility to monstrous saltation, then, makes sense if we assume that he was thinking in terms of 747 saltation—the sudden invention of new adaptive complexity. It is highly likely that that is what he was thinking of, because that is exactly what many of his opponents had in mind. Saltationists such as the Duke of Argyll (though presumably not Huxley) wanted to believe in 747 saltation, precisely because it did demand supernatural intervention. Darwin did not believe in it, for exactly the same reason. To quote Gillespie again (120): "...for Darwin, designed evolution, whether manifested in saltation, monstrous births, or manipulated variations, was but a disguised form of special creation."

I think this approach provides us with the only sensible reading of Darwin's well-known remark that "if it could be demonstrated that any complex organ existed, which could not possibly have been formed by numerous, successive, slight modifications, my theory would absolutely break down." That is not a plea for gradualism, as a modern palaeobiologist uses the term. Darwin's theory is falsifiable, but he was much too wise to make his theory that easy to falsify! Why on earth *should* Darwin have committed himself to such an arbitrarily restrictive version of evolution, a version that positively invites falsification? I think it is clear that he didn't. His use of the term "complex" seems to me to be clinching. Gould (1982) describes this passage from Darwin as "clearly invalid." So it is invalid if the alternative to slight modifications is seen as DC-8 saltation. But if the alternative is seen as 747 saltation, Darwin's remark is valid and very wise. Notwithstanding those whom Miller (1982) has unkindly called Darwin's more foolish critics, his theory is indeed falsifiable, and in the passage quoted he puts his finger on one way in which it might be falsified.

There are two kinds of imaginable saltation, then, DC-8 saltation and 747 saltation. DC-8 saltation is perfectly possible, undoubtedly happens in the laboratory and the farmyard, and may have made important contributions to evolution. 747 saltation is statistically ruled out unless there is supernatural intervention. In Darwin's own time, proponents and opponents of saltation often had 747 saltation in mind, because they believed in—or were arguing against—divine intervention. Darwin was hostile to (747) saltation, because he correctly saw natural selection as an *alternative* to the miraculous as an explanation for adaptive complexity. Nowadays saltation either means punctuation (which isn't saltation at all) or DC-8 saltation, neither of which Darwin would have had strong objections to in principle, merely doubts about the facts. In the modern context, therefore, I do not think Darwin should be labeled a strong gradualist. In the modern context, I suspect that he would be rather open-minded.

It is in the anti-747 sense that Darwin was a passionate gradualist, and it is in the same sense that we must all be gradualists, not just with respect to life on earth, but with respect to life all over the universe. Gradualism in this sense is essentially synonymous with evolution. The sense in which we may be nongradualists is a much less radical, although still quite interesting, sense. The theory of evolution by jerks has been hailed on television and elsewhere as radical and revolutionary, a paradigm shift. There is, indeed, an interpretation of it which is revolutionary, but that interpretation (the 747 macromutation version) is certainly wrong, and is apparently not held by its original proponents. The sense in which the theory might be right is not particularly revolutionary. In this field you may choose your jerks so as to be revolutionary, *or* so as to be correct, but not both.

Theory 5. Random evolution

Various members of this family of theories have been in vogue at various times. The "mutationists" of the early part of this century—De Vries, W. Bateson, and their colleagues—believed that selection served only to weed out deleterious freaks, and that the real driving force in evolution was mutation pressure. Unless you believe mutations are directed by some mysterious life force, it is sufficiently obvious that you can be a mutationist only if you forget about adaptive complexity—forget, in other words, most of the consequences of evolution that are of any interest! For historians there remains the baffling enigma of how such distinguished biologists as De Vries, W.

Bateson, and T. H. Morgan could rest satisfied with such a crassly inadequate theory. It is not enough to say that De Vries's view was blinkered by his working only on the evening primrose. He only had to look at the adaptive complexity in his own body to see that "mutationism" was not just a wrong theory: It was an obvious nonstarter.

These post-Darwinian mutationists were also saltationists and anti-gradualists, and Mayr treats them under that heading, but the aspect of their view that I am criticizing here is more fundamental. It appears that they actually thought that mutation, on its own without selection, was sufficient to explain evolution. This could not be so on any nonmystical view of mutation, whether gradualist or saltationist. If mutation is undirected, it is clearly unable to explain the adaptive directions of evolution. If mutation is directed in adaptive ways we are entitled to ask how this comes about. At least Lamarck's principle of use and disuse makes a valiant attempt at explaining how variation might be directed. The "mutationists" didn't even seem to see that there was a problem, possibly because they underrated the importance of adaptation—and they were not the last to do so. The irony with which we must now read W. Bateson's dismissal of Darwin is almost painful: "the transformation of masses of populations by imperceptible steps guided by selection is, as most of us now see, so inapplicable to the fact that we can only marvel . . . at the want of penetration displayed by the advocates of such a proposition . . ." (W. Bateson 1913; quoted in Mayr 1982).

Nowadays some population geneticists describe themselves as supporters of "non-Darwinian evolution." They believe that a substantial number of the gene replacements that occur in evolution are nonadaptive substitutions of alleles whose effects are indifferent relative to one another (Kimura 1968). This may well be true, if not in Israel (Nevo 1983) maybe somewhere in the universe. But it obviously has nothing whatever to contribute to solving the problem of the evolution of adaptive complexity. Modern advocates of neutralism admit that their theory cannot account for adaptation, but that doesn't seem to stop them from regarding the theory as interesting. Different people are interested in different things.

The phrase "random genetic drift" is often associated with the name of Sewall Wright, but Wright's conception of the relationship between random drift and adaptation is altogether subtler than the others I have mentioned

(Wright 1980). Wright does not belong in Mayr's fifth category, for he clearly sees selection as the driving force of adaptive evolution. Random drift may make it easier for selection to do its job by assisting the escape from local optima (Dawkins 1982a, 40), but it is still selection that is determining the rise of adaptive complexity.

Recently palaeontologists have come up with fascinating results when they perform computer simulations of "random phylogenies" (e.g., Raup 1977). These random walks through evolutionary time produce trends that look uncannily like real ones, and it is disquietingly easy, and tempting, to read into the random phylogenies apparently adaptive trends that, however, are not there. But this does not mean that we can admit random drift as an explanation of real adaptive trends. What it might mean is that some of us have been too facile and gullible in what we think are adaptive trends. This does not alter the fact that there are some trends that really are adaptive—even if we don't always identify them correctly in practice—and those real adaptive trends can't be produced by random drift. They must be produced by some nonrandom force, presumably selection.

So, finally, we arrive at the sixth of Mayr's theories of evolution.

Theory 6. Direction (order) imposed on random variation by natural selection

Darwinism—the nonrandom selection of randomly varying replicating entities by reason of their "phenotypic" effects—is the only force I know that can, in principle, guide evolution in the direction of adaptive complexity. It works on this planet. It doesn't suffer from any of the drawbacks that beset the other five classes of theory, and there is no reason to doubt its efficacy throughout the universe.

The ingredients in a general recipe for Darwinian evolution are replicating entities of some kind, exerting phenotypic "power" of some kind over their replication success. I have referred to these necessary entities as "active germ-line replicators" or "optimons" (Dawkins 1982a, chap. 5). It is important to keep their replication conceptually separate from their phenotypic effects, even though, on some planets, there may be a blurring in practice. Phenotypic adaptations can be seen as tools of replicator propagation.

Gould (1983) disparages the replicator's-eye view of evolution as preoccupied with "book-keeping." The metaphor is a superficially happy one: it is easy to see the genetic changes that accompany evolution as book-keeping entries, mere accountant's records of the really interesting phenotypic events going on in the outside world. Deeper consideration, however, shows that the truth is almost the exact opposite. It is central and essential to Darwinian (as opposed to Lamarckian) evolution that there shall be causal arrows flowing from genotype to phenotype, but not in the reverse direction. Changes in gene frequencies are not passive book-keeping records of phenotypic changes: it is precisely because (and to the extent that) they actively *cause* phenotypic changes that evolution of the phenotype can occur. Serious errors flow, both from a failure to understand the importance of this one-way flow (Dawkins 1982a, chapter 6), and from an overinterpretation of it as inflexible and undeviating "genetic determinism" (Dawkins 1982a, chap. 2).

The universal perspective leads me to emphasize a distinction between what may be called "one-off selection" and "cumulative selection." Order in the nonliving world may result from processes that can be portrayed as a rudimentary kind of selection. The pebbles on a seashore become sorted by the waves, so that larger pebbles come to lie in layers separate from smaller ones. We can regard this as an example of the selection of a stable configuration out of initially more random disorder. The same can be said of the "harmonious" orbital patterns of planets around stars, and electrons around nuclei, of the shapes of crystals, bubbles, and droplets, even, perhaps, of the dimensionality of the universe in which we find ourselves (Atkins 1981). But this is all one-off selection. It does not give rise to progressive evolution because there is no replication, no succession of generations. Complex adaptation requires many generations of cumulative selection, each generation's change building upon what has gone before. In one-off selection, a stable state develops and is then maintained. It does not multiply, does not have offspring.

In life the selection that goes on in *any one generation* is one-off selection, analogous to the sorting of pebbles on a beach. The peculiar feature of life is that successive generations of such selection build up, progressively and cumulatively, structures that are eventually complex enough to foster the strong illusion of design. One-off selection is a commonplace of physics and cannot give rise to adaptive complexity. Cumulative selection is the hallmark of biology and is, I believe, the force underlying all adaptive complexity.

OTHER TOPICS FOR A FUTURE SCIENCE OF UNIVERSAL DARWINISM

Active germ-line replicators together with their phenotypic consequences, then, constitute the general recipe for life, but the form of the system may vary greatly from planet to planet, both with respect to the replicating entities themselves, and with respect to the "phenotypic" means by which they ensure their survival. Indeed, the very distinction between "genotype" and "phenotype" may be blurred (L. Orgel, personal communication). The replicating entities do not have to be DNA or RNA. They do not have to be organic molecules at all. Even on this planet it is possible that DNA itself is a late usurper of the role, taking over from some earlier, inorganic crystalline replicator (Cairns-Smith 1982). It is also arguable that today selection operates on several levels, for instance, the levels of the gene and the species or lineage, and perhaps some unit of cultural transmission (Lewontin 1970).

A full science of Universal Darwinism might consider aspects of replicators transcending their detailed nature and the time-scale over which they are copied.

For instance, the extent to which they are "particulate" as opposed to "blending" probably has a more important bearing on evolution than their detailed molecular or physical nature. Similarly, a universe-wide classification of replicators might make more reference to their dimensionality and coding principles than to their size and structure. DNA is a digitally coded one-dimensional array. A "genetic" code in the form of a two-dimensional matrix is conceivable. Even a three-dimensional code is imaginable, although students of Universal Darwinism will probably worry about how such a code could be "read." (DNA is, of course, a molecule whose 3-dimensional structure determines how it is replicated and transcribed, but that doesn't make it a 3-dimensional code. DNA's meaning depends upon the 1-dimensional sequential arrangement of its symbols, not upon their 3-dimensional position relative to one another in the cell.) There might also be theoretical problems with analogue, as opposed to digital codes, similar to the theoretical problems that would be raised by a purely analogue nervous system (Rushton 1961).

As for the phenotypic levers of power by which replicators influence their survival, we are so used to their being bound up into discrete organisms or "vehicles" that we forget the possibility of a more diffuse extracorporeal or "extended" phenotype. Even on this Earth a large amount of interesting adaptation can be interpreted as part of the extended phenotype (Dawkins 1982a, chaps. 11, 12, and 13). There is, however, a general theoretical case that can be made in favor of the discrete organismal body, with its own recurrent life cycle, as a necessity in any process of evolution of advanced adaptive complexity (Dawkins 1982a, chap. 14), and this topic might have a place in a full account of Universal Darwinism.

Another candidate for full discussion might be what I shall call divergence, and convergence or recombination of replicator lineages. In the case of Earthbound DNA, "convergence" is provided by sex and related processes. Here the DNA "converges" within the species after having very recently "diverged." But suggestions are now being made that a different kind of convergence can occur among lineages that originally diverged an exceedingly long time ago. For instance, there is evidence of gene transfer between fish and bacteria (Jacob 1983). The replicating lineages on other planets may permit very varied kinds of recombination, on very different time-scales. On Earth the rivers of phylogeny are almost entirely divergent: if main tributaries ever recontact each other after branching apart it is only through the tiniest of trickling cross-streamlets, as in the fish/bacteria case. There is, of course, a richly anastomosing delta of divergence and convergence due to sexual recombination within the species, but only *within* the species. There may be planets on which the "genetic" system permits much more cross talk at all levels of the branching hierarchy, one huge fertile delta.

I have not thought enough about the fantasies of the previous paragraphs to evaluate their plausibility. My general point is that there is one limiting constraint upon all speculations about life in the universe. If a life-form displays adaptive complexity, it must possess an evolutionary mechanism capable of generating adaptive complexity. However diverse evolutionary mechanisms may be, if there is no other generalization that can be made about life all around the universe, I am betting it will always be recognizable as Darwinian life. The Darwinian Law (Eigen 1983) may be as universal as the great laws of physics.

NOTE

As usual I have benefited from discussions with many people, including especially Mark Ridley, who also criticized the manuscript, and Alan Grafen. Dr. F. J. Ayala called attention to an important error in the original spoken version of the paper.

REFERENCES

Asimov, I. *Extraterrestrial Civilizations*. London: Pan, 1979.

Atkins, P. W. *The Creation*. Oxford: W. H. Freeman, 1981.

Bateson, P. P. G. "Specificity and the Origins of Behavior." *Advances in the Study of Behavior* 6 (1976): 1–20.

———. "Rules for Changing the Rules." In *Evolution from Molecules to Men*, edited by D. S. Bendall, 483–501. Cambridge, MA: Cambridge University Press, 1983.

Billingham, J. *Life in the Universe*. Cambridge, MA: MIT Press, 1981.

Burnet, F. M. *Cellular immunology*. Melbourne: Melbourne University Press, 1969.

Cairns-Smith, A. G. *Genetic Takeover*. Cambridge, MA: Cambridge University Press, 1982.

Crick, F. H. C. *Life Itself*. London: Macdonald, 1982.

Danchin A. "Themes de La biologie: theories instructives et theories selectives." *Revue des Questions Scientifiques* 150 (1970): 151–64.

Darwin, C. R. *The Origin of Species*. 1859. Reprint, London: Penguin, 1968.

Dawkins, M. *Animal Suffering: The Science of Animal Welfare*. London: Chapman & Hall, 1980.

Dawkins, R. *The Extended Phenotype*. Oxford: W. H. Freeman, 1982a.

———. "The Necessity of Darwinism." *New Scientist* 94 (1982b): 130–32. Reprinted in *Darwin Up to Date*, edited by J. Cherfas. London: New Scientist, 1982.

Eigen, M. "Self-replication and Molecular Evolution." In S. D. S. Bendall, *Evolution from Molecules to Men*, 105–30. Cambridge, MA: Cambridge University Press, 1983.

Eldredge, N., and S.J. Gould. "Punctuated Equilibria: An Alternative to Phyletic Gradualism." In *Models in Paleobiology*, edited by T. J. M. Schopf. San Francisco: Freeman Cooper, 1972.

Fisher, R. A. *The Genetical Theory of Natural Selection*. New York: Dover, 1958.

Gillespie, N. C. *Charles Darwin and the Problem of Creation*. Chicago: University of Chicago Press, 1979.

Goldschmidt, R. *The Material Basis of Evolution.* New Haven, CT: Yale University Press, 1940.

Gould, S. J. *The Panda's Thumb.* New York: W. W. Norton, 1980.

———. "The Meaning of Punctuated Equilibrium and Its Role in Validating a Hierarchical Approach to Macroevolution." In S. R. Milkman, *Perspectives on Evolution,* 83–104. Sunderland, MA: Sinauer, 1982.

———. "Irrelevance, Submission and Partnership: The Changing Role of Palaeontology in Darwin's Three Centennials and a Modest Proposal for Macroevolution." In S. D. S. Bendall, *Evolution from Molecules to Men,* 387–402. Cambridge, MA: Cambridge University Press, 1983.

Ho, M. W., and P. T. Saunders. "Adaptation and Natural Selection: Mechanism and Teleology." In *Towards a Liberatory Biology,* edited by S. Rose, 85–102. 1982.

Hoyle, E., and N. C. Wickramasinghe. *Evolution from Space.* London: J. M. Dent, 1981.

Jacob, F. "Molecular Tinkering in Evolution." In *Evolution from Molecules to Men,* edited by D. S. Bendall, 131–44. Cambridge, MA: Cambridge University Press, 1983.

Kimura, M. "Evolutionary Rate at the Molecular Level." *Nature* 217 (1967): 624–26.

Koestler, A. *The Ghost in the Machine.* London: Hutchinson, 1967.

Lewin, R. "Evolutionary Theory Under Fire." *Science* 210 (1980): 883–87.

Lewontin, R. C. "The Units of Selection." *Annual Review of Ecology and Systematics* 1 (1970): 1–18.

———. "Sociobiology as an Adaptationist Program." *Behavioral Science* 24 (1979): 5–14.

Lorenz, K. *Evolution and Modification of Behavior.* London: Methuen, 1966.

Maynard Smith, J. "The Status of Neo-Darwinism." In *Towards a Theoretical Biology,* edited by C. H. Waddington. Edinburgh: Edinburgh University Press, 1969.

Mayr, E. *The Growth of Biological Thought.* Cambridge, MA: Harvard University Press, 1982.

Miller, J. *Darwin for Beginners.* London: Writers and Readers, 1982.

Nevo, E. "Population Genetics and Ecology: The Interface." In *Evolution from Molecules to Men,* edited by D. S. Bendall, 287–44. Cambridge, MA: Cambridge University Press, 1983.

Paley, W. *Natural Theology,* 2nd ed. Oxford: J. Vincent, 1828.

Pringle, J. W. S. "On the Parallel between Learning and Evolution." *Behaviour* 3 (1951): 90–110.

Raup, D. M. "Stochastic Models in Evolutionary Palaeontology." In S. A. Hallam, *Patterns of Evolution.* Amsterdam: Elsevier, 1977.

Ridley, M. "Coadaptation and the Inadequacy of Natural Selection." *British Journal for the History of Science* 15 (1982): 45–68.

Rushton, W. A. H. "Peripheral Coding in the Nervous System." In S. W. A. Rosenblith, *Sensory Communication.* Cambridge, MA: MIT Press, 1961.

Stanley, S. M. *The New Evolutionary Timetable.* New York: Basic Books, 1981.

Stebbins, G. L. *Darwin to DNA, Molecules to Humanity.* San Francisco: W. H. Freeman, 1982.

Steele, E. J. *Somatic Selection and Adaptive Evolution.* Toronto: Williams and Wallace, 1979.

Thorpe, W. H. *Learning and Instinct in Animals*, 2nd ed. London: Methuen, 1963.

Turner, J. R. G. "Review of R. J. Berry, *Neo-Darwinism.*" *New Scientist* 94 (1982): 160–62.

Wright, S. "Genic and Organismic Selection." *Evolution* 34 (1980): 825–43.

PART II

CREATION SCIENCE AND THE *McLEAN* CASE

INTRODUCTION TO PART II

What is creation science (also known as scientific creationism) and where does it come from? The second of these questions is answered in the superb essay by Ronald L. Numbers, "The Creationists." In this article, Numbers, a historian of science, is interested in outlining the history that led to creationism as it was seen in the *McLean* case and focuses just on Christian creationism rather than casting his net wider to include non-Christian creationist views. But Numbers does note that there are a wide variety of views even among Christian creationists. He first distinguishes between what he calls "strict creationists," who take the days of Genesis to be twenty-four-hour periods, and "progressive" creationists, who take the days of creation to be immense periods of time, but then points out that there are substantial differences even within these camps. (Readers should note that terminology has shifted somewhat since this piece was published in 1982. Numbers's two camps would now more commonly be referred to as "young-earth" and "old-earth" creationists; the latter group sees itself as interpreting scripture no less strictly, after all. Old-earthers do also believe in progressive creationism but this latter term is now used to refer to God's mode of creation—that is, by a progressive series of creative acts—rather than to the length of the Mosaic days per se.) Of course, this article makes no mention of intelligent design creationism, which would arise toward the end of the decade; its story will be told in the article by Nick Matzke in part III.

There is nothing more we need add to the content of Numbers's article, but do let us draw out one implication. However you may decide about the status of creationism—science, religion, or whatever—do not pretend to yourself that this, and this alone, represents accepted Christianity (or Judaism). Perhaps God is a creation scientist and a six-literal-day, young-earth interpretation of Genesis is actually true. The fact of the matter is that many Christians and Jews think otherwise. Nor is such dissent confined to liberal theologians from trendy New England divinity schools. No less a conservative on doctrinal matters than Pope John Paul II has said:

> The Bible itself speaks to us of the origin of the universe and its makeup, not in order to provide us with a scientific treatise but in order to state the correct relationships of man with God and with the universe. Sacred Scripture wishes simply to declare that the world was created by God, and in order to reach this truth it expresses itself in the terms of the cosmology in use at the time of the writer. The Sacred Book likewise wishes to tell men that the world was not created as the seat of the gods, as was taught by other cosmogonies and cosmologies, but was rather created for the service of man and the glory of God. Any other teaching about the origin and makeup of the universe is alien to the intentions of the Bible, which does not wish to teach how the heavens were made but how one goes to heaven. (Address to the Pontifical Academy of Science, October 1981)

To give the ideas of creation science, we next present an article by the leading scientific creationist, Duane T. Gish. We shall make no comment here about the content of Gish's discussion, except to point out that he at once explicitly moves the creation/evolution clash to the level of status and methodology. Is creationism scientific? Is evolutionism scientific? In some respects Gish seems to think that neither are scientific.

With the stage now set, we move on to the Arkansas case. The issue at stake was whether "scientific creationism" or, alternatively, "creation science" is genuine science, with, it must be admitted, the background question whether evolution (particularly Darwinian evolution) is likewise genuine science. The suit against the state challenged the constitutionality of Act 590, the so-called Balanced Treatment Act—which had been enacted into law in Arkansas in 1981. (The text of Act 590 is available online at http://www.antievolution.org/projects/mclean/new_site/legal/act_590.htm, and we recommend that readers refer to it as they read the trial testimony and opinion reproduced in chapters 13 and 14.) The act required that public schools give balanced treatment to creation science and "evolution science." It made no explicit reference to God or the Bible. Indeed, the act said that it did not permit instruction in any religious doctrine or materials, and it defined creation science as "scientific evidences for creation and inferences from those scientific evidences" followed by six specific theses it held, such as sudden creation of the universe and life from nothing, rejection of the Darwinian mechanism of random mutation and natural selection, fixed limits of change within originally created kinds of plants and animals, and so on. Scientists rejected scientific creationism as an oxymoron. Among the scientists who served as expert witnesses in *McLean* were geologist G. Brent Dalrymple, molecular biophysicist and biochemist Harold Morowitz, and paleontologist Steven Jay Gould. But it was not just scientists who objected to the bill. The first named plaintiff, William McLean, was a Presbyterian minister, and other plaintiffs in the case included bishops and clergy from Southern Baptist, United Methodist, Episcopalian, Roman Catholic, and Christian churches, as well as representatives from the American Jewish Congress and other Jewish groups.

We include here my (Ruse's) contribution to the case, where I, as a historian and philosopher of science, argued strongly that whereas Darwinian evolutionary theory is scientific, creation science is not. It is religion. I prepared three position papers for the American Civil Liberties Union lawyers. These covered my views on the science/religion relationship, together with articles I had published previously. From these papers was prepared my script for the trial, known informally as my "Questions and Answers," which I include here. This was put together by the lawyers, and was the basis for my direct testimony at the trial. As I remember, we stuck to it almost line by line.

Next, in chapter 14, we have Judge Overton's ruling. The most pertinent section for our question is 4 (C), though the issue of what counts or doesn't count as science appears indirectly in other sections as well. I (Ruse) might add that none of the other witnesses on either side spoke very much at all to philosophical issues. I might add also—with a little surprise, having reread the material following an interim of some years—how important the notion of falsifiability proved to be. (Apart from the fact that it was dwelt on at great length in cross-examination.) I believed, and still believe, that it is an important mark of the scientific, but forces other than I gave falsifiability major significance in *Arkansas*. In particular, for all that Popper has criticized in Darwinism and would (no doubt) criticize even more in creationism, both Darwinians and creationists had already staked their case on falsifiability. Francisco Ayala (also an *Arkansas* witness), for instance, tied himself to Popper's criterion (see the reference in my critique of Popper), and we saw Gish do just the same. For these historical reasons alone, falsifiability was practically bound to come to the fore, and it did. However, it is worth looking carefully at the rest of Overton's opinion as well, as many commentators overlook the range of arguments he gives. As you will see, he finds that creation science fails for a wide range of reasons, including one which will be important to keep in mind when comparing *McLean* to the *Kitzmiller* case, namely, that creationism is not science because it depends upon the concept of supernatural intervention.

Those who wish to delve into *McLean* in greater detail should consult the *McLean v. Arkansas* Documentation Project (http://www.antievolution.org/projects/mclean/new_site/index.htm), which has the most complete record of expert witness depositions, transcripts of oral testimony, and other material. But we now continue with a look into the philosophical aftermath of the trial and focus on an important exchange of articles with historian and philosopher of science Larry Laudan.

Laudan takes issue with the conceptual framework that he takes Judge Overton, following Ruse, to rely upon in drawing his conclusion that teaching creationism is unconstitutional. He is particularly hard on Popper's falsifiability criterion and with Popper's demarcation problem generally. The idea of searching for criteria that strictly demarcate science from nonscience, argues Laudan, is hopeless—for any criterion that has been offered, he says one may find counterexamples from the history of science that violate and thus undermine it. Indeed, the demarcation problem itself was already a dead issue in philosophy, he claimed, and Ruse misrepresented philosophy of science in suggesting otherwise. McLean was a hollow victory, he claimed, because it rested on an outmoded conception of how science works. Rather than try to dismiss creationism as nonscience, one should dismiss it as bad science. The specific claims that creation science makes, he claims, have been shown to be false, so just leave it at that.

In chapter 17, Ruse defends his five criteria and Overton's reasoning, which he says emerges unscathed from Laudan's critique. Moreover, he points out, Laudan's own suggestion would have been impotent for the legal purposes of the trial, as the Constitution bars religion, not weak science. Laudan was given the final word against Ruse in the exchange (chapter 18). However, philosopher Barry Gross (who was a philosophical consultant to the ACLU in the *McLean* case) also takes up the gauntlet against Laudan's critique and is far less tolerant of what he sees as Laudan's misguided arguments that mistook the courtroom for a philosophy seminar. *McLean*, he concludes contra Laudan, "was a triumph, not a disaster; cause for rejoicing, not tears. The right side won for the right reasons; the necessary standard of proof was met in the case at bar. . . . Disaster is much more likely to occur when scholars eminent in one field venture to apply inappropriate standards to another."

CHAPTER 11

THE CREATIONISTS

RONALD L. NUMBERS

Scarcely twenty years after the publication of Charles Darwin's *Origin of Species* in 1859 special creationists could name only two working naturalists in North America: John William Dawson (1820–1899) of Montreal and Arnold Guyot (1806–1884) of Princeton, who had not succumbed to some theory of organic evolution (Heifer 1974, 203; Gray 1963, 202–203). The situation in Great Britain looked equally bleak for creationists, and on both sides of the Atlantic liberal churchmen were beginning to follow their scientific colleagues into the evolutionist camp.[1] By the closing years of the nineteenth century evolution was infiltrating even the ranks of the evangelicals, and, in the opinion of many observers, belief in special creation seemed destined to go the way of the dinosaur. However, contrary to the hopes of liberals and the fears of conservatives, creationism did not become extinct. The majority of late-nineteenth-century Americans remained true to a traditional reading of Genesis, and as late as 1982 a public-opinion poll revealed that 44

From Ronald L. Numbers, "The Creationists," in David C. Lindberg and Ronald L. Numbers, eds., *God and Nature* (Berkeley: University of California Press, 1986). Originally published as "Creationism in 20th Century America," *Science* 218 (November 5, 1982): 538–44.

percent of Americans, nearly a fourth of whom were college graduates, continued to believe that "God created man pretty much in his present form at one time within the last 10,000 years" ("Poll" 1982, 22).[2]

Such surveys failed, however, to disclose the great diversity of opinion among those professing to be creationists. Risking oversimplification, we can divide creationists into two main camps: "strict creationists," who interpret the days of Genesis literally, and "progressive creationists," who construe the Mosaic days to be immense periods of time. Yet, even within these camps substantial differences exist. Among strict creationists, for example, some believe that God created all terrestrial life—past and present—less than ten thousand years ago, while others postulate one or more creations prior to the seven days of Genesis. Similarly, some progressive creationists believe in numerous creative acts, while others limit God's intervention to the creation of life and perhaps the human soul. Since this last species of creationism is practically indistinguishable from theistic evolutionism, this essay focuses on the strict creationists and the more conservative of the progressive creationists, particularly on the small number who claimed scientific expertise. Drawing on their writings, it traces the ideological development of creationism from the crusade to outlaw the teaching of evolution in the 1920s to the current battle for equal time. During this period the leading apologists for special creation shifted from an openly biblical defense of their views to one based largely on science. At the same time they grew less tolerant of notions of an old earth and symbolic days of creation, common among creationists early in the century, and more doctrinaire in their insistence on a recent creation in six literal days and on a universal flood.

THE LOYAL MAJORITY

The general acceptance of organic evolution by the intellectual elite of the late Victorian era has often obscured the fact that the majority of Americans remained loyal to the doctrine of special creation (Dillenberger and Welch 1954, 227). In addition to the masses who said nothing, there were many people who vocally rejected kinship with the apes and other, more reflective, persons who concurred with the Princeton theologian Charles Hodge (1797–1878) that

Darwinism was atheism. Among the most intransigent foes of organic evolution were the premiliennialists, whose predictions of Christ's imminent return depended on a literal reading of the scriptures (Whalen 1972, 219–29; Numbers 1975, 18–23). Because of their conviction that one error in the Bible invalidated the entire book, they had little patience with scientists who, as described by the evangelist Dwight L. Moody (1837–1899), "Dug up old carcasses... to make them testify against God" (McLoughlin 1959, 213).

Such an attitude did not, however, prevent many biblical literalists from agreeing with geologists that the earth was far older than six thousand years. They did so by identifying two separate creations in the first chapter of Genesis: the first, "in the beginning," perhaps millions of years ago, and the second, in six actual days, approximately four thousand years before the birth of Christ. According to the so-called gap theory, most fossils were relics of the first creation, destroyed by God prior to the Adamic restoration (Numbers 1977, 89–90; Ramm 1954, 195–98). In 1909 the Scofield Reference Bible, the most authoritative biblical guide in fundamentalist circles, sanctioned this view.[3]

Scientists like Guyot and Dawson, the last of the reputable nineteenth-century creationists, went still further to accommodate science by interpreting the days of Genesis as ages and by correlating them with successive epochs in the natural history of the world (O'Brien 1971; Numbers 1977, 91–100). Although they believed in special creative acts, especially of the first humans, they tended to minimize the number of supernatural interventions and to maximize the operation of natural law. During the late nineteenth century the theory of progressive creation circulated widely in the colleges and seminaries of America.[4]

The early Darwinian debate focused largely on the implications of evolution for natural theology (Moore 1979); and so long as these discussions remained confined to scholarly circles, those who objected to evolution on biblical grounds saw little reason to participate. However, when the debate spilled over into the public arena during the 1880s and 1890s, creationists grew alarmed, "When these vague speculations, scattered to the four winds by the million-tongued press, are caught up by ignorant and untrained men," declared one premillennialist in 1889, "it is time for earnest Christian men to call a halt" (Hastings 1889).

The questionable scientific status of Darwinism undoubtedly encouraged such critics to speak up ("Evolutionism in the Pulpit" 1910–1915; Bowler 1983). Although the overwhelming majority of scientists after 1880 accepted a long earth history and some form of organized evolution, many in the late nineteenth century were expressing serious reservations about the ability of Darwin's particular theory of natural selection to account for the origin of species. Their published criticisms of Darwinism led creationists mistakenly to conclude that scientists were in the midst of discarding evolution. The appearance of books with such titles as *The Collapse of Evolution* and *At the Death Bed of Darwinism* bolstered this belief and convinced antievolutionists that liberal Christians had capitulated to evolution too quickly. In view of this turn of events it seemed likely that those who had "abandoned the stronghold of faith out of sheer fright will soon be found scurrying back to the old and impregnable citadel, when they learn that 'the enemy is in full retreat'" (Young 1909, 41).

For the time being, however, those conservative Christians who would soon call themselves fundamentalists perceived a greater threat to orthodox faith than evolution—higher criticism, which treated the Bible more as a historical document than as God's inspired word. Their relative apathy toward evolution is evident in *The Fundamentals*, a mass-produced series of twelve booklets published between 1910 and 1915 to revitalize and reform Christianity around the world. Although one contributor identified evolution as the principal cause of disbelief in the Scriptures and another traced the roots of higher criticism to Darwin, the collection as a whole lacked the strident antievolutionism that would characterize the fundamentalist movement of the 1920s (Mauro 1910–1915; Reeve 1910–1915).

This is particularly true of the writings of George Frederick Wright (1838–1921), a Congregational minister and amateur geologist of international repute (Wright 1916). At first glance his selection to represent the fundamentalist point of view seems anomalous. As a prominent Christian Darwinist in the 1870s he had argued that the intended purpose of Genesis was to protest polytheism, not teach science (Wright 1898). By the 1890s, however, he had come to espouse the progressive creationism of Guyot and Dawson, partly, it seems, in reaction to the claims of higher critics regarding the accuracy of the Pentateuch (Wright 1902). Because of his standing as a scientific authority and his conservative view of the Scriptures, the editors of *The Fun-*

damentals selected him to address the question of the relationship between evolution and the Christian faith.

In an essay misleadingly titled "The Passing of Evolution" Wright attempted to steer a middle course between the theistic evolution of his early days and the traditional views of some special creationists. On the one hand, he argued that the Bible itself taught evolution, "an orderly progress from lower to higher forms of matter and life." On the other hand, he limited evolution to the origin of species, pointing out that even Darwin had postulated the supernatural creation of several forms of plants and animals, endowed by the Creator with a "marvelous capacity for variation." Furthermore, he argued that, despite the physical similarity between human beings and the higher animals, the former "came into existence as the Bible represents, by the special creation of a single pair, from whom all the varieties of the race have sprung" (Wright 1910–1915).[5]

Although Wright represented the left wing of fundamentalism, his moderate views on evolution contributed to the conciliatory tone that prevailed during the years leading up to World War I. Fundamentalists may not have liked evolution, but few, if any, at this time saw the necessity or desirability of launching a crusade to eradicate it from the schools and churches in America.

THE ANTIEVOLUTION CRUSADE

Early in 1922 William Jennings Bryan (1860–1925), Presbyterian layman and thrice-defeated Democratic candidate for the president of the United States, heard of an effort in Kentucky to ban the teaching of evolution in public schools. "The movement will sweep the country," he predicted hopefully, "and we will drive Darwinism from our schools" (Levine 1965, 277). His prophecy proved overly optimistic, but before the end of the decade more than twenty state legislatures did debate antievolution laws, and four—Oklahoma, Tennessee, Mississippi, and Arkansas—banned the teaching of evolution in public schools (Shipley 1927; 1930). At times the controversy became so tumultuous that it looked to some as though "America might go mad" (Nelson 1964, 319). Many persons shared responsibility for these events, but none more than Bryan. His entry into the fray had a catalytic effect (Szasz 1982, 107–16) and gave antievolutionists what they needed most: "a spokesman with a national reputation, immense prestige, and a loyal following" (Levine 1965, 272).

The development of Bryan's own attitude toward evolution closely paralleled that of the fundamentalist movement. Since early in the century he had occasionally alluded to the silliness of believing in monkey ancestors and to the ethical dangers of thinking that might makes right, but until the outbreak of World War I he saw little reason to quarrel with those who disagreed. The war, however, exposed the darkest side of human nature and shattered his illusions about the future of Christian society. Obviously something had gone awry, and Bryan soon traced the source of the trouble to the paralyzing influence of Darwinism on the human conscience. By substituting the law of the jungle for the teaching of Christ, it threatened the principles he valued most: democracy and Christianity. Two books in particular confirmed his suspicion. The first, Vernon Kellogg's *Headquarters Nights* in 1917, recounted firsthand conversations with German officers that revealed the role Darwin's biology had played in persuading the Germans to declare war. The second, Benjamin Kidd's *Science of Power* in 1918, purported to demonstrate the historical and philosophical links between Darwinism and German militarism (Levine 1965, 216–65).

About the time that Bryan discovered the Darwinian origins of the war, he also became aware, to his great distress, of unsettling effects the theory of evolution was having on America's own young people. From frequent visits to college campuses and from talks with parents, pastors, and Sunday school teachers, he heard about an epidemic of unbelief that was sweeping the country. Upon investigating the cause, his wife reported, "he became convinced that the teaching of Evolution as a fact instead of a theory caused the students to lose faith in the Bible, first, in the story of creation, and later in other doctrines, which underlie the Christian religion" (Williams 1936, 448). Again Bryan found confirming evidence in a recently published book, *Belief in God and Immortality*, authored in 1916 by the Bryn Mawr psychologist James H. Leuba, who demonstrated statistically that college attendance endangered traditional religious beliefs (Levine 1965, 266–67).

Armed with this information about the cause of the world's and the nation's moral decay, Bryan launched a nationwide crusade against the offending doctrine. In one of his most popular and influential lectures, "The Menace of

Darwinism," he summed up his case against evolution, arguing that it was both un-Christian and unscientific. Darwinism, he declared, was nothing but "guesses strung together," and poor guesses at that. Borrowing from a turn-of-the-century tract, he illustrated how the evolutionist explained the origin of the eye:

> The evolutionist guesses that there was a time when eyes were unknown—that is a necessary part of the hypothesis.... [A] piece of pigment, or, as some say, a freckle appeared upon the skin of an animal that had no eyes. This piece of pigment or freckle converged the rays of the sun upon that spot and when the little animal felt the heat on that spot it turned the spot to the sun to get more heat. The increased heat irritated the skin—so the evolutionists guess, and a nerve came there and out of the nerve came the eye! (Bryan 1922, 94, 97–98)[6]

"Can you beat it?" he asked incredulously—and that it happened not once but twice? As for himself, he would take one verse in Genesis over all that Darwin wrote.

Throughout his political career Bryan had placed his faith in the common people, and he resented the attempt of a few thousand scientists "to establish an oligarchy over the forty million American Christians," to dictate what should be taught in the schools (Coletta 1969, 230). To a democrat like Bryan it seemed preposterous that this "scientific soviet" (Levine 1965, 289) would not only demand to teach its insidious philosophy but impudently insist that society pay its salaries. Confident that nine-tenths of the Christian citizens agreed with him, he decided to appeal directly to them, as he had done so successfully in fighting the liquor interests.[7] "Commit your case to the people," he advised creationists. "Forget, if need be, the highbrows both in the political and college world, and carry this cause to the people. They are the final and efficiently corrective power" ("Progress" 1929, 13).

Who were the people who joined Bryan's crusade? As recent studies have shown, they came from all walks of life and from every region of the country. They lived in New York, Chicago, and Los Angeles as well as in small towns and in the country. Few possessed advanced degrees, but many were not without education. Nevertheless, Bryan undeniably found his staunchest supporters and won his greatest victories in the conservative and still largely rural

South, described hyperbolically by one fundamentalist journal as "the last stronghold of orthodoxy on the North American continent," a region where the "masses of the people in all denominations 'believe the Bible from lid to lid'" ("Fighting Evolution" 1925, 5).[8]

The strength of Bryan's following within the churches is perhaps more difficult to determine, because not all fundamentalists were creationists and many creationists refused to participate in the crusade against evolution. However, a 1929 survey of the theological beliefs of seven hundred Protestant ministers provides some valuable clues (Bets 1929, 26, 44). The question "Do you believe that the creation of the world occurred in the manner and time recorded in Genesis?" elicited the following positive responses:

Lutheran	89%
Baptist	63%
Evangelical	62%
Presbyterian	35%
Methodist	24%
Congregational	12%
Episcopalian	11%
Other	60%

Unfortunately, these statistics tell us nothing about the various ways respondents may have interpreted the phrase "in the manner and time recorded in Genesis," nor do they reveal anything about the level of political involvement in the campaign against evolution. Lutherans, for example, despite their overwhelming rejection of evolution, generally preferred education to legislation and tended to view legal action against evolution as "a dangerous mingling of church and state" (Rudnick 1966, 88–90; Szasz 1969, 279). Similarly, premillennialists, who saw the spread of evolution as one more sign of the world's impending end, sometimes lacked incentive to correct the evils around them (Sandeen 1971, 266–68).[9]

Baptists and Presbyterians, who dominated the fundamentalist movement, participated actively in the campaign against evolution. The Southern Baptist Convention, spiritual home of some of the most outspoken foes of evolution, lent encouragement to the creationist crusaders by voting unanimously in 1926 that "this Convention accepts Genesis as teaching that man was the special creation of God, and rejects every theory, evolution or other, which teaches that man originated in, or came by way of, a lower animal ancestry" (Dark 1952, 154; Thompson 1975–1976). The Presbyterian Church contributed Bryan and other leaders to the creationist cause but, as the above survey indicates, also harbored many evolutionists. In 1923 the General Assembly turned back an attempt by Bryan and his fundamentalist cohorts to cut off funds to any church school found teaching human evolution, approving instead a compromise measure that condemned only materialistic evolution (Loetscher 1954, 111). The other major Protestant bodies paid relatively little attention to the debate over evolution; and Catholics, though divided on the question of evolution, seldom favored restrictive legislation (Morrison 1953).[10]

Leadership of the antievolution movement came not from the organized churches of America but from individuals like Bryan and interdenominational organizations such as the World's Christian Fundamentals Association, a predominantly premillennialist body founded in 1919 by William Bell Riley (1861–1947), pastor of the First Baptist Church in Minneapolis.[11] Riley became active as an antievolutionist after discovering, to his apparent surprise, that evolutionists were teaching their views at the University of Minnesota. The early twentieth century witnessed an unprecedented expansion of public education; enrollment in public high schools nearly doubled between 1920 and 1930 (Bailey 1964, 72–73). Fundamentalists like Riley and Bryan wanted to make sure that students attending these institutions would not lose their faith. Thus they resolved to drive every evolutionist from the public school payroll. Those who lost their jobs as a result deserved little sympathy, for, as one rabble-rousing creationist put it, the German soldiers who killed Belgian and French children with poisoned candy were angels compared with the teachers and textbook writers who corrupted the souls of children and thereby sentenced them to eternal death (Martin 1923, 164–65).

The creationists, we should remember, did not always act without provocation. In many instances their opponents displayed equal intolerance and insensitivity. In fact, one contemporary observer blamed the creation-evolution controversy in part on the "intellectual flapperism" of irresponsible and poorly informed teachers who delighted in shocking naïve students with unsupportable statements about evolution. It was understandable, wrote an

Englishman, that American parents would resent sending their sons and daughters to public institutions that exposed them to "a multiple assault upon traditional faiths" (Beale 1936, 249–51).

CREATIONIST SCIENCE AND SCIENTISTS

In 1922 Riley outlined the reasons why fundamentalists opposed the teaching of evolution. "The first and most important reason for its elimination," he explained, "is the unquestioned fact that evolution is not a science; it is a hypothesis only, a speculation" ([Riley] 1922, 5). Bryan often made the same point, defining true science as "classified knowledge...the explanation of facts" (Bryan 1922, 94). Although creationists had far more compelling reasons for rejecting evolution than its alleged unscientific status, their insistence on this point was not merely an obscurantist ploy. Rather it stemmed from their commitment to a once-respected tradition, associated with the English philosopher Sir Francis Bacon (1561–1626), that emphasized the factual, nontheoretical nature of science (Marsden 1977, 214–15). By identifying with the Baconian tradition, creationists could label evolution as false science, could claim equality with scientific authorities in comprehending facts, and could deny the charge of being antiscience. "It is not 'science' that orthodox Christians oppose," a fundamentalist editor insisted defensively. "No! no! a thousand times, No! They are opposed only to the theory of evolution, which has not yet been proved, and therefore is not to be called by the sacred name of science" (K[eyser] 1925, 413).

Because of their conviction that evolution was unscientific, creationists assured themselves that the world's best scientists agreed with them. They received an important boost at the beginning of their campaign from an address by the distinguished British biologist William Bateson (1861–1926) in 1921, in which he declared that scientists had not discovered "the actual mode and process of evolution" (Bateson 1922).[12] Although he warned creationists against misinterpreting his statement as a rejection of evolution, they paid no more attention to that caveat than they did to the numerous pro-evolution resolutions passed by scientific societies (Shipley 1927, 384).

Unfortunately for the creationists, they could claim few legitimate scientists of their own: a couple of self-made men of science, one or two physicians, and a handful of teachers who, as one evolutionist described them, were "trying to hold down, not a chair, but a whole settee, of 'Natural Science' in some little institution."[13] Of this group the most influential were Harry Rimmer (1890–1952) and George McCready Price (1870–1963).

Rimmer, Presbyterian minister and self-styled "research scientist," obtained his limited exposure to science during a term or two at San Francisco's Hahnemann Medical College, a small homeopathic institution that required no more than a high school diploma for admission. As a medical student he picked up a vocabulary of "double-jointed, twelve cylinder, knee-action words" that later served to impress the uninitiated (Rimmer 1945, 14). After his brief stint in medical school he attended Whittier College and the Bible Institute of Los Angeles for a year each before entering full-time evangelistic work. About 1919 he settled in Los Angeles, where he set up a small laboratory at the rear of his house to conduct experiments in embryology and related sciences. Within a year or two he established the Research Science Bureau "to prove through findings in biology, paleontology, and anthropology that science and the literal Bible were not contradictory." The bureau staff—that is, Rimmer—apparently used income from the sale of memberships to finance anthropological field trips in the western United States, but Rimmer's dream of visiting Africa to prove the dissimilarity of gorillas and humans failed to materialize. By the late 1920s the bureau lay dormant, and Rimmer signed on with Riley's World's Christian Fundamentals Associations as a field secretary.[14]

Besides engaging in research, Rimmer delivered thousands of lectures, primarily to student groups, on the scientific accuracy of the Bible. Posing as a scientist, he attacked Darwinism and poked fun at the credulity of evolutionists. To attract attention, he repeatedly offered one hundred dollars to anyone who could discover a scientific error in the Scriptures; not surprisingly, the offer never cost him a dollar ("World Religious Digest" 1939, 215). He also, by his own reckoning, never lost a public debate. Following one encounter with an evolutionist in Philadelphia, he wrote home gleefully that "the debate was a simple walkover, a massacre—murder pure and simple. The eminent professor was simply scared stiff to advance any of the common arguments of the evolutionists, and he fizzled like a wet firecracker" (Edmondson 1969, 329–30, 333–34).

Price, a Seventh-Day Adventist geologist, was less skilled at debating than Rimmer but more influential scientifically. As a young man Price attended an Adventist college in Michigan for two years and later completed a teacher training course at the provincial normal school in his native New Brunswick. The turn of the century found him serving as principal of a small high school in an isolated part of eastern Canada, where one of his few companions was a local physician. During their many conversations, the doctor almost converted his fundamentalist friend to evolution, but each time Price wavered, he was saved by prayer and by reading the works of the Seventh-Day Adventist prophetess Ellen G. White (1827–1915), who claimed divine inspiration for her view that Noah's flood accounted for the fossil record on which evolutionists based their theory. As a result of these experiences, Price vowed to devote his life to promoting creationism of the strictest kind.[15]

By 1906 he was working as a handyman at an Adventist sanitarium in southern California. That year he published a slim volume titled *Illogical Geology: The Weakest Point in the Evolution Theory*, in which he brashly offered one thousand dollars "to any one who will, in the face of the facts here presented, show me how to prove that one kind of fossil is older than another." (Like Rimmer, he never had to pay.) According to Price's argument, Darwinism rested "logically and historically on the succession of life idea as taught by geology" and "if this succession of life is not an actual scientific fact, then Darwinism . . . is a most gigantic hoax" (Price 1906, 9).[16]

Although a few fundamentalists praised Price's polemic, David Starr Jordan (1851–1931), president of Stanford University and an authority on fossil fishes, warned him that he should not expect "any geologist to take [his work] seriously." Jordan conceded that the unknown author had written "a very clever book" but described it as "a sort of lawyer's plea, based on scattering mistakes, omissions and exceptions against general truths that anybody familiar with the facts in a general way cannot possibly dispute. It would be just as easy and just as plausible and just as convincing if one should take the facts of European history and attempt to show that all the various events were simultaneous."[17] As Jordan recognized, Price lacked any formal training or field experience in geology. He was, however, a voracious reader of geological literature, an armchair scientist who self-consciously minimized the importance of field experience.

During the next fifteen years Price occupied scientific settees in several Seventh-Day Adventist schools and authored six more books attacking evolution, particularly its geological foundation. Although not unknown outside his own church before the early 1920s, he did not attract national attention until then. Shortly after Bryan declared war on evolution, Price published in 1923 *The New Geology*, the most systematic and comprehensive of his many books. Uninhibited by false modesty, he presented his "great *law of conformable stratigraphic sequences*...by all odds the most important law ever formulated with reference to the order in which the strata occur." This law stated that "*any kind of fossiliferous beds whatever, 'young' or 'old,' may be found occurring conformably on any other fossiliferous beds, 'older' or 'younger'*" (Price 1923, 637–38).[18] To Price, so-called deceptive conformities (where strata seem to be missing) and thrust faults (where the strata are apparently in the wrong order) proved that there was no natural order to the fossil-bearing rocks, all of which he attributed to the Genesis flood.

A Yale geologist reviewing the book for *Science* accused Price of "harboring a geological nightmare" (Schuchert 1924). Despite such criticism from the scientific establishment—and the fact that his theory contradicted both the day-age and gap interpretations of Genesis—Price's reputation among fundamentalists rose dramatically. Rimmer, for example, hailed *The New Geology* as "a masterpiece of REAL science [that] explodes in a convincing manner some of the ancient fallacies of science 'falsely so called'" (Rimmer 1925, 28). By the mid-1920s Price's byline was appearing with increasing frequency in a broad spectrum of conservative religious periodicals, and the editor of *Science* could accurately describe him as "the principal scientific authority of the Fundamentalists" (*Science* 1926).

THE *SCOPES* TRIAL AND BEYOND

In the spring of 1925 John Thomas Scopes, a high school teacher in the small town of Dayton, Tennessee, confessed to having violated the state's recently passed law banning the teaching of human evolution in public schools. His subsequent trial focused international attention on the antievolution crusade and brought William Jennings Bryan to Dayton to assist the prosecution. In anticipation of arguing the scientific merits of evolution, Bryan sought out the best scientific minds in the creationist camp to serve as expert witnesses. The response to his inquiries could only have disappointed the aging crusader. Price, then teaching in England, sent his regrets—along with advice for Bryan to stay away from scientific topics (Numbers 1979, 24). Howard A. Kelly, a prominent Johns Hopkins physician who had contributed to *The Fundamentals*, confessed that, except for Adam and Eve, he believed in evolution. Louis T. More, a physicist who had just written a book in 1925 on *The Dogma of Evolution*, replied that he accepted evolution as a working hypothesis. Alfred W. McCann, author in 1922 of *God—or Gorilla*, took the opportunity to chide Bryan for supporting prohibition in the past and for now trying "to bottle-up the tendencies of men to think for themselves."[19]

At the trial itself things scarcely went better. When Bryan could name only Price and the deceased Wright as scientists for whom he had respect, the caustic Clarence Darrow (1857–1938), attorney for the defense, scoffed, "You mentioned Price because he is the only human being in the world so far as you know that signs his name as a geologist that believes like you do . . . every scientist in this country knows [he] is a mountebank and a pretender and not a geologist at all." Eventually Bryan conceded that the world was indeed far more than six thousand years old and that the six days of creation had probably been longer than twenty-four hours each—concessions that may have harmonized with the progressive creationism of Wright but hardly with the strict creationism of Price (Numbers 1979, 24; Levine 1965, 349).

Though one could scarcely have guessed it from some of his public pronouncements, Bryan had long been a progressive creationist. In fact, his beliefs regarding evolution diverged considerably from those of his more conservative supporters. Shortly before his trial he had confided to Dr. Kelly that he, too, had no objection to "evolution before man but for the fact that a concession as to the truth of evolution up to man furnishes our opponents with an argument which they are quick to use, namely, if evolution accounts for all the species up to man, does it not raise a presumption in behalf of evolution to include man?" Until biologists could actually demonstrate the evolution of one species into another, he thought it best to keep them on the defensive.[20]

Bryan's admission at Dayton spotlighted a serious and long-standing problem among antievolutionists: their failure to agree on a theory of creation. Even the most visible leaders could not reach a consensus. Riley, for example, followed Guyot and Dawson (and Bryan) in viewing the days of Genesis as ages, believing that the testimony of geology necessitated this interpretation.

Rimmer favored the gap theory, which involved two separate creations, in part because his scientific mind could not fathom how, given Riley's scheme, plants created on the third day could have survived thousands of years without sunshine, until the sun appeared on the fourth. According to the testimony of acquaintances, he also believed that the Bible taught a local rather than a universal flood (Culver 1955, 7). Price, who cared not a whit about the opinion of geologists, insisted on nothing less than a recent creation in six literal days and a worldwide deluge. He regarded the day-age theory as "the devil's counterfeit" and the gap theory as only slightly more acceptable (Price 1902, 125–27; 1954, 39). Rimmer and Riley, who preferred to minimize the differences among creationists, attempted the logically impossible, if ecumenically desirable, task of incorporating Price's "new geology" into their own schemes (Riley and Rimmer n.d.; Riley 1930, 45).

Although the court in Dayton found Scopes guilty as charged, creationists had little cause for rejoicing. The press had not treated them kindly, and the taxing ordeal no doubt contributed to Bryan's death a few days after the end of the trial. Nevertheless, the antievolutionists continued their crusade, winning victories in Mississippi in 1926 and in Arkansas two years later (Shipley 1930, 330–32). By the end of the decade, however, their legislative campaign had lost its steam. The presidential election of 1928, pitting a Protestant against a Catholic, offered fundamentalists a new cause, and the onset of the depression in 1929 further diverted their attention (Szasz 1981, 117–25).

Contrary to appearances, the creationists were simply changing tactics, not giving up. Instead of lobbying state legislatures, they shifted their attack to local communities, where they engaged in what one critic described as "the emasculation of textbooks, the 'purging' of libraries, and above all the continued hounding of teachers" (Shipley 1930, 330). Their new approach attracted less attention but paid off handsomely, as school boards, textbook publishers, and teachers in both urban and rural areas, North and South, bowed to their pressure. Darwinism virtually disappeared from high school texts, and for years many American teachers feared being identified as evolutionists (Beale 1936, 228–37; Gatewood 1969, 39; Grabiner and Miller 1974; Laba and Gross 1950).

CREATIONISM UNDERGROUND

During the heady days of the 1920s, when their activities made front-page headlines, creationists dreamed of converting the world; a decade later, forgotten and rejected by the establishment, they turned their energies inward and began creating an institutional base of their own. Deprived of the popular press and frustrated by their inability to publish their views in organs controlled by orthodox scientists, they determined to organize their own societies and edit their own journals (Carpenter 1980).[21] Their early efforts, however, encountered two problems: the absence of a critical mass of scientifically trained creationists and lack of internal agreement.

In 1935 Price, along with Dudley Joseph Whitney, a farm journalist, and L. Allen Higley, a Wheaton College science professor, formed a Religion and Science Association to create "a united front against the theory of evolution." Among those invited to participate in the association's first—and only—convention were representatives of the three major creationist parties, including Price himself, Rimmer, and one of Dawson's sons, who, like his father, advocated the day-age theory.[22] But as soon as the Price faction discovered that its associates had no intention of agreeing on a short earth history, it bolted the organization, leaving it a shambles.[23]

Shortly thereafter, in 1938, Price and some Seventh-Day Adventist friends in the Los Angeles area, several of them physicians associated with the College of Medical Evangelists (now part of Loma Linda University), organized their own Deluge Geology Society and, between 1941 and 1945, published a *Bulletin of Deluge Geology and Related Science.* As described by Price, the group consisted of "a very eminent set of men. In no other part of this round globe could anything like the number of scientifically educated believers in Creation and opponents of evolution be assembled, as here in Southern California" (Numbers 1979, 26). Perhaps the society's most notable achievement was its sponsorship in the early 1940s of a hush-hush project to study giant fossil footprints, believed to be human, discovered in rocks far older than the theory of evolution would allow. This find, the society announced excitedly, thus demolished that theory "at a single stroke" and promised to "*astound the scientific world!*" Yet despite such activity and the group's religious homogeneity, it, too, soon foundered on "the same rock," complained a disappointed member, that wrecked the Religion and Science Association, that is "*pre-Genesis time for the earth.*"[24]

By this time creationists were also beginning to face a new problem: the presence within their own ranks of young university-trained scientists who wanted to bring evangelical Christianity more into line with mainstream science. The encounter between the two generations often proved traumatic, as is illustrated by the case of Harold W. Clark (1891b). A former student of Price's, he had gone on to earn a master's degree in biology from the University of California and taken a position at a small Adventist college in northern California. By 1940 his training and field experience had convinced him that Price's *New Geology* was "entirely out of date and inadequate" as a text, especially in its rejection of the geological column. When Price learned of this, he angrily accused his former disciple of suffering from "the modern mental disease of university-itis" and of currying the favor of "tobacco-smoking, Sabbath-breaking, God-defying" evolutionists. Despite Clark's protests that he still believed in a literal six-day creation and universal flood, Price kept up his attack for the better part of a decade, at one point addressing a vitriolic pamphlet, *Theories of Satanic Origin*, to his erstwhile friend and fellow creationist (Numbers 1979, 25).

The inroads of secular scientific training also became apparent in the American Scientific Affiliation (ASA), created by evangelical scientists in 1941.[25] Although the society took no official stand on creation, strict creationists found the atmosphere congenial during the early years of the society. In the late 1940s, however, some of the more progressive members, led by J. Laurence Kulp, a young geochemist on the faculty of Columbia University, began criticizing Price and his followers for their allegedly unscientific effort to squeeze earth history into less than ten thousand years. Kulp, a Wheaton alumnus and member of the Plymouth Brethren, had acquired a doctorate in physical chemistry from Princeton University and gone on to complete all the requirements, except a dissertation, for a PhD in geology. Although initially suspicious of the conclusions of geology regarding the history and antiquity of the earth, he had come to accept them. As one of the first evangelicals professionally trained in geology, he felt a responsibility to warn his colleagues in the ASA about Price's work, which, he believed, had "infiltrated the greater portion of fundamental Christianity in America primarily due to the absence of trained Christian geologists." In what was apparently the first systematic critique of the "new geology" Kulp concluded that the "major propositions of the theory are contradicted by established physical and chemical laws" (Kulp 1950).[26] Conservatives within the ASA not unreasonably suspected that Kulp's exposure to "the orthodox, geological viewpoint" had severely undermined his faith in a literal interpretation of the Bible ("Comment" 1940, 2).

Before long it became evident that a growing number of ASA members, like Kulp, were drifting from strict to progressive creationism and sometimes on to theistic evolutionism. The transition for many involved immense personal stress, as revealed in the autobiographical testimony of another Wheaton alumnus, J. Frank Cassel:

> First to be overcome was the onus of dealing with a "verboten" term and in a "non-existent" area. Then, as each made an honest and objective consideration of the data, he was struck with the validity and undeniability of datum after datum. As he strove to incorporate each of these facts into his biblico-scientific frame of reference, he found that—while the frame became more complete and satisfying—he began to question first the feasibility and then the desirability of an effort to refute the total evolutionary concept, and finally he became impressed by its impossibility on the basis of existing data. This has been a heart-rending, soul-searching experience for the committed Christian as he has seen what he had long considered the raison d'être of God's call for his life endeavor fade away, and as he has struggled to release strongly held convictions as to the close limitations of creationism.

Cassel went on to note that the struggle was "made no easier by the lack of approbation (much less acceptance) of some of his less well-informed colleagues, some of whom seem to question motives or even to imply heresy" (Cassel 1959, 26–27).[27] Strict creationists, who suffered their own agonies, found it difficult not to conclude that their liberal colleagues were simply taking the easy way out. To both parties a split seemed inevitable.

CREATIONISM ABROAD

During the decades immediately following the crusade of the 1920s American antievolutionists were buoyed by reports of a creationist revival in Europe, especially in England, where creationism was thought to be all but dead. The Victoria Institute in London, a haven for English creationists in the nineteenth century, had by the 1920s become a stronghold of theistic evolution. When Price visited the institute in 1925 to receive its Langhorne-Orchard Prize for an essay on "Revelation and Evolution," several members protested his attempt to export the fundamentalist controversy to England. Even evangelicals refused to get caught up in the turmoil that engulfed the United States. As historian George Marsden has explained, English evangelicals, always a minority, had developed a stronger tradition of theological toleration than revivalist Americans, who until the twentieth century had never experienced minority status. This, while the displaced Americans fought to recover their lost position, English evangelicals adopted a nonmilitant live-and-let-live philosophy that stressed personal piety (Numbers 1975, 25; Marsden 1977; 1980, 222–26).

The sudden appearance of a small but vocal group of British creationists in the early 1930s caught nearly everyone by surprise. The central figure in this movement was Douglas Dewar (1875–1957), a Cambridge graduate and amateur ornithologist, who had served for decades as a lawyer in the Indian Civil Service. Originally an evolutionist, he had gradually become convinced of the necessity of adopting "a provisional hypothesis of special creation supplemented by a theory of evolution." This allowed him to accept unlimited development within biological families. His published views, unlike those of most American creationists, betrayed little biblical influence (Dewar 1931, 158; Lunn 1947, 1, 154; *Evolution Protest* 1965). His greatest intellectual debt was not to Moses but to a French zoologist, Louis Vialleton (1859–1929), who had attracted considerable attention in the 1920s for suggesting a theory of discontinuous evolution, which antievolutionists eagerly—but erroneously—equated with special creation (Paul 1979, 99–100).

Soon after announcing his conversion to creationism in 1931, Dewar submitted a short paper on mammalian fossils to the Zoological Society of London, of which he was a member. The secretary of the society subsequently rejected the piece, noting that a competent referee thought Dewar's evidence "led to no valuable conclusion." Such treatment infuriated Dewar and convinced him that evolution had become "a scientific creed." Those who questioned scientific orthodoxy, he complained, "are deemed unfit to hold scientific offices; their articles are rejected by newspapers or journals; their contributions are refused by scientific societies, and publishers decline to publish their books except at the author's expense. Thus the independents are today pretty effectually muzzled" (Dewar 1932, 142). Because of such experiences Dewar and other British dissidents in 1932 organized the Evolution Protest Movement, which after two decades claimed a membership of two hundred ("EPM" 1972).

HENRY M. MORRIS AND THE REVIVAL OF CREATIONISM

In 1964 one historian predicted that "a renaissance of the [creationist] movement is most unlikely" (Halliburton 1964, 283). And so it seemed. But even as these words were penned, a major revival was under way, led by a Texas engineer, Henry M. Morris (1918b). Raised a nominal Southern Baptist, and as such a believer in creation, Morris as a youth had drifted unthinkingly into evolutionism and religious indifference. A thorough study of the Bible following graduation from college convinced him of its absolute truth and prompted him to reevaluate his belief in evolution. After an intense period of soul-searching he concluded that creation had taken place in six literal days, because the Bible clearly said so and "God doesn't lie." Corroborating evidence came from the book of nature. While sitting in his office at Rice Institute, where he was teaching civil engineering, he would study the butterflies and wasps that flew in through the window; being familiar with structural design, he calculated the improbability of such complex creatures developing by chance. Nature as well as the Bible seemed to argue for creation.[2]

For assistance in answering the claims of evolutionists, he found little creationist literature of value apart from the writings of Rimmer and Price. Although he rejected Price's peculiar theology, he took an immediate liking to the Adventist's flood geology and in 1946 incorporated it into a little book, *That You Might Believe*, the first book, so far as he knew, "published since the *Scopes* trial in which a scientist from a secular university advocated recent special creation and a worldwide flood" (Morris 1978, 10). In the late 1940s he joined the American Scientific Affiliation—just in time to protest Kulp's attack on Price's geology. Yet his words fell largely on deaf ears. In 1953 when he presented some of his own views on the flood to the ASA, one of the few compliments came from a young theologian, John C. Whitcomb Jr., who belonged to the Grace Brethren. The two subsequently became friends and decided to collaborate on a major defense of the Noachian flood. By the time they finished their project, Morris had earned a PhD in hydraulic engineering from the University of Minnesota and was chairing the civil engineering department at Virginia Polytechnic Institute; Whitcomb was teaching Old Testament studies at Grace Theological Seminary in Indiana.[29]

In 1961 they brought out *The Genesis Flood*, the most impressive contribution to strict creationism since the publication of Price's *New Geology* in 1923. In many respects their book appeared to be simply "a reissue of C. M. Price's views, brought up to date," as one reader described it. Beginning with a testimony to their belief in "the verbal inerrancy of Scripture," Whitcomb and Morris went on to argue for a recent creation of the entire universe, a fall that triggered the second law of thermodynamics, and a worldwide flood that in one year laid down most of the geological strata. Given this history, they argued, "the last refuge of the case for evolution immediately vanishes away, and the record of the rocks becomes a tremendous witness... to the holiness and justice and power of the living God of Creation!"(Whitcomb and Morris 1961, xx, 451). Despite the book's lack of conceptual novelty, it provoked an intense debate among evangelicals. Progressive creationists denounced it as a travesty on geology that threatened to set back the cause of Christian science a generation, while strict creationists praised it for making biblical catastrophism intellectually respectable. Its appeal, suggested one critic, lay primarily in the fact that, unlike previous creationist works, it "looked legitimate as a scientific contribution," accompanied as it was by footnotes and other scholarly appurtenances. In responding to their detractors, Whitcomb and Morris repeatedly refused to be drawn into a scientific debate, arguing that "the real issue is not the correctness of the interpretation of various details of the geological data, but simply what God has revealed in His Word concerning these matters" (Morris and Whitcomb 1964, 60).[30]

Whatever its merits, *The Genesis Flood* unquestionably "brought about a stunning renaissance of flood geology" (Young 1977, 7), symbolized by the establishment in 1963 of the Creation Research Society. Shortly before the publication of his book Morris had sent the manuscript to Walter E. Lammerts (1904b), a Missouri-Synod Lutheran with a doctorate in genetics from the University of California. As an undergraduate at Berkeley Lammerts had discovered Price's *New Geology*, and during the early 1940s, while teaching at UCLA, he had worked with Price in the Creation-Deluge Society. After the mid-1940s, however, his interest in creationism had flagged—until awakened by reading the Whitcomb and Morris manuscript. Disgusted by the ASA's flirtation with evolution, he organized in the early 1960s a correspondence network with Morris and eight other strict creationists, dubbed the "team of ten." In 1963 seven of the ten met with a few other like-minded scientists at the home of a team member in Midland, Michigan, to form the Creation Research Society (CRS) (Lammerts 1974).

The society began with a carefully selected eighteen-man "inner-core steering committee," which included the original team of ten. The composition of this committee reflected, albeit imperfectly, the denominational, regional, and professional bases of the creationist revival. There were six Missouri-Synod Lutherans, five Baptists, two Seventh-Day Adventists, and one each from the Reformed Presbyterian Church, the Reformed Christian Church, the Church of the Brethren, and an independent Bible church. (Information about one member is not available.) Eleven lived in the Midwest, three in the South, and two in the Far West. The committee included six biologists but only one geologist, an independent consultant with a master's degree. Seven members taught in church-related colleges, five in state institutions; the others worked for industry or were self-employed.[31]

To avoid the creeping evolutionism that had infected the ASA and to ensure that the society remained loyal to the Price-Morris tradition, the CRS required members to sign a statement of belief accepting the inerrancy of the Bible, the special creation of "all basic types of living things," and a worldwide deluge (Creation Research 1964, [13]). It restricted membership to Christians only. (Although creationists liked to stress the scientific evidence for their position, one estimated that "only about five percent of evolutionists-turned-creationists did so on the basis of the overwhelming evidence for creation in the world of nature"; the remaining 95 percent became creationists because they believed in the Bible [Lang 1978, 2].)[32] To legitimate its claim to being a scientific society, the CRS published a quarterly journal and limited full membership to persons possessing a graduate degree in a scientific discipline.

At the end of its first decade the society claimed 450 regular members, plus 1,600 sustaining members, who failed to meet the scientific qualifications. Eschewing politics, the CRS devoted itself almost exclusively to education and research, funded "at very little expense, and . . . with no expenditure of public money" (Lammerts 1974, 63). CRS-related projects included expeditions to search for Noah's ark, studies of fossil human footprints and pollen grains found out of the predicted evolutionary order, experiments on radiation-produced mutations in plants, and theoretical studies in physics demonstrating a recent origin of the earth (Gish 1975). A number of members collaborated in preparing a biology textbook based on creationist principles (Moore and Slusher 1970). In view of the previous history of creation science, it was an auspicious beginning.

While the CRS catered to the needs of scientists, a second, predominantly lay organization carried creationism to the masses. Created in 1964 in the wake of interest generated by *The Genesis Flood*, the Bible-Science Association came to be identified by many with one man: Walter Lang, an ambitious Missouri-Synod pastor who self-consciously prized spiritual insight above scientific expertise. As editor of the widely circulated *Bible-Science Newsletter* he vigorously promoted the Price-Morris line—and occasionally provided a platform for individuals on the fringes of the creationist movement, such as those who questioned the heliocentric theory and who believed that Albert Einstein's theory of relativity "was invented in order to circumvent the evidence that the earth is at rest." Needless to say, the pastor's broad-mindedness greatly embarrassed creationists seeking scientific respectability, who feared that such bizarre behavior would tarnish the entire movement (Lang 1977a, 4–5; 1977b, 2–3; 1978b, 1–3; Wheeler 1976, 101–102).

SCIENTIFIC CREATIONISM

The creationist revival of the 1960s attracted little public attention until late in the decade, when fundamentalists became aroused about the federally funded *Biological Sciences Curriculum Study* texts (Skoog 1979; "A Critique" 1966, 1), which featured evolution, and the California State Board of Education voted to require public school textbooks to include creation along with evolution. This decision resulted in large part from the efforts of two southern California housewives, Nell Segraves and Jean Sumrall, associates of both the Bible-Science Association and the CRS. In 1961 Segraves learned of the US Supreme Court's ruling in the Madalyn Murray case protecting atheist students from required prayers in public schools. Murray's ability to shield her child from religious exposure suggested to Segraves that creationist parents like herself "were entitled to protect our children from the influence of beliefs that would be offensive to our religious beliefs." It was this line of argument that finally persuaded the Board of Education to grant creationists equal rights (Bates 1975, 58; "Fifteen Years" 1979, 2; Wade 1972; see also Moore 1974; and Nelkin 1982).

Flushed with victory, Segraves and her son Kelly in 1970 joined an effort to organize a Creation Science Research Center (CSRC), affiliated with Christian Heritage College in San Diego, to prepare creationist literature suitable for adoption in public schools. Associated with them in this enterprise was Morris, who resigned his position at Virginia Polytechnic Institute to help establish a center for creation research. Because of differences in personalities and objectives, the Segraveses in 1972 left the college, taking the CSRC with them; Morris thereupon set up a new research division at the college, the Institute for Creation Research (ICR), which, he announced with obvious relief, would be "controlled and operated by scientists" and would engage in research and education, not political action. During the 1970s Morris added five scientists to his staff and, funded largely by small gifts and royalties from institute publications, turned the ICR into the world's leading center for the propagation of strict creationism (Morris 1972).[33] Meanwhile, the CSRC continued campaigning for the legal recognition of special creation, often citing a direct relationship between the acceptance of evolution and the breakdown of law and order. Its own research, the CSRC announced, proved that evolution fostered "the moral decay of spiritual values which contribute to the destruction of mental health and... [the prevalence of] divorce, abortion, and rampant venereal disease" (Segraves 1977, 17; "Fifteen Years" 1979, 2–3).

The 1970s witnessed a major shift in creationist tactics. Instead of trying to outlaw evolution, as they had done in the 1920s, antievolutionists now fought to give creation equal time. And instead of appealing to the authority of the Bible, as Morris and Whitcomb had done as recently as 1961, they consciously downplayed the Genesis story in favor of what they called "scientific creationism." Several factors no doubt contributed to this shift. One sociologist has suggested that creationists began stressing the scientific legitimacy of their enterprise because "their theological legitimation of reality was no longer sufficient for maintaining their world and passing on their worldview to their children" (Bates 1976, 98). However, there were also practical considerations. In 1968 the US Supreme Court declared the Arkansas antievolution law unconstitutional, giving creationists reason to suspect that legislation requiring the teaching of biblical creationism would meet a similar fate. They also feared that requiring the biblical account "would open the door to a wide variety of interpretations of Genesis" and produce demands for the inclusion of non-Christian versions of creation (Morris 1974a, 2; see also Larson 1984).

In view of such potential hazards, Morris recommended that creationists ask public schools to teach "only the scientific aspects of creationism" (Morris 1974a, 2), which in practice meant leaving out all references to the six days of Genesis and Noah's ark and focusing instead on evidence for a recent worldwide catastrophe and on arguments against evolution. Thus the product remained virtually the same; only the packaging changed. The 1974 ICR textbook *Scientific Creationism*, for example, came in two editions: one for public schools, containing no references to the Bible, and another for use in Christian schools that included a chapter on "Creation According to Scripture" (Morris 1974b).

In defending creation as a scientific alternative to evolution, creationists relied less on Francis Bacon and his conception of science and more on two new philosopher-heroes: Karl Popper and Thomas Kuhn. Popper required all scientific theories to be falsifiable; since evolution could not be falsified, reasoned the creationists, it was by definition not science. Kuhn described scientific progress in terms of competing models or paradigms rather than the accumulation of objective knowledge.[34] Thus creationists saw no reason why their flood-geology model should not be allowed to compete on an equal scientific basis with the evolution model. In selling this two-model approach to school boards, creationists were advised: "Sell more SCIENCE.... Who can object to teaching more science? What is controversial about that? . . . Do not use the word 'creationism.' Speak only of science. Explain that withholding scientific information contradicting evolution amounts to 'censorship' and smacks of getting into the province of religious dogma. . . . Use the 'censorship' label as one who is against censoring science. YOU are for science; anyone else who wants to censor scientific data is an old fogey and too doctrinaire to consider" (Leitch 1980, 2). This tactic proved extremely effective, at least initially. Two state legislatures, in Arkansas and Louisiana, and various school boards adopted the two-model approach, and an informal poll of school board members in 1980 showed that only 25 percent favored teaching nothing but evolution ("Finding" 1980, 52; Segraves 1977, 24). In 1982, however, a federal judge declared the Arkansas law, requiring a "balanced treatment" of creation and evolution, to be unconstitutional ("Creationism in Schools" 1982). Three years later a similar decision was reached regarding the Louisiana law.

Except for the battle to get scientific creationism into public schools, nothing brought more attention to the creationists than their public debates with prominent evolutionists, usually held on college campuses. During the 1970s the ICR staff alone participated in more than a hundred of these contests and, according to their own reckoning, never lost one. Although Morris preferred delivering straight lectures—and likened debates to the bloody confrontations between Christians and lions in ancient Rome—he recognized their value in carrying the creationist message to "more non-Christians and non-creationists than almost any other method" (Moths 1981, iii; 1974d, 2). Fortunately for him, an associate, Duane T. Gish, holder of a doctorate in biochemistry from the University of California, relished such confrontations. If the mild-mannered, professorial Morris was the Darwin of the creationist movement, then the bumptious Gish was its T. H. Huxley. He "hits the floor running" just like a bulldog, observed an admiring colleague; and "I go for the jugular vein," added Gish himself. Such enthusiasm helped draw crowds of up to five thousand.[35]

Early in 1981 the ICR announced the fulfillment of a recurring dream among creationists: a program offering graduate degrees in various creation-oriented sciences ("ICR Schedules" 1981). Besides hoping to fill an anticipated demand for teachers trained in scientific creationism, the ICR wished to provide an academic setting where creationist students would be free from discrimination. Over the years a number of creationists had reportedly been kicked out of secular universities because of their heterodox views, prompting leaders to warn graduate students to keep silent, "because if you don't, in almost 99 percent of the cases you will be asked to leave." To avoid anticipated harassment, several graduate students took to using pseudonyms when writing for creationist publications.[36]

Creationists also feared—with good reason—the possibility of defections while their students studied under evolutionists. Since the late 1950s the Seventh-Day Adventist Church had invested hundreds of thousands of dollars to staff its Geoscience Research Institute with well-trained young scientists, only to discover that in several instances exposure to orthodox science had destroyed belief in strict creationism. To reduce the incidence of apostasy, the church established its own graduate programs at Loma Linda University, where Price had once taught (Numbers 1979, 27–28; Couperus 1980).

TO ALL THE WORLD

It is still too early to assess the full impact of the creationist revival sparked by Whitcomb and Morris, but its influence, especially among evangelical Christians, seems to have been immense. Not least, it has elevated the strict creationism of Price and Morris to a position of apparent orthodoxy. It has also endowed creationism with a measure of scientific respectability unknown since the deaths of Guyot and Dawson. Yet it is impossible to determine how much of the creationists' success stemmed from converting evolutionists as opposed to mobilizing the already converted, and how much it owed to widespread disillusionment with established science. A sociological survey of church members in northern California in 1963 revealed that over a fourth of those polled—30 percent of Protestants and 28 percent of Catholics—were already opposed to evolution when the creationist revival began (Bainbridge and Stark 1980, 20). Broken down by denomination, it showed

Liberal Protestants (Congregationalists, Methodists, Episcopalians, Disciples)	11%
Moderate Protestants (Presbyterians, American Lutherans, American Baptists)	29%
Church of God	57%
Missouri-Synod Lutherans	64%
Southern Baptists	72%
Church of Christ	78%
Nazarenes	80%
Assemblies of God	91%
Seventh-Day Adventists	94%

Thus the creationists launched their crusade having a large reservoir of potential support.

Has belief in creationism increased since the early 1960s? The scanty evidence available suggests that it has. A nationwide Gallup poll in 1982, cited at the beginning of this [chapter], showed that nearly as many Americans (44 percent) believed in a recent special creation as accepted theistic (38 percent) or nontheistic (9 percent) evolution ("Poll" 1982, 22). These figures, when compared with the roughly 30 percent of northern California church members who opposed evolution in 1963, suggest, in a grossly imprecise way, a substantial gain in the actual number of American creationists. Bits and pieces of additional evidence lend credence to this conclusion. For example, in 1935 only 36 percent of the students at Brigham Young University, a Mormon school, rejected human evolution; in 1973 the percentage had climbed to 81 (Christensen and Cannon 1978). Also, during the 1970s both the Missouri-Synod Lutheran and Seventh-Day Adventist churches, traditional bastions of strict creationism, took strong measures to reverse a trend toward greater toleration of progressive creationism ("Return to Conservatism" 1973, 1; Numbers 1979, 27–28). In at least these instances, strict creationism did seem to be gaining ground.

Unlike the antievolution crusade of the 1920s, which remained confined mainly to North America, the revival of the 1960s rapidly spread overseas as American creationists and their books circled the globe. Partly as a result of stimulation from America, including the publication of a British edition of *The Genesis Flood* in 1969, the lethargic Evolution Protest Movement in Great Britain was revitalized; and two new creationist organizations, the Newton

Scientific Association and the Biblical Creation Society, sprang into existence (Barker 1979; [Clark] 1972–1973; 1977; "British Scientists" 1973; "EPM" 1972). On the Continent the Dutch assumed the lead in promoting creationism, encouraged by the translation of books on flood geology and by visits from ICR scientists (Ouweneel 1978). Similar developments occurred elsewhere in Europe, as well as in Australia, Asia, and South America. By 1980 Morris's books alone had been translated into Chinese, Czech, Dutch, French, German, Japanese, Korean, Portuguese, Russian, and Spanish. Strict creationism had become an international phenomenon.[38]

NOTES

1. Michael Ruse (1979) argues that most British biologists were evolutionists by the mid-1860s, while David L Hull, Peter D. Tessner, and Arthur M. Diamond (1978, 721) point out that more than a quarter of British scientists continued to reject the evolution of species as late as 1869. On the acceptance of evolution among religious leaders see, e.g., Frank Hugh Foster (1939, 38–58) and Owen Chadwick (1972, 23–24).

2. According to the poll, 9 percent of the respondents favored an evolutionary process in which God played no part, 38 percent believed God directed the evolutionary process, and 9 percent had no opinion.

3. On the influence of the *Scofleld Reference Bible* see Ernest R.. Sandeen (1971, 222).

4. On the popularity of the Guyot-Dawson view, also associated with the geologist James Dwight Dana, see William North Rice (1904, 101).

5. The Scottish theologian James Orr contributed an equally tolerant essay in *The Fundamentals* (Orr 1910–15).

6. "The Menace of Darwinism" appears in Bryan's book, *In His Image*, as chapter 4, "The Origin of Man." Bryan apparently borrowed his account of the evolution of the eye from Patterson (1902, 32–33).

7. Bryan gives the estimate of nine-tenths in a letter to W. A. McRae, April 5, 1924 (Bryan Papers, box 29).

8. The best state histories of the antievolution crusade are Bailey (1950); Gatewood (1966); and Gray (1970). Szasz (1969, 351) stresses the urban dimension of the crusade.

9. For examples of prominent fundamentalists who stayed aloof from the antievolution controversy see Stonehouse (1954, 401–402) and Lewis (1963, 86–88).

10. Furness (1954) includes chapter-by-chapter surveys of seven denominations.

11. On Riley see Riley (1938, 101–102) and Sansa (1980, 89–91). Maraden (1980, 167–70) stresses the interdenominational character of the antievolution crusade.

12. The creationists' use of Bateson provoked the evolutionist Henry Fairfield Osborn into repudiating the British scientist (Osborn 1926, 29).

13. Heber D. Curtis to W. J. Bryan, May 22, 1923 (Bryan Papers, box 37). Two physicians, Arthur I. Brown of Vancouver and Howard A. Kelly of Johns Hopkins, achieved prominence in the fundamentalist movement, but Kelly leaned toward theistic evolution.

14. See Edmondson (1969, 276–336); Cole (1931, 264–65); B[oyer] (1939, 6–7); and "Two Great Field Secretaries" (1926, 11).

15. This and the following paragraphs on Price closely follow the account in Numbers (1979, 22–24).

16. Price's first antievolution book was published four years earlier (Price 1902).

17. David Star Jordan to G. M. Price, May 5, 1911 (Price Papers).

18. The discovery of Price's law was first announced in Price (1913, 119).

19. Howard A. Kelly to W. J. Bryan, June 15, 1925; Louis T. More to W. I. Bryan, July 7, 1925; and Alfred W. McCann to W. J. Bryan, June 30, 1925 (Bryan Papers, box 47).

20. W. J. Bryan to Howard A. Kelly, June 22, 1925 (Bryan Papers, box 47). In a letter to the editor of the *Forum*, Bryan (1923) asserted that he had never taught that the world was made in six literal days. I am indebted to Paul M. Waggoner for bringing this document to my attention.

21. For a typical statement of creationist frustration see Price (1935). The title for this section comes from Morris (1974, 13).

22. See "Announcement of the Religion and Science Association" (Price Papers); "The Religion and Science Association" (1936, 159–60); "Meeting of the Religion and Science Association" (1936, 209); Clark (1977, 168).

23. On the attitude of the Price faction see Harold W. Clark to G. M. Price, September 12, 1937 (Price Papers).

24. Ben F. Allen to the Board of Directors of the Creation-Deluge Society, August 12, 1945 (courtesy of Molleurus Couperus). Regarding the fossil footprints, see the *Newsletters of the Creation-Deluge Society* for August 19, 1944, and February 17, 1945.

25. On the early years of the American Scientific Affiliation see Everest (1951).

26. Kulp (1949, 20) mentions his initial skepticism of geology.

27. For a fuller discussion see Numbers (1984).

28. Interviews with Henry M. Morris, October 26, 1980, and January 6, 1981. See also the autobiographical material in Morris (1984).

29. Interviews with Morris.

30. The statement regarding the appearance of the book comes from Walter Hearn, quoted in Bates (1976, 52). See also Roberts (1964); Van de Fliert (1969); and Lammerts (1964). Among Missouri-Synod Lutherans, John W. Klotz (1955) may have had an even greater influence than Morris and Whitcomb.

31. Names, academic fields, and institutional affiliations are given in *Creation Research Society Quarterly* (1964 [113]); for additional information I am indebted to Duane T. Gish, John N. Moore, Henry M. Moths, Harold Slusher, and William J. Tinkle.

32. Other creationists have disputed the 5 percent estimate.

33. Information also obtained from the interview with Morris, January 6, 1981.

34. On Popper's influence see, e.g., Roth (1977). In a letter to the editor of *New Scientist*, Popper (1980) affirmed that the evolution of life on earth was testable and, therefore, scientific. On Kuhn's influence see, e.g, Roth (1975); Brand (1974); and Wheeler (1975, 192–230).

35. The reference to Gish comes from an interview with Harold Slusher and Duane T. Gish, January 6, 1981.

36. Evidence for alleged discrimination and the use of pseudonyms comes from: "Grand Canyon Presents Problems for Long Ages" (1980); interview with Ervil D. Clark, January 9, 1981; interview with Steven A. Austin, January 6, 1981; and interview with Duane T. Gish, October 26, 1980, the source of the quotation.

37. Barker greatly underestimates the size of the E. P. M. in 1966.

38. Notices regarding the spread of creationism overseas appeared frequently in *Bible-Science Newsletter* and *Acts & Facts.* On translations see "ICR Books Available in Many Languages" (1980, 2, 7).

REFERENCES

Bailey, Kenneth K. "The Enactment of Tennessee's Antievolution Law." *Journal of Southern History* 16 (1950): 472–510.

———. *Southern White Protestantism in the Twentieth Century.* New York: Harper & Row, 1964.

Bainbridge, William Sims, and Rodney Stark. "Superstitions Old and New." *Skeptical Inquirer* 4 (Summer 1980).

Barker, Eileen. "In the Beginning: The Battle for Creationist Science against Evolutionism." In *On the Margins of Science: The Social Construction of Rejected Knowledge*, edited by Roy Wallis, 197–200. Sociological Review Monograph, no. 27. Keele: University of Keele, 1979.

Bates, Vernon Lee. "Christian Fundamentalism and the Theory of Evolution in Public School Education: A Study of the Creation Science Movement." PhD diss., University of California, Davis, 1976.

Bateson, William. "Evolutionary Faith and Modern Doubts." *Science* 55 (1922): 55–61.

Beale, Howard K. *Are American Teachers Free? An Analysis of Restraints upon the Freedom of Teaching in American Schools.* New York: Charles Scribner's Sons, 1936.

Betts, George Herbert. 1929. *The Beliefs of 700 Ministers and Their Meaning for Religious Education.* New York: Abingdon Press, 1929.

Bowler, Peter J. *'The Eclipse of Darwinism.' Anti-Darwinian Evolution Theories in the Decades around 1900.* Baltimore, MD: Johns Hopkins University Press, 1983.

B[oyer], F. J. "Harry Rimmer, D.D." *Christian Faith and Life* 45 (1939).

Brand, Leonard R. "A Philosophic Rationale for a Creation-Flood Model." *Origins* 1 (1974): 73–83.

"British Scientists Form Creationist Organization." *Acts & Facts* 2 (November–December 1973): 3.

Bryan, William Jennings. *In His Image.* New York: Fleming H. Revell, 1922.

———. Letter to the editor, *Forum* 70 (1923): 1852–53.

Bryan Papers. N.d. Library of Congress.

Carpenter, Joel A. "Fundamentalist Institutions and the Rise of Evangelical Protestantism, 1929–1942." *Church History* 49 (1980): 62–75.

Cassel, J. Frank. "The Evolution of Evangelical Thinking on Evolution." *Journal of the American Scientific Affiliation* 11 (December 1959): 26–27.

Chadwick, Owen. *The Victorian Church, Part 2*, 2nd ed. London: Adam and Charles Black, 1972.

Christensen, Harold T., and Kenneth L. Cannon. "The Fundamental Emphasis at Brigham Young University: 1935–1973." *Journal for the Scientific Study of Religion* 17 (1978): 53–57.

Clark, Edward Lassiter. "The Southern Baptist Reaction to the Darwinian Theory of Evolution." PhD diss., Southwestern Baptist Theological Seminary, Fort Worth, 1952.

Clark, Harold W. *The Battle over Genesis.* Washington: Review and Herald Publishing Association, 1977.

[Clark, Robert E. D.] "Evolution: Polarization of Views." *Faith and Thought* 100 (1972–1973): 227–29.

———. "American and English Creationists." *Faith and Thought* 104 (1977): 6–8.

Cole, Steward G. *The History of Fundamentalism.* New York: Richard R. Smith, 1931.

Coletta, Paolo E. *William Jennings Bryan*, vol. 3. *Political Puritan*, Lincoln: University of Nebraska Press, 1915–1925.

"Comment on the 'Deluge Geology' Paper of J. L. Kulp." *Journal of the American Scientific Affiliation* 2 (June 1950).

Couperus, Molleurus. "Tensions between Religion and Science." *Spectrum* 10 (March 1980): 74–78.

"Creationism in Schools: The Decision in MacLean versus the Arkansas Board of Education." *Science* 215 (1982): 934–43.

Creation Research Society Quarterly 1 (July 1964).

"A Critique of BSCS Biology Texts." *Bible-Science Newsletter* 4 (March 15, 1966).

Culver, Robert D. "An Evaluation of the Christian View of Science and Scripture by Bernard Ramm from the Standpoint of Christian Theology." *Journal of the American Scientific Affiliation* 7 (December 1955).

Dewar, Douglas. *The Difficulties of the Evolution Theory.* London: Edward Arnold and Co., 1931.

———. "The Limitations of Organic Evolution," *Journal of the Victoria Institute* 64 (1932).

Dillenberger, John, and Claude Welch. *Protestant Christianity Interpreted Through Its Development.* New York: Charles Scribner's Sons, 1954.

Edmondson, William D. "Fundamentalist Sects of Los Angeles, 1900–1930." PhD diss., Claremont Graduate School, Claremont, 1969.

"EPM—40 Years On; Evolution 114 Years Off." Supplement to *Creation* 1 (May 1972).

Everest, Alton. "The American Scientific Affiliation—The First Decade," *Journal of the American Scientific Affiliation* 3 (September 1951): 31–38.

"Evolutionism in the Pulpit." In *The Fundamentals*, vol. 8, 28–30. Chicago: Testimony, 1910–1915.

Evolution Protest Movement Pamphlet no. 125 (April 1965).

"Fifteen Years of Creationism." *Five Minutes with the Bible and Science.* Supplement to *Bible-Science Newsletter* 17 (May 1979): 2.

"Fighting Evolution at the Fundamentals Convention." *Christian Fundamentals in School and Church* 7 (July–September 1925).

"Finding: Let Kids Decide How We Got Here." *American School Board Journal* 167 (March 1980).

Foster, Frank Hugh. *The Modern Movement in American Theology: Sketches in the History of American Protestant Thought from the Civil War to the World War.* New York: Fleming H. Revel Co., 1939.

The Fundamentals. 12 vols. Chicago: Testimony, 1910–1915.

Furniss, Norman K. *The Fundamentalist Controversy, 1918–1931.* New Haven, CT: Yale University Press, 1954.

Gatewood, William J., Jr. *Preachers, Pedagogues and Politicians: The Evolution Controversy in North Carolina, 1920–1927.* Chapel Hill: University of North Carolina Press, 1966.

———, ed. *Controversy in the Twenties: Fundamentalism, Modernism, and Evolution.* Nashville, TN: Vanderbilt University Press, 1969.

Gish, Duane T. "A Decade of Creationist Research." *Creation Research Society Quarterly* 12 (June 1975): 34–36.

Grabiner, Judith V., and Peter D. Miller. "Effects of the Scopes Trial." *Science* 185 (1974): 832–37.

"Grand Canyon Presents Problems for Long Ages." *Five Minutes with the Bible and Science.* Supplement to *Bible-Science Newsletter* 18 (June 1980): 1–2.

Gray, Asa. *Darwinism: Essays and Reviews Pertaining to Darwinism*, edited by A. Hunter Dupree. Cambridge, MA: Harvard University Press, 1963.

Gray, Virginia. "Antievolution Sentiment and Behavior: The Case of Arkansas." *Journal of American History* 57 (1970): 352–66.

Halliburton, R., Jr. "The Adoption of Arkansas' Antievolution Law." *Arkansas Historical Quarterly* 23 (1964).

Hastings, H. L "Preface." In Robert Patterson, *The Errors of Evolution: An Examination of the Nebular Theory, Geological Evolution, the Origin of Life, and Darwinism*, 3rd ed. Boston: Scriptural Tract Repository, 1889.

Hull, David L., Peter D. Tessner, and Arthur M. Diamond. "Planck's Principle." *Science* 202 (1978).

"ICR Books Available in Many Languages." *Acts & Facts* 9 (February 1980).

"ICR Schedules M. S. Programs." *Acts & Facts* 10 (February 1981): 1–2.

K[eyser], L. S. "No War against Science—Never!" *Bible Champion* 31 (1925).

Klotz, John W. *Genes, Genesis, and Evolution.* St. Louis, MO: Concordia Publishing House, 1955.

Kulp, J. Laurence. "Some Presuppositions of Evolutionary Thinking." *Journal of the American Scientific Affiliation* 1 (June 1949).

———. "Deluge Geology." *Journal of the American Scientific Affiliation* 2 (March 1950): 1–15.

Laba, Estelle R., and Eugene W. Gross. "Evolution Slighted in High-School Biology." *Clearing House* 24 (1950): 396–99.

Lammerts, Walter E. "Introduction." *Annual Creation Research Society* (1964).

———. "The Creationist Movement in the United States: A Personal Account." *Journal of Christian Reconstruction* (Summer 1974): 49–63.

Lang, Walter. "A Naturalistic Cosmology vs. a Biblical Cosmology." *Bible-Science Newsletter* 15 (January–February 1977a).

———. "Editorial Comments." *Bible-Science Newsletter* 15 (March 1977b).

———. "Editorial Comments." *Bible-Science Newsletter* 16 (June 1978a).

———. "Fifteen Years of Creationism." *Bible-Science Newsletter* 16 (October 1978b).

Larson, Edward J. "Public Science vs. Popular Opinion: The Creation-Evolution Legal Controversy." PhD diss., University of Wisconsin, Madison, 1984.

Leitch, Russell H. "Mistakes Creationists Make." *Bible-Science Newsletter* 18 (March 1980).

Levine, Lawrence W. *Defender of the Faith—William Jennings Bryan: The Last Decade, 1915–25.* New York: Oxford University Press, 1965.

Lewis, William Bryant. "The Role of Harold Paul Sloan and His Methodist League for Faith and Life in the Fundamentalist-Modernist Controversy of the Methodist Episcopal Church." PhD diss., Vanderbilt University, Nashville, 1963.

Loetscher, Lefferts A. *The Broadening Church: A Study of Theological Issues in the Presbyterian Church Since 1869.* Philadelphia: University of Pennsylvania Press, 1954.

Lunn, Arnold, ed. *Is Evolution Proved? A Debate between Douglas Dewar and H. S. Shelton.* London: Hollis and Carter, 1947.

Marsden, George M. "Fundamentalism as an American Phenomenon: A Comparison with English Evangelicalism." *Church History* 46 (1977): 215–32.

———. *Fundamentalism and American Culture: The Shaping of Twentieth Century Evangelicalism, 1870–1925.* New York: Oxford University Press, 1980.

Martin, T. T. *Hell and the High School Christ or Evolution, Which?* Kansas City, MO: Western Baptist Publishing Co., 1923.

Mauro, Philip. "Modern Philosophy." In *The Fundamentals*, vol. 2, pp. 85–305. Chicago: Testimony, 1910–1915.

McLoughlin, William G., Jr. *Modern Revivalism: Charles Grandison Finney to Billy Graham.* New York: Ronald Press, 1959.

"Meeting of the Religion and Science Association." *Christian Faith and Life* 42 (1936).

Moore, James R. *The Post-Darwinian Controversies: A Study of the Protestant Struggle to Come to Terms with Darwin in Great Britain and America, 1870–1900.* Cambridge, MA: Cambridge University Press, 1979.

Moore, John A. "Creationism in California." *Daedalus* 103 (1974): 173–89.

Moore, John N., and Harold Schultz Slusher, eds. *Biology: A Search for Order in Complexity.* Grand Rapids, MI: Zondervan Publishing House, 1970.

Morris, Henry M. "Director's Column." *Acts & Facts* 1 (June–July 1972).

———. "Director's Column." *Acts & Facts* 3 (March 1974a).

———. "Director's Column." *Acts & Facts* 3 (September 1974b).
———, ed. *Scientific Creationism.* San Diego, CA: Creation-Life Publishers, 1974c.
———. *The Troubled Waters of Evolution.* San Diego, CA: Creation-Life Publishers, 1974d.
———. *That You Might Believe.* San Diego, CA: Creation-Life Publishers, 1978.
———. "Director's Column." *Acts & Facts* 1 (June–July 1981a).
———. "Two Decades of Creation: Past and Future." *Impact.* Supplement to *Acts & Facts* 10 (January 1981b).
———. *History of Modern Creationism.* San Diego, CA: Master Book Publishers, 1984.
Morris, Henry M., and John C. Whitcomb Jr. "Reply to Reviews in the March 1964 Issue." *Journal of the American Scientific Affiliation* 16 (June 1964).
Morrison, John L. "American Catholics and the Crusade against Evolution." *Records of American Catholic Historical Society of Philadelphia* 64 (1953): 59–71.
Nelkin, Dorothy. *The Creation Controversy: Science or Scriptures in the Schools.* New York: W. W. Norton, 1982.
Nelson, Roland T. "Fundamentalism and the Northern Baptist Convention." PhD diss., University of Chicago, Chicago, 1964.
Numbers, Ronald L. "Science Falsely So-Called: Evolution and the Adventists in the Nineteenth Century." *Journal of the American Scientific Affiliation* 27 (March 1975).
———. *Creation by Natural Law: Laplace's Nebular Hypothesis in American Thought.* Seattle: University of Washington Press, 1977.
———. "Sciences of Satanic Origin: Adventist Attitudes toward Evolutionary Biology and Geology." *Spectrum* 9 (January 1979).
———. "The Dilemma of Evangelical Scientists." In *Evangelism and Modern America,* edited by George M. Marsden, 150–60. Grand Rapids, MI: William B. Eerdmans, 1984.
O'Brien, Charles F. *Sir William Dawson: A Life in Science and Religion.* Philadelphia: American Philosophical Society, 1971.
Orr, James. "Science and Christian Faith." In *The Fundamentals,* vol. 4, 91–104. Chicago: Testimony, 1910–1915.
Osborn, Henry Fairfield. *Evolution and Religion in Education: Polemics of the Fundamentalist Controversy of 1922 to 1926.* New York: Charles Scribner's Sons, 1926.
Ouweneel, W. J. "Creationism in the Netherlands." *Impact.* Supplement to *Acts & Facts* 9 (February 1978): i–iv.
Patterson, Alexander. *The Other Side of Evolution: An Examination of Its Evidences.* Chicago: Winona Publishing Co., 1902.
Patterson, Robert. *The Errors of Evolution: An Examination of the Nebular Theory, Geolog-*

ical Evolution, the Origin of Life, and Darwinism, 3rd ed. Boston: Scriptural Tract Repository, 1893.

Paul, Harry W. *The Edge of Contingency: French Catholic Reaction to Scientific Change from Darwin to Duhem.* Gainesville: University Presses of Florida, 1979.

Pfeifer, Edward J. "United States." In *The Comparative Reception of Darwinism*, edited by Thomas F. Glick. Austin: University of Texas Press, 1974.

"Poll Finds Americans Split on Creation Idea." *New York Times*, August 29, 1982.

Popper, Karl. "Letter to the editor." *New Scientist* 87 (August 21, 1980): 611.

Price, George McCready. *Outlines of Modern Science and Modern Christianity.* Oakland, CA: Pacific Press, 1901.

———. *Illogical Geology: The Weakest Point in the Evolution Theory.* Los Angeles, CA: Modern Heretic Co., 1906.

———. *The Fundamentals of Geology and Their Bearings on the Doctrine of a Literal Creation.* Mountain View, CA: Pacific Press, 1913.

———. *The New Geology.* Mountain View, CA: Pacific Press, 1923.

———. "Guarding the Sacred Cow." *Christian Faith and Life* 41 (1935): 124–27.

———. *The Story of the Fossils.* Mountain View, CA: Pacific Press, 1954.

Price Papers. N.d. Andrews University.

"Progress of Antievolution." *Christian Fundamentalist* 2 (1929).

Ramm, Bernard. *The Christian View of Science and Scripture.* Grand Rapids, MI: William B. Eerdmans, 1954.

Reeve, J. J. "My Personal Experience with the Higher Criticism." In *The Fundamentals*, vol. 3, 98–118. Chicago: Testimony, 1910–1935.

"The Religion and Science Association." *Christian Faith and Life* 42 (1936): 159–60.

"Return to Conservatism." *Bible-Science Newsletter* 11 (August 1973).

Rice, William North. *Christian Faith in an Age of Science*, 2nd ed. New York: A. C. Armstrong and Son, 1904.

Riley, Marie Acomb. *The Dynamic of a Dream: The Life Story of Dr. William B. Riley.* Grand Rapids, MI: William B. Eerdmans, 1938.

[Riley, William B.] "The Evolution Controversy." *Christian Fundamentals in School and Church* 4 (April–May 1922).

———. "The Creative Week." *Christian Fundamentalist* 4 (1930).

Riley, William B., and Harry Rimmer. N.d. *A Debate Resolved. That the Creative Days in Genesis Were Aeons, Not Solar Days.* Pamphlet.

Rimmer, Harry. *The Harmony of Science and Scripture*, 11th ed. Grand Rapids, MI: William B. Eerdmans, 1945.

Roberts, Frank H. Review of *The Genesis Flood*, by Henry M. Morris and John C. Whitcomb Jr. *Journal of the American Scientific Affiliation* 16 (March 1964): 28–29.

Roth, Ariel A. "The Pervasiveness of the Paradigm." *Origins* 2 (1975): 55–57.

———. "Does Evolution Qualify as a Scientific Principle?" *Origins* 4 (1977): 4–10.

Rudnick, Milton L. *Fundamentalism and the Missouri Synod: A Historical Study of Their Interaction and Mutual Influence.* St. Louis, MO: Concordia Publishing House, 1966.

Ruse, Michael. *The Darwinian Revolution: Science Red in Tooth and Claw.* Chicago: University of Chicago Press, 1979.

Sandeen, Ernest R. *The Roots of Fundamentalism: British and American Millenarianism, 1800–1930.* Chicago: University of Chicago Press, 1971.

Schuchert, Charles. Review of *The New Geology*, by George McCready Price. *Science* 59 (1924): 486–87.

Science 63 (1926): 259.

Segraves, Nell J. *The Creation Report.* San Diego, CA: Creation Science Research Center, 1977.

Shipley, Maynard. *The War on Modern Science: A Short History of the Fundamentalist Attacks on Evolution and Modernism.* New York: Alfred A. Knopf, 1927.

———. "Growth of the Antievolution Movement." *Current History* 32 (1930).

Skoog, Gerald. "Topic of Evolution in Secondary School Biology Textbooks: 1900–1977." *Science Education* 63 (1979): 621–40.

Stonehouse, Ned B. *J. Gresham Machen: A Biographical Memoir.* Grand Rapids, MI: William B. Eerdmans, 1954.

Szasz, Ferenc Morton. "Three Fundamentalist Leaders: The Roles of William Bell Riley, John Roach Straton, and William Jennings Bryan in the Fundamentalist-Modernist Controversy." PhD diss., University of Rochester, 1969.

———. *The Divided Mind of Protestant America, 1889–1930.* Tuscaloosa: University of Alabama Press, 1982.

Thompson, James J. Jr. "Southern Baptists and the Antievolution Controversy of the 1920s." *Mississippi Quarterly* 29 (1975–1976): 65–81.

"Two Great Field Secretaries—Harry Rimmer and Dr. Arthur I. Brown." *Christian Fundamentals in School and Church* 8 (July–September 1926).

Van de Fliert, J. R. "Fundamentalism and the Fundamentals of Geology." *Journal of the American Scientific Association* 21 (September 1969): 69–81.

Wade, Nicholas. "Creationists and Evolutionists: Confrontation in California." *Science* 178 (1972): 724–29.

Whalen, Robert D. "Millenarianism and Millennialism in America, 1790–1880." PhD diss., State University of New York at Stony Brook, 1972.

Wheeler, Gerald. *The Two-Taled Dinosaur: Why Science and Religion Conflict over the Origin of Life.* Nashville, TN: Southern Publishing Association, 1975.

———. "The Third National Creation Science Conference." *Origins* 3 (1976).

Whitcomb, John C., Jr., and Henry M. Moths. *The Genesis Flood: The Biblical Record and Its Scientific Implications.* Philadelphia: Presbyterian and Reformed Publishing Co., 1961.

Whitney, Dudley Joseph. "'What Theory of Earth History Shall We Adopt?" *Bible Champion* 34 (1928).

Williams, Wayne C. *William Jennings Bryan.* New York: G. P. Putnam, 1936.

"World Religious Digest." *Christian Faith and Life* 45 (1939).

Wright, G. Frederick. "The First Chapter of Genesis and Modern Science." *Homiletic Review* 35 (1898): 392–99.

———. "Introduction." In Alexander Patterson, *The Other Side of Evolution An Examination of Its Evidences*, xvii–xix. Chicago: Winona Publishing Co., 1902.

———. "The Passing of Evolution." In *The Fundamentals*, vol. 7, 5–20. Chicago: Testimony, 1910–1915.

———. *Story of My Life and Work.* Oberlin, OH: Bibliotheca Sacra Co., 1916.

Young, Davis A. *Creation and the Flood: An Alternative to Flood Geology and Theistic Evolution.* Grand Rapids, MI: Baker Book House, 1977.

Young, G. L. "Relation of Evolution and Darwinism to the Question of Origins." *Bible Student and Teacher* 11 (July 1909).

CHAPTER 12

CREATION, EVOLUTION, AND THE HISTORICAL EVIDENCE

DUANE T. GISH

For a clear understanding of the issues to be discussed in this [chapter], I must begin by defining evolution and creation. When the term evolution is used it will refer to the general theory of organic evolution, or the molecules-to-man theory of evolution. According to this theory all living things have arisen by naturalistic, mechanistic, evolutionary processes from a single living source, which itself had arisen by similar processes from inanimate matter. These processes are attributable solely to properties inherent in matter and are, therefore, still operative today. Creation theory postulates, on the other hand, that all basic animal and plant types (the created kinds) were brought into being by the acts of a preexisting Being by means of special processes that are not operative today. The variation that has occurred since creation has been restricted within the limits of each created kind.

Evolutionists adamantly insist that special creation be excluded from any consideration as a possible explanation for origins, because it does not qualify as a scientific theory. The proponents of evolution theory at the same time would view as unthinkable the consideration of evolution as anything less

From Duane T. Gish, "Creation, Evolution, and the Historical Evidence," *American Biology Teacher* (March 1973): 132–40. Reprinted by permission of the National Association of Biology Teachers.

than pure science: and indeed most of them insist that evolution must no longer be thought of as a theory, but must be considered to be a fact.

WHAT IS THEORY? WHAT IS FACT?

What criteria must be met for a theory to be considered scientific in the usually accepted sense? George Gaylord Simpson (1964) has stated, "It is inherent in any definition of science that statements that cannot be checked by observation are not really about anything... or at the very least, they are not *science*." A definition of science in the *Oxford English Dictionary* is "a branch of study which is concerned either with a connected body of *demonstrated truths* or with *observed facts* systematically classified and more or less colligated by being brought under general laws, and which includes trustworthy methods for the discovery of new truth within its own domain" (emphasis added).

Thus, for a theory to qualify as a scientific theory, it must be supported by events or processes that can be observed to occur, and the theory must be useful in predicting the outcome of future natural phenomena or laboratory experiments. An additional limitation usually imposed is that the theory must be capable of falsification; that is, one must be able to conceive some experiment the failure of which would disprove the theory. It is on the basis of such criteria that most evolutionists insist that creation be refused consideration as a possible explanation for origins. Creation has not been witnessed by human observers, it cannot be tested scientifically, and as a theory it is nonfalsifiable.

The general theory of evolution (molecules-to-man theory) also fails to meet all three of these criteria, however. Dobzhansky (1958), while seeking to affirm the factuality of evolution, actually admits that it does not meet the criteria of a scientific theory, when he says, "The occurrence of the evolution of life in the history of the earth is established about as well as events *not witnessed by human observers* can be" (emphasis added).

Goldschmidt, who has insisted that evolution is a fact for which no further proof is needed, also reveals its failure to meet the usually accepted criteria for a scientific theory. After outlining his postulated systemic-mutation, or "hopeful monster," mechanism for evolution, Goldschmidt (1952, 94) states, "Such an assumption is violently opposed by the majority of geneticists, who claim that the facts found on the sub-specific level must apply also to the higher categories. Incessant repetition of this *unproved claim*, glossing lightly over the difficulties, and the assumption of an arrogant attitude toward those who are not so easily swayed by fashions in science, are considered to afford scientific proof of the doctrine. It is true that nobody thus far has produced a new species or genus, etc., by macromutation. It is equally true that nobody has produced even a species by the selection of micromutations" (emphasis added). Later in the same paper (97) he says, "Neither has anyone witnessed the production of a new specimen of a higher taxonomic category by selection of micromutants." Goldschmidt has thus affirmed that, in the molecules-to-man context, only the most trivial change, or that at the subspecies level, has actually ever been observed.

Furthermore, the architects of the modern synthetic theory of evolution have so skillfully constructed their theory that it is not capable of falsification. The theory is so plastic that it is capable of explaining anything. This is the complaint of Olson (1960, 580) and of several participants in the Wistar Symposium on Mathematical Challenges to the Neo-Darwinian Interpretation of Evolution (Moorhead and Kaplan, 1967)—even including Ernst Mayr, a leading exponent of the theory. Eden (1967, 71), one of the mathematicians, puts it this way, with reference to falsifiability, "This cannot be done in evolution, taking it in its broad sense, and this is really all I meant when I called it tautologous in the first place. It can, indeed, explain anything. You may be ingenious or not in proposing a mechanism which looks plausible to human beings and mechanisms which are consistent with other mechanisms which you have discovered, but it is still an unfalsifiable theory."

A RISING TIDE OF CRITICISM

In addition to scientists who are creationists, a growing number of other scientists have expressed doubts that modern evolution theory could explain more than trivial change. Eden (1967, 109) is so discouraged, after a thorough consideration of the modern theory from a probabilistic point of view, that he proclaims, "an adequate scientific theory of evolution must await the discovery and elucidation of new laws—physical, physico-chemical, and biological." Salisbury (1969; 1971) similarly expresses doubts based on probabilistic considerations.

The attack on the theory by French scientists has been intense in recent years. In a review of the French situation Litynski (1961) says, "This year saw the controversy rapidly growing, until recently it culminated in the title 'Should We Burn Darwin?' spread over two pages of the magazine *Science et Vie.* The article, by the science writer Aimé Michel, was based on the author's interviews with such specialists as Mrs. Andrêe Tetry, professor at the famous Ecole des Hautes Etudes and a world authority on problems of evolution, Professor Renô Chauvin and other noted French biologists, and on his thorough study of some 600 pages of biological data collected, in collaboration with Mrs. Tetry, by the late Michael Cuenot, a biologist of international fame. Aimé Michel's conclusion is significant: the classical theory of evolution in its strict sense belongs to the past. Even if they do not publicly take a definite stand, almost all French specialists hold today strong mental reservations as to the validity of natural selection."

E. C. Olson (1960, 523), one of the speakers at the Darwinian Centennial Celebration at Chicago, made the following statement on that occasion, "There exists, as well, a generally silent group of students engaged in biological pursuits who tend to disagree with much of the current thought but say and write little because they are not particularly interested, do not see that controversy over evolution is of any particular importance, or are so strongly in disagreement that it seems futile to undertake the monumental task of controverting the immense body of information and theory that exists in the formulation of modern thinking. It is, of course, difficult to judge the size and composition of this silent segment, but there is no doubt that the numbers are not inconsiderable."

Fothergill (1961) refers to what he calls "the paucity of evolutionary theory as a whole." Ehrlich and Holm (1962) have stated their reservations in the following way: "Finally, consider the third question posed earlier: 'What accounts for the observed patterns in nature?' It has become fashionable to regard modern evolutionary theory as the only possible explanation of these patterns rather than just the best explanation that has been developed so far. It is conceivable, even likely, that what one might facetiously call a non-Euclidean theory of evolution lies over the horizon. Perpetuation of today's theory as dogma will not encourage progress toward more satisfactory explanations of observed phenomena."

Sometimes the attacks are openly critical. Such is Danson's letter that appeared recently in *New Scientist.* He states in part, "The Theory of Evolution is no longer with us, because neo-Darwinism is now acknowledged as being unable to explain anything more than trivial changes and in default of some other theory we have none... despite the hostility of the witness provided by the fossil record, despite the innumerable difficulties, and despite the lack of even a credible theory, evolution survives.... Can there be any other area of science, for instance, in which a concept as intellectually barren as embryonic recapitulation could be used as evidence for a theory?" (Danson 1971).

Macbeth (1971) has provided an especially incisive criticism of evolution theory. He points out that although evolutionists have abandoned classical Darwinism, the modern synthetic theory they have proposed as a substitute is equally inadequate to explain progressive change as the result of natural selection; as a matter of fact, they cannot even define natural selection in nontautologous terms. Inadequacies of the present theory and failure of the fossil record to substantiate its predictions leave macroevolution, and even microevolution, intractable mysteries, according to Macbeth. He suggests that no theory at all may be preferable to the existing one.

In view of the above considerations, it is incredible that leading scientists, including several who addressed the NABT convention in San Francisco, dogmatically insist that the molecules-to-man evolution theory be taught as a fact to the exclusion of all other postulates. Evolution in this broad sense is unproven and unprovable and thus cannot be considered to be fact. It is not subject to test by the ordinary methods of experimental science: observation and falsification. It thus does not, in a strict sense, even qualify as a scientific theory. It is a postulate, and it may serve as a model within which attempts may be made to explain and correlate the evidence from the historical record—that is, the fossil record—and to make predictions concerning the nature of future discoveries.

Creation is, of course, unproven and unprovable by the methods of experimental science. Neither can it qualify, according to the above criteria, as a scientific theory, because creation would have been unobservable and, as a theory, would be nonfalsifiable. Creation is therefore (like evolution) a postulate that may serve as a model to explain and correlate the evidence relating to origins. Creation is, in this sense, no more religious or less scientific than evolution. In fact, to many well-informed scientists creation seems to be far superior to the evolution model as an explanation for origins.

I strongly suspect that the dogmatic acceptance of evolution is not due, primarily, to the nature of the evidence but the philosophic bias peculiar to our times. Watson (1929), for example, has referred to the theory of evolution as "a theory universally accepted not because it can be proved by logically coherent evidence to be true but because the only alternative, special creation, is clearly incredible."

That this is the philosophy held by most biologists has been recently emphasized by Dobzhansky. In his review of Monod's book *Chance and Necessity* Dobzhansky (1972) says, "He has stated with admirable clarity, and eloquence often verging on pathos, the mechanistic materialist philosophy shared by most of the present 'establishment' in the biological sciences."

TWO MODELS TO BE TESTED

The exclusion of creation from science teaching as a credible explanation of origins is unwarranted and undesirable on both philosophic and scientific grounds. Under the present system, whereby evolution is taught as an established fact to the exclusion of creation, the student is being indoctrinated in a philosophy of secular humanism rather than benefiting from an objective presentation of the evidence.

This situation could be remedied by (i) presenting creation and evolution in the form of models, (ii) making predictions based on each model, and (iii) comparing the actual scientific evidence with the predictions of the models. The students would then be able to make up their minds on the basis of this objective presentation. This is what I would like to do in the remainder of this [chapter]. I will restrict myself to an examination of the fossil record.

Although various scientific fields could be investigated in attempts to determine which model appears to be the more plausible of the two, the fossil record offers the only source of scientific evidence that would allow a determination of whether living organisms actually did arise by creation or by evolutionary process. The case is well stated by Le Gros Clark (1955) when he says, "That evolution actually did occur can only be scientifically established by the discovery of the fossilized remains of representative samples of those intermediate types which have been postulated on the basis of the indirect evidence. In other words, the really crucial evidence for evolution must be provided by the paleontologist whose business it is to study the evidence of the fossil record." Gavin de Beer (1964) echoes this view when he states, "The last word on the credibility and course of evolution lies with the paleontologists."

In his revolutionary work *The Origin of Species*, Darwin (1859) says, "the number of intermediate and transitional links, between all living and extinct species, must have been inconceivably great." This conclusion seems inescapable, whether it be based either on the concepts of classical Darwinism or on those of the modern synthetic theory. Because the number of transitional forms predicted by evolution theory is inconceivably great, the number of such forms that would have become fossilized, according to this theory, would have been very great indeed, even though only a very minute fraction of all plants and animals that ever existed had become fossilized.

Sampling of the fossil record has now been so thorough that appeals to the imperfections in the record are no longer valid. George (1960, 1) has stated, "There is no need to apologize any longer for the poverty of the fossil record. In some ways it has become almost unmanageably rich and discovery is outpacing integration." It seems clear, then, that after 150 years of intense searching, a very large number of obvious transitional forms would have been discovered if the predictions of evolution theory are valid.

On the basis of the creation model, on the other hand, the visual absence of apparent transitional forms between the higher categories or created kinds would be predicted. The presence of apparent transitional forms could not be rigidly excluded, however, for two reasons: (i) tremendous diversity is exhibited within each major type of plant and animal and (ii) possession of similar modes of existence or activities would require similar structures or functions. On the basis of the creation model such pseudotransitional forms should be rare and would not be connected by intermediate types. Gaps in the fossil record, therefore, should be systematic and nearly universal between the higher categories or created kinds. The fossil record should permit a clear choice between the two models.

The two models may thus be constructed as follows:

Creation model	*Evolution model*
By acts of a Creator	By naturalistic, mechanistic processes due to properties inherent in inanimate matter
Creation of basic plant and animal kinds with ordinal characteristics complete in first representatives	Origin of all living things from a single living source, which itself arose from inanimate matter. Origin of each kind from an ancestral form by slow, gradual change
Variation and speciation limited within each kind	Unlimited variation. All forms genetically related

These two models would permit the following predictions to be made about the fossil record:

Creation model	*Evolution model*
Sudden appearance in great variety of highly complex forms	Gradual change of simplest forms into more and more complex forms
Sudden appearance of each created kind with ordinal characteristics complete. Sharp boundaries separating major taxonomic groups. No transitional forms between higher categories	Transitional arteries linking all categories. No systematic gaps

Let us now compare the known facts of the fossil record with the predictions of the two models.

ADVENT OF LIFE IN THE CAMBRIAN

The oldest rocks in which indisputable fossils are found are those of the Cambrian Period. In these sedimentary deposits are found billions and billions of fossils of highly complex forms of life. These include sponges, corals, jellyfish, worms, mollusks, and crustaceans; in fact, every one of the major invertebrate forms of life has been found in Cambrian rocks. These animals were so highly complex that, it is conservatively estimated, they would have required 1.5 billion years to evolve.

What do we find in rocks older than the Cambrian? Not a single, indisputable multicellular fossil has ever been found in Precambrian rocks. Certainly it can be said without fear of contradiction that the evolutionary ancestors of the Cambrian fauna, if they ever existed, have never been found (Simpson 1960, 143; Cloud 1968; Axelrod 1968).

Concerning this problem, Axelrod (1968) has stated, "One of the major unsolved problems of geology and evolution is the occurrence of diversified, multicellular marine invertebrates in Lower Cambrian rocks on all the continents and their absence in rocks of greater age." After discussing the varied types that are found in the Cambrian, Axelrod goes on to say, "However, when we turn to examine the Precambrian rocks for the forerunners of these Early Cambrian fossils they are nowhere to be found. Many thick (over 5,000 feet) sections of sedimentary rock are now known to lie in unbroken succession below strata containing the earliest Cambrian fossils. These sediments apparently were suitable for the preservation of fossils because they are often identical with overlying rocks which are fossiliferous, yet no fossils are found in them."

From all appearances, then, based on the known facts of the historical record, there occurred a sudden great outburst of life at a high level of complexity. The fossil record gives no evidence that these Cambrian animals were derived from preceding, ancestral forms. Furthermore, not a single fossil has been found that can be considered to be a transitional form between the major groups, or phyla. At their earliest appearance these major invertebrate types were just as clearly and distinctly set apart as they are today.

How do these facts compare with the predictions of the evolution model? They are in clear contradiction to such predictions. This has been admitted, for instance, by George (1960, 5), who states, "Granted an evolutionary origin of the main groups of animals and not an act of special creation, the absence of any record whatsoever of a single member of any of the phyla in the Precambrian rocks remains as inexplicable on orthodox grounds as it was to Darwin." Simpson has struggled valiantly but not fruitfully with this problem and has been forced to concede (1949, 18) that the absence of Precambrian fossils (other than alleged fossil microorganisms) is the "major mystery of the history of life."

These facts, however, are in full agreement with the predictions of the creation model. The fossil record *does* reveal (i) a sudden appearance, in great variety, of highly complex forms with no evolutionary ancestors and (ii) the absence of transitional forms between the major taxonomic groups, just as postulated on the basis of creation. Most emphatically, the known facts of the fossil record from the very outset support the predictions of the creation model but unquestionably contradict the predictions of the evolution model.

DISCRETE NATURE OF VERTEBRATE CLASSES

The remainder of the history of life reveals a remarkable absence of the many transitional forms demanded by the theory. There is, in fact, a systematic deficiency of transitional forms between the higher categories, just as predicted by the creation model.

The idea that the vertebrates are derived from the invertebrates is purely an assumption that cannot be documented from the fossil record. In the history of the study of the comparative anatomy and embryology of living forms almost every invertebrate group has been proposed, at one time or another, as the ancestor of the vertebrates (E. G. Conklin, as quoted in Allen 1969; Romer 1966, 12). The transition from invertebrate to vertebrate supposedly passed through a simple chordate stage. Does the fossil record provide evidence for such a transition? Not at all. Ommaney (1964) has stated, "How this earliest chordate stock evolved, what stages of development it went through to eventually give rise to truly fishlike creatures we do not know. Between the Cambrian when it probably originated, and the Ordovician when the first fossils of animals with really fishlike characteristics appeared, there is a gap of perhaps 100 million years which we will probably never be able to fill."

Incredible! One hundred million years of evolution and no transitional forms! All hypotheses combined, no matter how ingeniously, could never pretend, on the basis of evolution theory, to account for a gap of such magnitude. Such facts, on the other hand, are in perfect accord with the predictions of the creation model.

A careful reading of Romer's *Vertebrate Paleontology* (1966) seems to allow no other conclusion than that the major classes of fish are clearly and distinctly set apart from one another with no transitional forms linking them.

The fossil record has not produced ancestors or transitional forms for these classes. Hypothetic ancestors and the required transitional forms must, on the basis of the known record, be merely the products of speculation. How then can it be argued that the evolution model's explanation of such evidence is more scientific than that of the creation model?

The fossil record has been diligently searched for a transitional series linking fish to amphibian, but as yet no such series has been found. The closest link that has been proposed is that allegedly existing between rhipidistian crossopterygian fish and the amphibians of the genus *Ichthyoszega*, of the labyrinthodont family Ichthyostegidae. There is a tremendous gap, however, between the crossopterygians and the ichthyostegids—a gap that would have spanned many millions of years, during which innumerable transitional forms should have existed. These transitional forms should reveal a slow, gradual change of the pectoral and pelvic fins of the crossopterygian fish into the feet and legs of the amphibian, along with loss of other fins, and the accomplishments of other transformations required for adaptation to a terrestrial habitat.

What is the fact? Not a single transitional form has ever been found showing an intermediate stage between the fin of the crossopterygian and the foot of the ichthyostegid. The limb and the limb-girdle of *Ichthyostega* is already of the basic amphibian type, showing no vestige of a fin ancestry.

The extremely broad gap between fish and amphibians, as observed between the rhipidistian crossoptenjgians and the ichthyostegids; the sudden appearance, in fact, of all Paleozoic amphibian orders with diverse ordinal characteristics complete in the first representatives; the absence of any transitional forms between these Paleozoic orders; and the absence of transitional forms between the Paleozoic orders and the three living orders—all these conditions are contradictory to the predictions of the evolution model. These facts, however, are just as predicted by the creation model.

It is at the amphibian-reptilian and the reptilian-mammalian boundaries that the strongest claims have been advanced for transitional types bridging classes. But these are just those classes that are most closely similar in skeletal features; that is, the parts that are preserved in the fossil record.

The conversion of an invertebrate into a vertebrate, a fish into a tetrapod with feet and legs, and a nonflying animal into a flying animal are a few examples of changes that would require a revolution in structure. Such transformations should provide readily recognizable transitional series in the fossil record if they occurred through evolutionary processes. On the other hand, if the creation model is the true model, it is at just such boundaries that the absence of transitional forms would be most evident.

The opposite is true at the amphibian-reptilian and reptilian-mammalian boundaries—particularly the former. Although it is feasible to distinguish between living reptiles and amphibians on the basis of skeletal features, they are much more readily distinguishable by means of their soft parts; and, in fact, the major definitive characteristic that separates reptiles from amphibians is the possession by the reptile, in contrast with the amphibian, of an amniote egg.

Many of the diagnostic features of mammals, of course, reside in their soft anatomy or their physiology. These include their mode of reproduction, warm-bloodedness, mode of breathing due to possession of a diaphragm, suckling of the young, and possession of hair.

The two most easily distinguishable osteologic differences between reptiles and mammals, however, have never been bridged by transitional series. All mammals, living or fossil, have a single bone, the dentary, on each side of the lower jaw; and all mammals, living or fossil, have three auditory ossicles, or ear bones: the malleus, incus, and stapes. In some fossil reptiles the number and size of the bones of the lower jaw are reduced, by comparison with those of living reptiles. Every reptile, living or fossil, however, has at least four bones in the lower jaw and only one auditory ossicle, the stapes. There are no transitional forms showing, for instance, three or two jaw bones or two ear bones. No one has explained yet, for that matter, how the transitional form would have managed to chew while its jaw was being unhinged and rearticulated or how it would hear while dragging two of its jaw bones up into its ear.

SPECIAL FEATURES OF FLYING ANIMALS

The origin of flight should provide excellent case histories for testing the evolution model versus the creation model. Almost every structure in a nonflying animal would require modification for flight, and resultant transitional forms should be easily detectable in the fossil record. Flight is supposed to have evolved four times, separately and independently: in insects, birds, mammals (bats), and reptiles (pterosaurs, now extinct). In each case the origin of flight is supposed to have required many millions of years, and almost innumerable transitional forms would have been required in each case. Yet not in a single case can anything even approaching a transitional series be produced.

E. C. Olson, an evolutionist and geologist, in his book *The Evolution of Life* (1965) states that "[a]s far as flight is concerned there are some very big gaps in the record" (180). Concerning insects he says, "There is almost nothing to give any information about the history of the origin of flight in insects" (180). Concerning flying reptiles, Olson reports that "true flight is first recorded among the reptiles by the pterosaurs in the Jurassic Period. Although the earliest of these were rather less specialized for flight than in the later ones, there is absolutely no sign of intermediate stages" (181). As for birds: Olson refers to Archaeopteryx as "reptile-like" but says that in possession of feathers "it shows itself to be a bird" (182). Finally, with reference to mammals Olson states that "[t]he first evidence of flight in mammals is in *fully developed* bats of the Eocene epoch" (182; emphasis added).

Thus, in not a single investigation of the origin of flight has a transitional series been documented. In the case of *Archaeopteryx*—a so-called intermediate —all paleontologists now acknowledge that it was a true bird. It had wings; it was completely feathered; it flew. It was not a halfway bird; it was a bird. No transitional form with part-wings and part-feathers has ever been found.

The alleged reptilian features of *Archaeopteryx* consist of the clawlike appendages on the leading edges of its wings and the possession of teeth and of vertebrae that extend out along the tail. It is believed to have been a poor flier, with a small keel on the sternum. Although such features might be expected if birds had evolved from reptiles, in no sense of the word do they constitute proof that *Archaeopteryx* was an intermediate between reptile and bird. For example, there is a bird living today in South America—the hoatzin,

Opisthocomus hoazin—which in the juvenile stage possesses two claws. Furthermore, it is a poor flier, with an astonishingly small keel (Grimmer 1962). This bird is unquestionably 100 percent bird, yet it possesses two of the characteristics that are used to impute a reptilian ancestry to *Archaeopteryx*.

Modern birds do not possess teeth; but certain ancient birds, unquestionably 100 percent birds, possessed teeth. Does the possession of teeth denote a reptilian ancestry for birds, or does it simply prove that some ancient birds had teeth and others did not? Some reptiles have teeth and some do not; some amphibians have teeth and some do not. In fact, this is true throughout the entire range of the vertebrate subphylum. On the principle that toothed birds are primitive and that toothless birds are more advanced, the Monotremata (the duck-billed platypus and the spiny anteater), which are mammals that do not possess teeth, should be considered more "advanced" than humans. Yet in every other respect these egg-laying mammals are considered to be the most primitive of all mammals (although they are among the last mammals to appear in the fossil record). Just what phylogenetic value, then, can be assigned to the possession or absence of teeth?

Concerning the status of *Archaeopteryx*, Lecomte du Noüy (1947, 58) has stated, "Unfortunately, the greater part of the fundamental types in the animal realm are disconnected from a paleontological point of view. In spite of the fact that it is undeniably related to the two classes of reptiles and birds (a relation which the anatomy and physiology of actually living specimens demonstrates), we are not even authorized to consider the exceptional case of the *Archaeopteryx* as a true link. By link, we mean a necessary stage of transition between classes such as reptiles and birds, or between smaller groups. An animal displaying characters belonging to two different groups cannot be treated as a true link as long as the intermediary stages have not been found, and as long as the mechanisms of transition remain unknown."

What seems to be the most reasonable conclusion? I believe that the fossil record would permit no better assessment of the facts than that voiced by Swinton (1960): "The origin of birds is largely a matter of deduction. There is no fossil of the stages through which the remarkable change from reptile to bird was achieved."

The absence of any indication whatsoever from the fossil record that feathers gradually evolved is usually excused by the allegation that such delicate structures are not likely to be preserved in fossils. No such explanation is admissible, however, in the case of flying reptiles and the bats.

There are many significant differences between nonflying reptiles and flying reptiles. Again I refer to Romer's *Vertebrate Paleontology*. On p. 140 is shown a reconstruction of *Saltoposuchus* (fig. 214), which was a representative of the Triassic thecodonts—a group that Romer believes gave rise to flying reptiles (pterosaurs), dinosaurs, and birds. Comparison of this form with reconstructions of the earliest representatives among the two suborders of pterosaurs (pp. 144 and 146) reveals the vast gulf between them—a gulf not bridged by fossil intermediates. A similar gulf also exists, of course, between this creature and *Archaeopteryx*.

Almost every structure in *Rhaniphorhynchus*, a long-tailed pterosaur (fig. 222, 144), was unique to this creature. Especially obvious (as in all pterosaurs) is the enormous length of the fourth finger, in contrast with the other three fingers possessed by this reptile. This fourth finger provided support for the wing membrane. It is certainly not a delicate structure; and if the pterosaurs evolved from the thecodonts or some other earth-bound reptile, transitional forms should have been found showing a gradual lengthening of this fourth finger. Not even a hint of such a transitional form has ever been discovered.

Even more unusual was the pterodactyloid group of pterosaurs (Romer, fig. 225, 146). Pteranodon not only had a large, toothless beak and a long, rearward-extending bony crest, but its fourth fingers supported a wingspan of twenty-five feet. Where are the transitional forms documenting an evolutionary origin of these and other structures unique to the pterosaurs?

The bat is presumed to have evolved from nonflying insectivores—although, as stated earlier, the oldest-known bat to appear in the fossil record is 100 percent bat, and no trace of a transitional form can be found (Jepsen 1966). In the bat four of the five fingers support the membrane of the wing and are extremely long, compared with the normal hand. These and other unique structures are solid bone and are anything but delicate structures. Transitional forms, if they ever existed, should certainly have been preserved. The absence of such forms leaves unanswered, on the basis of the evolution model, such questions as when, from what, where, and how bats originated.

Now let me ask this question: concerning the origin of flight, does the creation model or the evolution model have greater support from the fossil record?

To me the answer seems obvious. Not a single fact contradicts the predictions of the creation model; but the actual evidence fails miserably to support the predictions of the evolution model. Here, where transitional forms should be the most obvious and easiest to find if evolution really accounts for the origin of these highly adapted and unique creatures, *none* is found. Could the fossil record really be that cruel and capricious to evolutionary paleontologists? The historical record inscribed in the rocks literally cries "Creation!"

SYSTEMIC DISCONTINUITY IS PERVASIVE

The examples cited in this [chapter] are in no way exceptional; rather, they serve to illustrate what is characteristic of the fossil record. Although transitions at the subspecies level are observable and those at the species level may be inferred, the absence of transitional forms between higher categories (the created kinds of the creation model) is regular and systematic.

Simpson, in his book *Tempo and Mode in Evolution* (1944), under the heading "Major Systematic Discontinuities of Record," states that nowhere in the world is there any trace of a fossil that would close the considerable gap between *Hyracotherium* and its supposed ancestral order, *Condylarthra.* He then goes on to say, "This is true of all the thirty-two orders of mammals.... The earliest and most primitive known members of every order already have the basic ordinal characters, and in no case is an approximately continuous sequence from one order to another known. In most cases the break is so sharp and the gap so large that the origin of the order is speculative and much disputed" (106). Later, Simpson states, "this regular absence of transitional forms is not confined to mammals, but is an almost universal phenomenon, as has long been noted by paleontologists. It is true of almost all orders of all classes of animals, both vertebrate and invertebrate. A fortiori, it is also true of the classes themselves, and of the major animal phyla, and it is apparently also true of analogous categories of plants" (107).

In his book *The Meaning of Evolution* (1949) Simpson, with reference to the appearance of new phyla, classes, and other major groups, states, "The process by which such *radical events* occur in evolution is the subject of one of the most serious remaining disputes among qualified professional students of evolution.

The question is whether such major events take place *instantaneously*, by some process essentially unlike those involved in lesser or more gradual evolutionary change, or whether all of evolution, including these major changes, is explained by the same principles and processes throughout, their results being greater or less according to the time involved, the relative intensity of selection, and other material variables in any given situation" (231). He continues: "Possibility for such dispute exists because transitions between major grades of organization are seldom well recorded by fossils. There is in this respect a tendency toward *systematic deficiency* in the record of the history of life. It is thus possible to claim that such transitions are not recorded because they did not exist, that the changes were not by transition but by sudden leaps in evolution" (emphasis added).

If phyla, classes, orders, and other major groups were connected by transitional forms rather than appearing suddenly in the fossil record with basic characteristics complete, it would not be necessary, of course, to refer to their appearances in the fossil record as "radical events." Furthermore, it cannot be emphasized too strongly that even evolutionists are arguing among themselves as to whether these major categories appeared *instantaneously* or not.

It is precisely the argument of creationists that these forms *did* arise instantaneously and that the transitional forms are not recorded because they never existed. Creationists thus would reword Simpson's statement to read, "It is thus possible to claim that such transitions are not recorded because they did not exist—that these major types arose by creation rather than by a process of gradual evolution."

In a more recent work, Simpson (1960, 149) says, "It is a feature of the known fossil record that most taxa appear abruptly." In the same paragraph he states further, "Gaps among known species are sporadic and often small. Gaps among known orders, classes, and phyla are systematic and almost always large."

It would hardly be necessary to document further the nature of the fossil record. It seems obvious that if the above statements of Simpson were stripped of all presuppositions and presumed evolutionary mechanisms to leave the bare record, they would describe exactly what is required by the creation model. This record is woefully deficient, however, in the light of the predictions of the evolution model.

No one has devoted himself more wholeheartedly than Simpson to what Dobzhansky (1972) has called the "mechanistic materialist philosophy shared by most of the present 'establishment' in the biological sciences." Simpson (1953, 360) therefore asserts that most paleontologists "find it logical, if not scientifically required, to assume that the sudden appearance of a new systematic group is not evidence for creation..." He has expended considerable effort (1944, 105–24; 1953, 360–76; 1960, 140–52) in attempts to bend and twist every facet of evolution theory to explain away the deficiencies of the fossil record. One needs to be reminded, however, that if evolution is adopted as an a priori principle, it is always possible to imagine auxiliary hypotheses—unproved and by nature unprovable—to make it work in any specific case. By this process biological evolution degenerates into what Thorpe (1969) calls one of his "four pillars of unwisdom": mental evolution that is the result of random tries preserved by reinforcements.

Concerning the plant kingdom, the following remark of E. J. H. Corner (1961), of the Cambridge University botany school, is refreshingly candid: "Much evidence can be adduced in favor of the theory of evolution—from biology, biogeography and paleontology, but I still think that to the unprejudiced, the fossil record of plants is in favor of special creation."

Even in the famous horse "series," which has been so highly touted as proof of evolution within an order, transitional forms between major types are missing. Lecomte du Nouy (1947, 74) has stated in reference to horses, "But each one of these intermediaries seems to have appeared 'suddenly,' and it has not yet been possible, because of the lack of fossils, to reconstitute the passage between those intermediaries. Yet it must have existed. The known forms remain separated like the piers of a ruined bridge. We know that the bridge has been built, but only vestiges of the stable props remain. The continuity we surmise may never be established by facts." Goldschmidt (1952, 97) has said, "Moreover, within the slowly evolving series, like the famous horse series, the decisive steps are abrupt."

THE "HOPEFUL MONSTER" ALTERNATIVE

Goldschmidt (1940; 1952, 84–98), in contrast with Simpson and the majority of evolutionists, accepts the discontinuities in the fossil record at face value. He rejects the neo-Darwinian interpretation of evolution (the modern synthesis), which is accepted by almost all evolutionists, at least among those who accept any theory concerning mechanisms at all. Goldschmidt instead has proposed that major categories (phyla, classes, orders, families) arose instantaneously by major saltations or systemic mutations. Goldschmidt terms this the "hopeful monster" mechanism. He has proposed, for instance, that at one time a reptile laid an egg and a bird was hatched from the egg. All major gaps in the fossil record are accounted for, according to Goldschmidt, by similar events: something laid an egg, and something else got born. Neo-Darwinists prefer to believe that Goldschmidt is the one who laid the egg; they maintain that there is not a shred of evidence to support his "hopeful monster" mechanism. Goldschmidt insists just as strongly that there is no evidence for the postulated neo-Darwinian mechanism (major transformations by the accumulation of micromutations). Creationists agree with both the neo-Darwinists and Goldschmidt: they are both wrong. However, Goldschmidt's publications do offer cogent arguments against the neo-Darwinian view of evolution, from both genetics and paleontology.

No one was more wholly committed to evolutionary philosophy than was Goldschmidt. If anybody wanted to find transitional forms, he did. If anybody would have admitted that a transitional form was a transitional form, if indeed that's what it was, he would have. But, concerning the fossil record, this is what Goldschmidt (1952, 97) says: "The facts of greatest general importance are the following. When a new phylum, class, or order appears, there follows a quick, explosive (in terms of geological time) diversification so that practically all orders or families known appear suddenly and without any apparent transitions."

Now, creationists ask what better description of the fossil record could one expect, based on the predictions of the creation model? On the other hand, unless one accepts Goldschmidt's "hopeful monster" mechanism of evolution, this description contradicts the most critical prediction of the evolution model: the presence in the fossil record of the intermediates demanded by the theory.

AGAINST AUTHORITARIAN MATERIALISM

Kerkut (1960), although not a creationist, wrote a notable little volume to expose the weaknesses and fallacies in the usual evidence used to support evolution theory. In the concluding paragraph of the book this author states that "there is the theory that all the living forms in the world have arisen from a single source which itself came from an inorganic form. This theory can be called the 'General Theory of Evolution' and the evidence that supports it is not sufficiently strong to allow us to consider it as anything more than a working hypothesis." There is a world of difference, of course, between a working hypothesis and established scientific fact. If one's philosophic presuppositions lead him to accept evolution as his working hypothesis, he should restrict it to that use, rather than force it on others as an established fact.

If, without the philosophic presuppositions of either the materialist or the theist, creation and evolution are used as models to predict the nature of the historical evidence, it can be seen that the creation model is just as credible as the evolution model (and, I believe, much more credible). And I reiterate: the one model is no more religious or any less scientific than the other.

No less convinced an evolutionist than Thomas H. Huxley (as quoted in L. Huxley 1903) acknowledged that "'creation', in the ordinary sense of the word is perfectly conceivable. I find no difficulty in conceiving that, at some former period, this universe was not in existence, and that it made its appearance in six days (or instantaneously, if that is preferred), in consequence of the volition of some pre-existing Being. Then, as now, the so-called *a priori* arguments against Theism and, given a Deity, against the possibility of creative acts, appeared to me to be devoid of reasonable foundation."

The majority in the scientific community and educational circles are using the cloak of "science" to force the teaching of their view of life upon all. The authoritarianism of the medieval church has been replaced by the authoritarianism of rationalistic materialism. Constitutional guarantees are violated and free scientific inquiry is stifled under this blanket of dogmatism. It is time for a change.

REFERENCES

Allen, G. E. "T. H. Morgan and the Emergence of a New American Biology." *Quarterly Review of Biology* 44 (1969): 168–88.

Axelrod, D. I. "Early Cambrian Marine Fauna." *Science* 123 (1956): 7–9.

Cloud, P. E. "Significance of the Gunflint (Precambrian) Microflora." *Science* 143 (1955): 27–35.

Corner, E.J. H. "Evolution." In *Contemporary Botantical Thought*, edited by A. M. McLeod and L. S. Cobley, 93–114. Chicago: Quadrangle Books, 1961.

Danson, R. "Evolution." *New Scientist* 49 (1971): 25.

Darwin, C. *Origin of Species*. Reprinted London: J. M. Dent & Sons.

de Beer, G. "The World of an Evolutionist." *Science* 143 (1964): 1911–17.

Dobzhansky, T. "Evolution at Work." *Science* 137 (1936): 1891–1903.

———. "A Biologist's World View." *Science* 175 (1972): 49.

Eden, M. "Inadequacies of Neo-Darwinian Evolution as a Scientific Theory." In *Mathematical Challenges to the Neo-Darwinian Interpretation of Evolution*, edited by P. S. Moorhead and M. M. Kaplan. Philadelphia: Wistar Institute Press, 1967.

Ehrlich, P. R., and R. W. Holm. "Patterns and Populations." *Science* 137 (1962): 655.

Fothergill, P. G. "Issues in Evolution." *Nature* 189: 425.

George, T. N. "Fossils in Evolutionary Perspective." *Science Progress* 48 (1960): 1–5.

Goldschmidt, R. R. *The Material Basis of Evolution*. New Haven, CT: Yale University Press, 1940.

———. "Evolution as Viewed by One Geneticist." *American Scientist* 40 (1952): 84–98.

Grimmer, J. L. "Strange Little World of the Hoatzin." *National Geographic* 122 (1962): 391–400.

Huxley, L., ed. *Life and letters of Thomas Henry Huxley*, vol. 2. New York: D. Appleton & Co., 1903.

Jepsen, G. L. "Early Eocene Bat from Wyoming." *Science* 154 (1966): 1333–39.

Kerkut, G. A. *Implications of Evolution*. New York: Pergamon Press, 1960.

Lecomte du Noüy, P. *Human Destiny*. New York: New American Library, 1947.

Le Gros Clark, W. E. *Discovery* (January 1955): 7.

Litynski, L. "Should We Burn Darwin?" *Science Digest* 50 (1961): 61–63.

Macbeth, N. *Darwin Retried*. Boston: Gambit, Inc., 1971.

Moorhead, P. S., and M. M. Kaplan, eds. *Mathematical Challenges to the Neo-Darwinian Interpretation of Evolution*. Philadelphia: Wistar Institute Press, 1937.

Olson, E. C. "Morphology, Paleontology, and Evolution." In *Evolution after Darwin, Volume 1: The Evolution of life*, edited by Sol Tax. Chicago: University of Chicago Press, 1960.

———. *The Evolution of Life.* New York: New American Library, 1965.

Ommaney, F. D. *The Fishes.* New York: Life Nature Library, Time-Life, Inc., 1964.

Romer, A. S. *Vertebrate Paleontology,* 3rd ed. Chicago: University of Chicago Press, 1966.

Salisbury, F. B. "Natural Selection and the Complexity of the Gene." *Nature* 224 (1969): 342–43.

———. "Doubts about the Modern Synthetic Theory of Evolution." *American Biology Teacher* 33 (1971): 335–38.

Simpson, G. G. *Tempo and Mode in Evolution.* New York: Columbia University Press, 1944.

———. *The Meaning of Evolution.* New Haven, CT: Yale University Press, 1949.

———. *The Major Features of Evolution.* New York: Columbia University Press, 1953.

———. "The History of Life." In *Evolution after Darwin, Volume 1: The Evolution of Life,* edited by Sol Tax. Chicago: University of Chicago Press, 1960.

———. "The Non-prevalence of Humanoids." *Science* 143 (1964): 709.

Swinton, W. E. "The Origin of Birds." In *Biology and Comparative Physiology of Birds*, vol. 1, edited by A. J. Marshall. New York: Academia Press, 1980.

Thorpe, W. "Reductionism v. Organicism." *New Scientist* 43 (1969): 635–38.

Watson, D. M. S. "Adaptation." *Nature* 194 (1929): 233.

CHAPTER 13

WITNESS TESTIMONY SHEET: *McLEAN V. ARKANSAS BOARD OF EDUCATION*

MICHAEL RUSE

Question: Please state your name.
Answer: Michael Escott Ruse.

***

Q: Are you a Canadian citizen?
A: Yes.

Q: What is your occupation?
A: I am a professor of philosophy at the University of Guelph in Ontario, Canada.

Q: What is your educational background?
A: [Attached vita.]

Q: Are you a member of any professional organizations?
A: [Attached vita.]

From *McLean v. Arkansas Board of Education*, 529 F. Supp. 1255, 1258–64 (ED Ark. 1982).

Q: Have you authored any books or articles?

A: Yes, I have authored the following books: *The Philosophy of Biology*; *Sociobiology: Sense or Nonsense?*; *The Darwinian Revolution: Science Red in Tooth and Claw*; *Is Science Sexist? And Other Problems in the Biomedical Sciences*; and *Darwinism Defended*, which is now in manuscript form and about to be published. In addition I have published well over seventy articles in professional journals. [See vita.]

Q: Professor Ruse, will you please describe your academic specialties at the present time.

A: My major academic interest at this time is the history and philosophy of science, in particular the history and philosophy of biology. I teach courses in the philosophy of science, ethics, logic, introductory philosophy, and the history of science. I also teach a course in the philosophy of religion.

THE HAPPENINGS AND MECHANISMS OF EVOLUTION

Q: Dr. Ruse, you mentioned that your latest book is titled *Darwinism Defended*. Does the title of that book suggest that evolution is in question and in need of defense?

A: No. That evolution happened is not questioned by any credible scientist. My book is, in large part, a defense of the Darwinian theory of about *how* evolution happened.

Q: Would you please explain that distinction.

A: Yes, that distinction is very important because creation scientists often confuse the two ideas. What is properly known as "evolution" has two major components. The first is the *happening* of evolution. That is, did one or a few organisms develop by naturalistic processes through a succession of forms which changed over long periods of time into the organisms alive today. I do not know of a single credible scientist, other than the people who call themselves "creation scientists," who questions that evolution actually happened.

The second component of evolution is *how* evolution happened. When scientists today speak of "evolutionary theory," they are usually referring to a theory explaining the mechanics of evolution; that is, the "how" of evolution.

Q: You say that scientists agree that evolution happened. Why is that?

A: Because the evidence is absolutely overwhelming. It convinces the unbiased observer beyond any reasonable doubt.

Q: What is the history of that consensus? When did scientists come to agree about the happening of evolution?

A: Scientists have been considering ideas about evolution since ancient Greece—although those early notions were rather primitive. In modern times, evolution has been seriously considered by scientists since the scientific revolution in the sixteenth and seventeenth centuries. Apart from the pressure that Copernicanism put upon literal readings of the Bible, the heliocentric worldview inspired all sorts of thoughts about inorganic evolution. In the eighteenth century, an idea which enjoyed much popularity was the "nebular hypothesis," formulated by Kant, Laplace, and William Herschel, supposing that universes are formed out of gaseous clouds. It was not long before people started to think analogically about the organic world. Developments in geology had started to convince people that the earth is far older than the traditionally supposed, biblically based 6000 years.

The first great organic evolutionist was the French scientist Jean Baptiste de Lamarck, whose major work *Philosophie Zoologique* was published in 1809. The next important figure, certainly in influence, was the then-anonymous Scottish publisher Robert Chambers, whose *Vestiges of the Natural History of Creation* appeared in 1844. And this leads to Charles Darwin and his *Origin of Species*, arriving on the scene in 1859. (In fact, Darwin's seminal work was performed in the 1830s, but he did not publish for twenty years.) The publication of *Origin* was the conclusive scholarship necessary to convince the science community of the day that evolution happened. That consensus has never been threatened since because every new piece of data, from every relevant discipline of science, confirms that evolution happened.

Of course, Darwin also proposed a theory about *how* evolution happened. That theory—natural selection—has been far more controversial.

Q: Do scientists generally agree now about how evolution happened?

A: No, not at all. With respect to this issue of how evolution happened there is still much debate.

Darwin suggested *natural selection* as the most important, but not the only, mechanism of evolution. For the most part, I think he was right and that accounts for the title of my book *Darwinism Defended.* However, other scientists think that natural selection may not be the most important element of evolutionary change. Some emphasize speciation. Others emphasize *genetic drift. Pure chance* may be another important factor. Still other scientists, such as Stephen Gould (who I understand will also be a witness in this trial), propose a theory of *punctuated equilibria*—that evolution happened more abruptly than the slow, gradual change suggested by Darwinian theory.

In other words, though scientists are all agreed that evolution happened, there is a vigorous and, I believe, a healthy debate among proponents of very many different ideas about how evolution happened.

Q: Dr. Ruse, you testified earlier that creation scientists often confuse the difference between the *happening* of evolution and the *how* of evolution. Would you please explain what you meant?

A: Creation scientists frequently extract quotes from scientific literature which seem to challenge, or question the sufficiency of, evolution. They use those quotes to advance the creationist theory that evolution did not happen. However, in virtually every instance I have examined, the authors of those quotations were discussing or criticizing a particular theory about *how* evolution happened, not whether evolution happened. Indeed, the scientists quoted are often among the foremost evolutionists. However, by taking their words out of context, the creation scientists camouflage the fact that those same authorities are unanimously in agreement that evolution actually happened.

Q: Do you know of any specific examples of such out of context quotations?

A: Yes. [Insert appropriate examples such as Theodosius Dobzhansky, Ernst Mayr, or Stephen Gould.]

CREDENTIALS AS EXPERT IN CREATION SCIENCE

Q: In your book *Darwinism Defended* do you also say anything about creation science?

A: Yes. I devote two chapters to creation science. In the first of those chapters, chapter 14 in the manuscript (which will be chapter 13 in the published edition), I describe the creation science arguments in some detail. In the next chapter I analyze those contentions and conclude that philosophically and methodologically the creationists do not act like scientists, and that substantively the creationists' contentions are without scientific merit.

Q: Have you read much creation science literature?

A: Yes.

Q: Would you please describe the kinds of material you are familiar with.

A: In *Darwinism Defended* I analyzed in great detail a book titled *Scientific Creationism, Public School Edition.* The book is published by the Institute for Creation Research, a leading creation science organization. The book is edited by Henry Morris and was written by twenty-three creation scientists on the staff of ICR. Inasmuch as it is denoted a public school edition, it purports to be the best attempt to eliminate religion from creation science. I consider *Scientific Creationism* to be an authoritative statement of creation science doctrine.

In addition to *Scientific Creationism* I have read [Ruse offers list].

CONNECTION: ACT 590 AND CREATION SCIENCE GENERALLY

Q: Dr. Ruse, I would like to show you a copy of Act 590. I direct your attention to the references to creation science, especially in Section 4(a) that defines that term. As a philosopher and historian of science, and someone who has read extensively in creation science literature, what are your thoughts about how Act 590 relates to creation science generally?

A: When I read Act 590, and Section 4 in particular, I am struck by the similarities between the act and the principal creation science claims found throughout all of the creation science literature that I have read and that I know about.[1] Of course, some creation science books focus on specific issues. However, taken as a whole, the elements of creation science found in Section 4(a) track the major themes of creation science generally.

Q: What are the similarities that you see?

A: First of all, the act establishes the so-called dual-model approach to the teaching of what creation scientists call "origins."

Secondly, the six elements of creation science listed in 4(a) are the same claims repeatedly made throughout creation science literature.

Finally, creation scientists are the only ones I know of who talk about evolution as a single body of knowledge relating to everything from the origin of the universe, through the origin of the world, the origin of life, and the origin of human life, to flood geology. That typical creationist distortion of evolutionary theory is also incorporated into the statute.

ACT 590: DUAL MODEL

Q: Dr. Ruse, I would like to explore each of those elements with you one at a time.

First you mentioned the dual-model approach to teaching origins. Would you please describe what you meant by that?

A: Creation scientists like to say that there are only two models of origins: evolution and creation. This so-called dual-model approach to origins is a critical ingredient of the creation science model for two reasons. First, it enables creation scientists to argue that if evolution is taught, then only creation need also be taught—without having to include other possible explanations of what they call "origins." In other words, the "dual model" is an attempt to distinguish creation from other unconventional theories about how life on this planet began. Second, creation scientists use the dual model as a rationale for arguing that disproof of evolution is proof of creation.

Q: What are the various theories about how life began?

A: Of course, there is the creation theory and the theory of abiogenesis—that life developed from nonlife. Another theory is that life on this planet began as a result of the intelligent intervention of beings from elsewhere in the universe. That theory has been proposed by Von Daniken, in *Chariots of the Gods*, and more recently by Francis Crick.

Still another theory is that life began on this planet as a result of passing through a cloud of organic material, some of which took root here. In fact, I think one of the defendants' witnesses in this case, Chandra Wickramasinghe, is a proponent of this theory.

Q: As a philosopher of science, what is your professional opinion about the logic of the creation science contention that disproof of Darwinian evolution is proof of creation?

A: The contention is fallacious because, as I just demonstrated, creation and abiogenesis are not the only alternatives. The creation scientists try to invoke the following proposition of logic: If A or B, then "not A" equals B. That statement is true—but only if A and B are the only alternatives. If the statement read: If A, B, or C, then "not A" would not equal B. In fact, "not A" would not tell you anything about the truth value of B or C.

Similarly, there are many theories about how life began on this planet. Disproving one cannot prove any other. Also keep in mind that I am only speaking about theories of how life began on this planet. Creation science is also a theory about how the universe began; how the earth originated, how human life developed, and flood geology. And, in regard to each of those subjects scientists have proposed a great many alternative theories.

Q: Dr. Ruse, Act 590 does not contain the words "dual model." Would you tell us precisely what in the statute you are referring to?

A: The dual-model approach is incorporated throughout Act 590. For example, in the very first sentences of Section 1 the statute talks about creation science and evolution science as the two models that must be given balanced treatment in the classroom. Similarly, Section 3 talks about discrimination with respect to "either model," meaning the two models of evolution science and creation science. However, the clearest example of the two so-called models of origin is apparent in Section 4. Section 4 defines creation science and evolution science. As you can see from looking at that section, the definitions have been written in a way to establish a stark contrast between every element of creation science and corresponding elements of evolution science.

Q: What is your understanding of creation theory?

A: That the universe, the earth, all life, and especially human life, were created by a supernatural Creator—spelled with a capital C—and that the Creator also inundated the world with a great flood.

ACT 590: CREATION THEORY

Q: Is the creation theory part of Act 590?

A: Yes, Act 590 cannot be understood, in my opinion, unless you realize that the explanations of science are being contrasted with the explanations of supernatural creation.

Q: Where do you see that contrast?

A: In the first sentence of 4(a) and the first sentence of 4(b).

Looking first at 4(b) ["Evolution-science means the scientific evidences for evolution and inferences from those scientific evidences"]. Scientific evidences and inferences from evidences do not mean anything independent of a theory that joins them together. In Section 4(b) the unifying theory for those evidences and the inferences is evolution.

Now look at 4(a) ["Creation science means the scientific evidences for creation and inferences from those scientific evidences"]. Section 4(a) is identical to 4(b) word for word except that where 4(b) says "evolution," 4(a) speaks of "creation." Thus, in 4(a) the various evidences and inferences are meant to support the theory of creation. And that theory is that the Creator created the universe, the world, life, human life, and a worldwide flood.

Q: Do you see the word Creator in Act 590?

A: No.

Q: Why then do you refer to the intervention of a creator in analyzing the statute?

A: It is true that the act does not mention "Creator." It is also true that the act even tries to conceal that creation science is the theory of creation. And the theory of creation is that the Creator did it.

For example, look at sections 4(a)(1) and 4(b)(l), Section 4(b)(l) ["Emergence by naturalistic processes of the Universe from disordered matter and emergence of life from non-life"], which is supposed to be the evolution-science explanation of the origin of matter and life, refers to emergence by *naturalistic* processes. Now I do not like the word "emergence"—it is not scientific—and I do not think that evolutionary theory has anything to say about the origin or matter of life. However, the important point of comparison here is that 4(b)(1) pertains to "naturalistic processes."

Contrast that with 4(a)(l) ["Sudden creation of the universe, energy and life from nothing"]. In the statute, "sudden creation" in 4(a)(1) is contrasted with "naturalistic processes" in 4(b)(1). Therefore, "sudden creation" must mean by nonnaturalistic processes. And since we know that the unifying theory of 4(a) is the theory of creation, it follows that the nonnaturalistic process referred to in 4(a)(1) must be the intervention of a Creator.

Similarly, compare 4(b)(3) with 4(a)(3). 4(b)(3) ["Emergence by mutation and natural selection of present living kinds from simple earlier kinds"] refers to "mutation and natural selection." Both are naturalistic processes that contribute to the evolution of present living organisms from simpler earlier organisms.

In contrast, 4(a)(3) ["Changes only within fixed limits of originally created kinds of plants and animals"] refers to "originally created kinds of plants and animals." Since "originally created" must mean something other than "by naturalistic processes," it can only be understood with reference to the theory of creation—which is that the Creator created these original kinds of plants and animals.

I might add that there is another reason to believe that the reference to "originally created kinds" refers to the acts of a Creator. The word "kinds" is not a scientific word. The science of taxonomy—which involves classifying organisms—divides living organisms into categories of species, genus, family, order, class, and phylum. There is no taxonomic classification of "kind." In fact, the only place I know where the word "kind" is used in this sense is in the Bible. In the very first chapter of Genesis, describing the creation, the Bible says, "and God made the beast of the earth after its kind and the cattle after their kind and everything that creepeth upon the ground after its kind and God saw that it was good."

ACT 590: SIMILARITIES WITH CREATION SCIENCE GENERALLY

Q: Dr. Ruse, I believe you testified earlier that the six elements of creation science identified in sections 4(a)(1) through 4(a)(6) were similar to the elements of creation science as explained in creation science literature, is that so?

A: Yes.

Q: Would you please elaborate a bit on those similarities.

A: In my reading of creation science literature, I find that it constantly emphasizes each of the elements that are also incorporated into Section 4(a) of Act 590.

A good example of this is clear from chapter 14 of my book *Darwinism Defended.* In that chapter, as I mentioned earlier, I analyzed the creation science model as articulated in a basic text, *Scientific Creationism.* That chapter was written long before I ever saw Act 590; yet the elements of creation science as articulated in Section 4(a) are precisely the same as the elements of creation science articulated in *Scientific Creationism.*

In *Scientific Creationism* the authors refer to a "cause-and-effect" analysis and a "purpose" analysis to conclude that an omnipotent Creator is the first cause of matter, energy, and life (see *Darwinism Defended*, p. 430), just as in 4(a)(1).

The book *Scientific Creationism* then argues at some length that mutation and selection are not adequate explanations for the variable life-forms found today (*Darwinism Defended*, p. 433), which is directly comparable to the contention in 4(a)(2).

Next the book *Scientific Creationism* discusses homologies—which are structural similarities of different organisms—and concludes that homologies are not evidence of common ancestry (*Darwinism Defended*, p. 434). The authors similarly dismiss embryological similarities and behavorial similarities as evidences of common ancestry. That creationist position is incorporated into the statute in sections 4(a)(2) and 4(a)(3).

Creation scientists generally also refer to the absence of transitional forms in the fossil record to suggest the lack of common ancestry generally and, more specifically, that man did not descend from apes (*Darwinism Defended*, pp. 435, 441). These creationists' contentions are incorporated in 4(a)(3) and 4(a)(4).

Continuing with the analysis of parallels, *Scientific Creationism* next argues that a catastrophic worldwide flood once enveloped the earth, destroyed all living things, and profoundly affected earth's geology. Of course, this notion of a worldwide flood is incorporated into 4(a)(5) of Act 590.

Finally, *Scientific Creationism* devotes a whole chapter to its contention that the earth is relatively young (*Darwinism Defended*, p. 438). And that creation science is also in Act 590 in Section 4(a)(6).

ACT 590: EVOLUTION SCIENCE

Q: I would now like to direct your attention to Section 4(b) of the statute, which purports to define evolution science. What is your professional opinion of the definition of evolution science?

A: The definition does not make sense in any way that scientists talk about evolution.

Q: Would you please explain?

A: The statute adopts a common and quite distorted creation science characterization of evolution. Creation science attempts to create the impression that evolution pertains to the same things as creation science. That is not so. There is no scientific discipline known as evolution science. One cannot get a degree in evolution science.

In fact, the body of knowledge implicated by creation science includes a dozen different scientific disciplines including cosmology, astronomy, geology, biology, and all of its many subdisciplines: paleontology, chemistry, physics, botany, and the technologies of radiocarbon and radiometric dating.

Therefore, I find that incorporating many aspects of all these disciplines under the rubric of "evolution science" is both typical of creation science and thoroughly nonsensical.

Q: What then do scientists usually mean by "evolution"?

A: Scientists commonly think of evolutionary theory as attempting to explain how life developed after it was formed. Evolutionary theory does not focus on how life began, but only on what happened to life *after* it began. For example, one of the defendants' own experts—Chandra Wickramasinghe—believes that life on this planet took hold after earth passed through a cloud of organic material. Even if that were so, evolutionary theory would begin at that point and describe what happened to that life after it started to grow on the planet.

Q: Understanding that scientists do not generally recognize this thing called "evolution science," let me nonetheless direct your attention to Section 4(b)(1). What is your professional judgment of that provision as a scientific statement?

A: Of course, I do not know, and scientists generally do not know, how the universe originated or even how life on this planet originated.

Section 4(b)(l) may or may not be right. It is not the only naturalistic explanation of how the universe began. For example, the universe might not have ever "begun." Though it is a very hard concept to grasp, the universe may have always existed. Moreover, even if it began at some point, the so-called big bang may not have happened from "disordered matter," but rather from the most highly ordered matter imaginable: all of the energy and matter of the universe compressed into a single mass for an instant before explosion. Similarly, the theory of abiogenesis—that life developed through naturalistic processes from nonlife—is only a working hypothesis that scientists are presently attempting to confirm or deny. It will probably be quite some time before science has any clear understanding of how life began.

Thus the elements of 4(b)(1) are by no means the only scientific theories being examined.

The most important thing about 4(b)(1), however, is its reliance on naturalistic processes. I believe that the first and the most important characteristic of science is that it relies exclusively on blind, undirected natural laws and naturalistic processes. Therefore, whether Section 4(b)(l) is right or wrong, at least it is science.

Q: What is your professional assessment of 4(b)(2)?

A: Section 4(b)(2) is a good example of the confusion and distortion that results from the dual-model comparison of evolution science with creation science.

Section 4(b)(2) ["The sufficiency of mutation and natural selection in bringing about development of present living kinds from single earlier kinds"] is probably not accepted by any scientist as an accurate statement about evolution. In other words, scientists do not believe that mutation and natural selection are sufficient to bring about the development of present living kinds from simple earlier kinds. Other additional factors are surely also at work. For example, in addition to mutation and natural selection scientists recognize the importance of genetic drift, speciation, and pure chance in the development of present living organisms from simple earlier organisms. Therefore, Section 4(a)(2), which pertains to the *in*sufficiency of mutation and natural selection, is an accurate reflection of current scientific learning about evolution.

However, as used in the statute, 4(a)(2) is supposed to be an indication for creation. In that context, the section obviously means that mutation and natural selection are insufficient, not because other naturalistic processes are also operative, but because the Creator brought about the development of all living kinds.

Q: Dr. Ruse, do you understand the meaning of Section 4(b)(3)?

A: Standing alone, 4(b)(3) does not seem to mean very much in addition to 4(b)(2). There is the same problem with limiting the operative mechanisms of evolution to mutation and natural selection. However, the point of 4(b)(3) becomes evident by reference to 4(a)(3).

Section 4(a)(3) adopts the general creation science concession that some evolutionary changes have occurred. Creation science contends however that evolution has occurred only within kinds and not between kinds.

The reason creation science has to admit the happening of some evolution is that some evolutionary changes cannot be denied. To use an oft-cited example, [use moths, flies, Darwin's finches, etc.]. In order to account for these evolutionary changes creation science resorts to the 4(b)(3) contention: that the Creator created an original moth kind; although that moth may experience evolutionary adaptations to deal with changing environmental circumstances, it cannot change into any other kind.

By contrast, Section 4(b)(3) must mean that evolutionary changes do not know any fixed limits of original kinds, and that organisms existing today have descended from other organisms of very different species. Of course, 4(b)(4)—that man descended from apes—is the most troublesome example of that evolutionary change for the creation scientists.

Q: What is your professional opinion about the contrast between unrestrained evolutionary change and evolutionary changes only within fixed limits of created kinds?

A: I believe that these subsections are quite revealing about the creation science/evolution-science dual model. In section 4(b), although subsections (3) and (4) are incorrectly limited to mutation and natural selection, they reflect a reliance on natural processes that are not constrained by any supernatural limitations.

On the other hand, sections 4(a)(3) and (a)(4) impose the limitation of "originally created kinds." However, creation scientists offer no explanation of why admittedly ongoing evolutionary changes should stop at originally created kinds. In the creation model the only explanation is that the Creator fixed those limits of originally created kinds and that His will cannot be altered.

Q: Dr. Ruse, do you understand the use of the words "catastrophism" and "uniformitarianism" in subsection 5 of sections 4(a) and (b)?

A: No. These two words are used in the statute very loosely, without any sensitivity to the historical meanings of these terms.

In Section 4(a)(5) the reference to a worldwide flood suggests that catastrophism is used to mean that earth geology can be explained by catastrophes of a different order of magnitude than we know today. However, by the time the word "catastrophism" was introduced in the 1830s, it was meant to apply to naturalistic phenomena and its proponents, such as Whewell, specifically denied that it applied to a worldwide flood as described in the Bible. Indeed, one element of catastrophism was that the earth was very old and that the process of cooling from its molten state caused these catastrophes.

Q: And what about "uniformitarianism" as used in Section 4(b)(5)?

A: Standing alone, 4(b)(5) gives very little clue to the meaning of the word "uniformitarianism" as used in the statute. Once again, its content is understood only by reference to 4(b)(5). In that regard, it seems to me that uniformitarianism was intended to mean that the processes acting on the world's geology have been subject to the same natural laws in the past as in the present. Of course, conditions in the past may have been different from what they are today, and that would mean that the operations of the same laws might have different effects in the past from those they do today. However, the laws of nature would remain the same.

If that is the definition, then I think most scientists would agree with it. However, uniformitarianism then also does not say very much because all science is premised on the same natural laws operating in the past as they do today. Moreover, that meaning of uniformitarianism takes absolutely no account of the varied meanings attached to that word throughout the last hundred years.

Q: Dr. Ruse, do you find much reference to these words "uniformitarianism" and "catastrophism" in the creationist literature generally?

A: Yes. Indeed, it is quite striking that the statute embodies the same loose and historically meaningless use of the two words.

Q: What is your professional opinion about the significance of the worldwide flood contention as it relates to creation science?

A: In my opinion, the worldwide flood idea further highlights the nonscience of creation science. As used in 590, "uniformitarianism" refers to scientific reliance on blind, unchanging natural laws. Contrarily, since catastrophism is cited as an element of *creation* theory, and as an alternative to the naturalistic processes that are presumed to govern evolution theory, the word can only mean that catastrophes occurred in the past in a way that we do not see today because a Creator intervened and disrupted the natural processes. The most prominent example of that supernatural intervention is the worldwide flood described in Genesis: the Noachian flood.

I think the worldwide flood is especially revealing because it has nothing to do with "origins"—which is supposed to be the subject matter of creation science. To use the words of the statute, the worldwide flood is not directly related to the origin of man, life, the earth, or the universe. A cataclysmic worldwide flood is, of course, a biblical notion. Creation science generally has incorporated the flood into the creation model. That the worldwide flood found its way into the statute, even though it does not have anything to do with origins, further highlights the direct connection between the existing body of literature defining creation science and the creation science defined in the statute.

Q: Do you have an opinion about why creation science incorporates reference to the flood?

A: Yes. I think there are two reasons.

First, because it is an essential element of fundamentalist religious belief, and creation science is a manifestation of those religious views.

Second, I think creation scientists grasped at the flood to explain away a very difficult problem for them—the worldwide uniform ordering of the fossil record. In other words, if as the creationists contend, the Creator created all organisms at once, then the fossil remains of man and dinosaurs and fish, etc., should be found uniformly throughout the geologic column. Of course, as we know, that is not so. The creation scientists fabricate a farcically ridiculous explanation based on the flood. According to creation science this catastrophic catastrophe that destroyed the earth also had the side effect of neatly ordering the fossil record.

Q: And finally, Dr. Ruse, do you have any professional observations with respect to subsection 6?

A: Section 4(b) is rather vague for a scientific statement: the earth was created "*several* billion years ago" and "*life was created somewhat* later." However, at least the point of this provision is clear: the earth and life on the earth are very old.

In contrast, 4(a)(6), though similarly vague, identifies a young age for youth and life on the earth. In the creation science literature I've read, a standard tenet of the theory is that the world is only six to ten thousand years old.

Interestingly, a creation theory—that a Creator created the earth and life on the earth—does not necessarily require a young earth. The Creator could have acted many billions of years ago. Indeed, many of the government witnesses at this trial apparently believe that the earth is very old. Since a young earth is not a necessary prerequisite for creation by a Creator, it must have some other explanation. Obviously, that explanation is the Bible. Creation science attempts to prove that the world is six to ten thousand years old because the Bible, if literally read, would have the earth only that old.

Q: Dr. Ruse, having reviewed the definition and elements of creation science and evolution science, the dual models, do you have a professional opinion about the dual-model approach adopted in the statute?

A: Yes. The central premise of the dual-model approach is that an argument against evolution science must be an argument in support of creation science. That premise is without any logical basis whatsoever. The reason is that the so-called scientific evidences against evolution are not really against evolution at all. At very best, each scientific evidence cited by creation science suggests no more than some uncertainty about some small part of one discipline in an interrelated scheme of about a dozen scientific disciplines that comprise the content of what is defined in the statute as evolution science.

Q: Can you give us any example of a creation science argument that illustrates your point?

A: Yes. I think the issue of abiogenesis is a good example because science does not yet know how life on this planet began. Abiogenesis is one working theory that some scientists are considering

Creation scientists take great comfort in the present inability of science to fabricate life in a laboratory. They dismiss the whole effort for its present failure, conclude that naturalistic processes were not sufficient to create life, and cite that present failure as scientific evidence in support of creation.

The silliness of that conclusion is apparent. First, that scientists cannot *now* create life in the laboratory from naturalistic processes does not mean that, as knowledge and technology advances, life will not be so created in the future. And even if life could never be created in the laboratory, that does not mean that creation is the only alternative. Even today there are other theories. And advanced scientific learning may suggest still more other alternative explanations through naturalistic mechanisms for the origin of life. Science does not know the *right* answer to most things. But that ignorance does not prove creation, rather it proves only that current knowledge is incomplete.

Q: Dr. Ruse, does your learning in the history of science suggest any parallels with the example you have been discussing about the scientific attempt to understand the origins of life?

A: Yes, absolutely. We can see throughout the history of science that when people do not understand the world around them, some always attribute the unknown to the workings of a god.

The ancient Greeks' ignorance of celestial mechanics led them to believe that the sun was pulled across the sky by a god. And indeed that all natural phenomena were directed by gods in the sky at the mere mortals that inhabited the earth.

Similarly, for thousands of years, up until only about a century ago, people believed that disease was inflicted by God: divine retribution for sin.

Virtually all scientific learning—astronomy, biology, chemistry, physics, of course, psychology—was at one time profoundly influenced by the belief that what could not be understood must reflect the intervention of God. Consequently, perhaps the single most important element of our modern understanding of science is that science is limited to naturalistic processes that do not rely on, or permit, the intervention of supernatural forces. If something is not now understood, then that means only that more work must be done. When faced with the unknown, scientists today do not just throw up their hands and say the Creator must have done it.

WHAT IS SCIENCE?

Q: Dr. Ruse, you refer to a modern understanding of what science is. Will you please describe to the court your understanding, as a historian and philosopher of science, of what science is today?

A: Mr. Novik, that is a misleading question, in part because science is not one thing. For thousands of years, beginning with the Greek philosophers, people have been seriously thinking about that question. Needless to say, many different notions have been posited, considered, and criticized. Therefore, I cannot give you one single definition of science. Instead, I can describe what philosophers of science generally consider some of the attributes that characterize scientific thinking and methodology today.

Q: Would you please do that for the court?

A: To begin with, as I have mentioned earlier, the most important characteristic of modern science is that it depends entirely on the operation of blind, unchanging regularities in nature. We call those regularities "natural laws." Thus, scientists seek to understand the empirical world by reference to natural law and naturalistic processes. Therefore, any reliance on a supernatural force, a Creator intervening in a natural world by supernatural processes, is necessarily not science.

Q: If a theory involving the supernatural intervention of a Creator is not science, then what is it?

A: It is religion. In my opinion, reliance on the acts of a Creator is inherently religious. It is not necessarily wrong. It is just a different perspective. It has its place, just as science has its place, but it is not science.

Q: In addition to exclusive reliance on natural law, are there any other characteristics commonly recognized among philosophers of science as distinguishing science and scientific thought?

A: Yes, in my opinion science must also be at least explanatory, testable, and tentative.

Q: Tell the court what you mean in saying that science must be explanatory.

A: Science attempts to explain the empirical world in terms of natural law and naturalistic processes. A scientific explanation may try to explain how one phenomenon follows in a tight and definite way, as a result of the working of natural law. A scientific explanation may also attempt to explain how two different phenomena are related to each other by the operation of natural law.

For example, a scientist might attempt to explain the path of a basketball from player to hoop. A scientific explanation would focus on the physical laws of motion as they apply to the variables in the situation: in this case the force and direction of the throw as well as the resistance of gravity and friction. A scientific explanation would also involve a consistency that would transcend the one-of-a-kind throw of a basketball and relate that one throw to other basketball shots as well as to football passes and the flight of a rocket.

Q: Please explain what you mean by testable.

A: Testability is related to explanation.

Genuine science must lay itself open to testing against empirical reality. A scientific explanation leads to certain inferences—predictions, if you will—about what to expect the next time the explanation is applied to the empirical evidence. That ability to compare the evidence against the proposed explanation is the essence of scientific testability.

This characteristic of testability also suggests another consideration that does not usually come up but is peculiarly relevant to creation science. If a theory is to be taken seriously as science, it must have some positive confirming evidence in support of it that can be tested. If a theory does not have any empirical evidence in support of it, it cannot be tested and, therefore, cannot be science.

Q: In connection with your explanation of testable, can you explain the meaning of falsifiability?

A: Falsifiability is another way of looking at what I have called testability.

Falsifiability means that a theory is scientific only if there is some fact or observation that, if true, would tend to disprove, or falsify, the theory. In other words, science must be subject to falsification.

Falsifiability was a philosophical analysis developed by Karl Popper. The reason Popper focused on falsification, rather than verification, is that it is theoretically never possible to verify a theory with absolute certainty. In order for a theory to be absolutely verified it would have to be proven in every possible circumstance. Inasmuch as there are an infinite number of circumstances in which any theory could be tested, it would be impossible ever to verify a theory fully and unequivocally.

However, if a theory were false in any one circumstance, the theory would be falsified. (Of course, the theory might be modified or the theory might be discarded in favor of a new theory, but the original theory tested would be false.) Therefore, Popper argued that the true method of scientific inquiry is the attempt to falsify. And unless something was subject to falsification it was not scientific.

Q: Can you give us an example of how falsification works?

A: Yes, using an example in paleontology, evolutionary theory predicts an ordering of fossils in fossil record: less complex to more complex. If the fossil record contained evidences of all species—from trilobites to dinosaurs to man—at every level of the geologic column, that finding would tend to falsify evolution. That has not been the case, and therefore evolution has not been falsified. But it is subject to falsification.

Similarly, if modern technological advances in microbiology had resulted in the discovery that there were absolutely no similarities between the genetic material of different organisms, that finding would tend to falsify the evolutionary theory of common descent. Once again, that theory of common descent has not been falsified by the evidence. But the theory is nonetheless falsifiable.

Q: How does falsifiability relate to testability?

A: In attempting to falsify a theory, scientists test that theory in every imaginable circumstance. If such tests are negative, then the theory is falsified. But if the tests are positive then the theory is confirmed, at least in the circumstance of that test. And, as a theory is confirmed in more and more circumstances, it is given increasing weight by science. Of course, as I explained earlier, as a theoretical matter such verification will never prove the theory with absolute certainty. However, as a practical matter such confirming evidence may be so overwhelming, the theory may have been tested and validated under so many different circumstances, that scientists generally accept it as true. Such is the state of the theory that evolution has actually happened. There is not yet such a consensus with respect to the theories about how evolution has happened.

Q: The last characteristic you mentioned was that science is tentative. Would you please explain that concept?

A: The tentativeness of science follows directly from its testability. Science knows no ultimate truth not subject to revision. If a theory fails the test, then it is rejected.

Q: You testified earlier with respect to testability that sometimes a theory is confirmed to such a degree that it is accepted as true, for practical purposes. If that is so, how can science be tentative?

A: By saying that science is tentative, I do not mean that scientists must start from the beginning every Monday morning. For purposes of continuing their research and expanding their knowledge, scientists regularly accept many theories that have been proven sufficiently valid to warrant day-to-day reliance on their validity. That is not to say that such accepted theories can never be challenged. If accumulated evidence cannot be explained by the theory, then scientists regularly reassess their initial underlying theoretical assumptions.

For example, up until the early part of this century, Newtonian physics enjoyed virtually universal acceptance. Within a very short period of time, however, it was displaced by Einstein's theories, which better explained the empirical world.

Of course, a scientist does not have to discard the whole theoretical framework of his discipline merely because he finds some aberration. First, the peculiarity will be studied and explored and exposed and questioned, and certainly an effort will be made to fix it within the existing theoretical framework. Only then, if it is found hopelessly inconsistent with existing theory, will new ideas be formulated to take account of the discrepancy. That is precisely what happened with Newton's laws of motion. In certain respects they were seen to be inadequate, they were not able to explain the empirical evidence. Scientists did not immediately throw out Newtonian theory. They spent a considerable amount of time attempting to fit the facts within that theory. It was only after Einstein formulated a new theory, which was an alternative to Newtonian physics and which did explain the evidence, that scientists accepted the displacement of Newton's laws.

I like to think of tentativeness in science as analogous to the tentativeness in the judicial system. When a person is accused of crime, he is assumed innocent until proven guilty. The judicial system spends a considerable amount of time and energy examining and questioning the evidence and being very tentative about the guilt of the defendant. At some point, however, the case is presented to the jury, the evidence is assessed, and if the jury is convinced beyond a reasonable doubt, the defendant is convicted. At that point the judicial system ceases to look at the evidence; it does not reassess guilt every day. Of course, for sufficient cause, if the evidence is compelling, the conviction can be reexamined even though the jury was convinced beyond a reasonable doubt. But the extent to which the jury's initial judgment is subject to reexamination is a function of the strength of the contrary evidence. If a person is convicted, and then the very next week an identical crime is committed by another person actually caught doing it who happens to look just like the person convicted for the first crime, that is much stronger reason to question the determination than a bit of tangential evidence found twenty years after the fact.

You change your mind much more quickly in one case than the other. Thus, "tentative" means that a working conclusion or a theory is always open to change, depending upon the strength of the contrary evidence. Science must strive to be an honest, rigorous, and objective enterprise that constantly self-corrects to overcome personal bias and misconception. In pursuit of those goals, scientists search out new facts and generate new information to test and challenge their theories. Scientists subject their work to the logical and critical assessment of their colleagues. In other words, science rejects reliance on a priori assumptions that are not subject to scientific investigation.

Q: Dr. Ruse, having examined the creationist literature at great length, do you have a professional opinion about whether creation science measures up to the standards and characteristics of science that you have just been describing?

A: Yes, I do.

In my opinion creation science does not have those attributes that distinguish science from other endeavors.

Q: Would you please explain why you think it does not?

A: First, and most importantly, creation science necessarily looks to the supernatural acts of a Creator. According to creation science theory, the Creator has intervened in supernatural ways using supernatural forces. Moreover, because the supernatural forces are the acts of a Creator, that is, the acts of God, they are not subject to scientific investigation or understanding. This nonscientific aspect of creation science emerges quite clearly from the creation science literature I have read. [Appropriate quotes.]

Q: Do you think that creation science is explanatory?

A: No, because creation science relies on the acts of a Creator, whose purposes cannot be understood by us mortals, creation science has no explanatory power.

Let me give you an example in the evolution/creation controversy. The example pertains to homologies. Homologies are structural similarities between different organisms of widely different species. For example [as can be seen in these two pictures], the wing of a bat and the arm of a man are strikingly similar. These similarities cannot be explained by function: obviously, a man's arm and a bat's wing have different functions. Evolutionary theory gives an explanation for homologies. That explanation is: Organisms with homologies are descended from common ancestors. Creationism, on the other hand, does not give any explanation for homologies. The theory of creation is that a Creator created the bat's wing and man's arm independently. But that theory does not give any explanation of why they are similar. Nor can any explanation be provided because the purpose of the Creator can never be understood. The Creator could have made them similar; He could have made them different. Creation theory does not explain why He chose similarity rather than dissimilarity.

Q: But doesn't creation science theory explain some things? For example, the eye is a remarkable structure. Creation science would say it was made by the Creator. Isn't that an explanation?

A: No, not in the scientific sense. In science an explanation must explain more than that for which it was invented. Saying God created the eye of a dog, for example, does not explain anything about why it is structured the way it is, why it works the way it does, and why it is similar to the eye of a cat in some respects and different in others.

Q: Do you think that creation science is testable?

A: Creation science is neither testable nor tentative. Indeed, an attribute of creation science that distinguishes it quite clearly from science is that it is absolutely certain about all of the answers. And considering the magnitude of the questions it addresses—the origins of man, life, the earth, and the universe—that certainty is all the more revealing. Whatever the contrary evidence, creation science never accepts that its theory is falsified. This is just the opposite of tentativeness and makes a mockery of testing. [Quotes to illustrate.]

Q: Do you find that creation science measures up to the methodological considerations you described?

A: Creation science is woefully lacking in this regard as well. First of all, the dogmatic certainty of creation science I have described earlier necessarily precludes the requisite objectivity. For example, members of the Creation Research Society take an oath that is an outright admission that their religious views supersede scientific objectivity. [Read oath.]

Second, and most regrettably, I have found innumerable instances of outright dishonesty, deception, and distortion used to advance creation science arguments. [Quotes to illustrate.]

Q: Dr. Ruse, do you have an opinion to a reasonable degree of professional certainty about whether creation science is science?

A: Yes.

Q: What is your opinion?

A: In my opinion creation science is not science.

Q: What do you think it is?

A: As someone also trained in the philosophy of religion, in my opinion creation science is religion.

NOTE

1. In order to provide a constant reminder of the connection, all our witnesses should refer *only* to creation science and creation scientists. I think we make up what we lose in propaganda by enhancing the relevance of creation science literature. If the witness is uncomfortable, a general disclaimer at the outset ("just because they call it 'science' does not mean I concede it is") should suffice.

CHAPTER 14

UNITED STATES DISTRICT COURT OPINION: *McLEAN V. ARKANSAS BOARD OF EDUCATION*

JUDGE WILLIAM R. OVERTON

JUDGMENT

Pursuant to the court's memorandum opinion filed this date, judgment is hereby entered in favor of the plaintiffs and against the defendants. The relief prayed for is granted.

Dated this January 5, 1982.

INJUNCTION

Pursuant to the court's memorandum opinion filed this date, the defendants and each of them and all their servants and employees are hereby permanently enjoined from implementing in any manner Act 590 of the Acts of Arkansas of 1981.

It is so ordered this January 5, 1982.

From *McLean v. Arkansas Board of Education*, 529 F. Supp. 1255, 1258–64 (ED Ark. 1982)

MEMORANDUM OPINION

Introduction

On March 19, 1981, the governor of Arkansas signed into law Act 590 of 1981, entitled "Balanced Treatment for Creation Science and Evolution-Science Act." The act is codified as Ark. Stat. Ann. §80–1663, *et seq.* (1981 Suppl.). Its essential mandate is stated in its first sentence: "Public schools within this State shall give balanced treatment to creation science and to evolution-science." On May 27, 1981, this suit was filed[1] challenging the constitutional validity of Act 590 on three distinct grounds.

First, it is contended that Act 590 constitutes an establishment of religion prohibited by the First Amendment to the Constitution, which is made applicable to the states by the Fourteenth Amendment. Second, the plaintiffs argue the act violates a right to academic freedom which they say is guaranteed to students and teachers by the Free Speech Clause of the First Amendment. Third, plaintiffs allege the act is impermissibly vague and thereby violates the Due Process Clause of the Fourteenth Amendment.

The individual plaintiffs include the resident Arkansas Bishops of the United Methodist, Episcopal, Roman Catholic, and African Methodist Episcopal Churches, the principal official of the Presbyterian Churches in Arkansas, other United Methodist, Southern Baptist, and Presbyterian clergy, as well as several persons who sue as parents and next friends of minor children attending Arkansas public schools. One plaintiff is a high school biology teacher. All are also Arkansas taxpayers. Among the organizational plaintiffs are the American Jewish Congress, the Union of American Hebrew Congregations, the American Jewish Committee, the Arkansas Education Association, the National Association of Biology Teachers, and the national Coalition for Public Education and Religious Liberty, all of which sue on behalf of members living in Arkansas.[2]

The defendants include the Arkansas Board of Education and its members, the director of the Department of Education, and the State Textbooks and Instructional Materials Selecting Committee.[3] The Pulaski County Special School District and its directors and superintendent were voluntarily dismissed by the plaintiffs at the pretrial conference held October 1, 1981.

The trial commenced December 7, 1981, and continued through December 17, 1981. This memorandum opinion constitutes the court's findings of fact and conclusions of law. Further orders and judgments will be in conformity with this opinion.

I

There is no controversy over the legal standards under which the Establishment Clause portion of this case must be judged. The Supreme Court has on a number of occasions expounded on the meaning of the clause, and the pronouncements are clear. Often the issue has arisen in the context of public education, as it has here. In *Everson v. Board of Education*, 330 U.S. 1, 15–16 (1947), Justice Black stated:

> The "establishment of religion" clause of the First Amendment means at least this: Neither a state nor the Federal Government can set up a church. Neither can pass laws which aid one religion, aid all religions, or prefer one religion over another. Neither can force nor influence a person to go to or to remain away from church against his will or force him to profess a belief or disbelief in any religion. No person can be punished for entertaining or professing religious beliefs or disbeliefs, for church attendance or non-attendance. No tax, large or small, can be levied to support any religious activities or institutions, whatever they may be called, or whatever form they may adopt to teach or practice religion. Neither a state nor the Federal Government can, openly or secretly, participate in the affairs of any religious organizations or groups and vice versa. In the words of Jefferson, the clause ... was intended to erect "a wall of separation between church and State."

The Establishment Clause thus enshrines two central values: voluntarism and pluralism. And it is in the area of the public schools that these values must be guarded most vigilantly.

> Designed to serve as perhaps the most powerful agency for promoting cohesion among a heterogeneous democratic people, the public school must keep scrupulously free from entanglement in the strife of sects. The preservation of the community from divisive conflicts, of Government from irreconcil-

> able pressures by religious groups, of religion from censorship and coercion however subtly exercised, requires strict confinement of the State to instruction other than religious, leaving to the individual's church and home, indoctrination in the faith of his choice. [*McCollum v. Board of Education*, 333 U.S. 203, 216–217 (1948) (Opinion of Justice Frankfurter, joined by Justices Jackson, Burton, and Rutledge)]

The specific formulation of the establishment prohibition has been refined over the years, but its meaning has not varied from the principles articulated by Justice Black in *Everson*. In *Abbington School District v. Schempp*, 374 U.S. 203, 222 (1963), Justice Clark stated that "to withstand the strictures of the Establishment Clause there must be a secular legislative purposed and a primary effect that neither advances nor inhibits religion." The Court found it quite clear that the First Amendment does not permit a state to require the daily reading of the Bible in public schools, for "[s]urely the place of the Bible as an instrument of religion cannot be gainsaid" (ibid., 224). Similarly, in *Engel v. Vitale*, 370 U.S. 421 (1962), the Court held that the First Amendment prohibited the New York Board of Regents from requiring the daily recitation of a certain prayer in the schools. With characteristic succinctness, Justice Black wrote: "Under [the First] Amendment's prohibition against governmental establishment of religion, as reinforced by the provisions of the Fourteenth Amendment, government in this country, be it state or federal, is without power to prescribe by law any particular form of prayer which is to be used as an official prayer in carrying on any program of governmentally sponsored religious activity" (ibid., 430). Black also identified the objective at which the Establishment Clause was aimed: "its first and most immediate purpose rested on the belief that a union of government and religion tends to destroy government and to degrade religion" (ibid., 431).

Most recently, the Supreme Court has held that the clause prohibits a state from requiring the posting of the Ten Commandments in public school classrooms for the same reasons that officially imposed daily Bible reading is prohibited [*Stone v. Graham*, 449 U.S. 39 (1980)]. The opinion in *Stone* relies on the most recent formulation of the Establishment Clause test, that of *Lemon v. Kurtzman*, 403 U.S. 602, 612–613 (1971):

> First, the statute must have a secular legislative purpose; second, its principal or primary effect must be one that neither advances nor inhibits religion; . . . finally, the statute must not foster "an excessive government entanglement with religion." [*Stone v. Graham*, 449 U.S. 40]

It is under this three-part test that the evidence in this case must be judged. Failure on any of these grounds is fatal to the enactment.

II

The religious movement known as fundamentalism began in nineteenth-century America as part of evangelical Protestantism's response to social changes, new religious thought, and Darwinism. Fundamentalists viewed these developments as attacks on the Bible and as responsible for a decline in traditional values.

The various manifestations of fundamentalism have had a number of common characteristics,[4] but a central premise has always been a literal interpretation of the Bible and a belief in the inerrancy of the scriptures. Following World War I, there was again a perceived decline in traditional morality, and fundamentalism focused on evolution as responsible for the decline. One aspect of their efforts, particularly in the South, was the promotion of statutes prohibiting the teaching of evolution in public schools. In Arkansas, this resulted in the adoption of Initiated Act 1 of 1929.[5]

Between the 1920s and early 1960s, antievolutionary sentiment had a subtle but pervasive influence on the teaching of biology in public schools. Generally, textbooks avoided the topic of evolution and did not mention the name of Darwin. Following the launch of the *Sputnik* satellite by the Soviet Union in 1957, the National Science Foundation funded several programs designed to modernize the teaching of science in the nation's schools. The Biological Sciences Curriculum Study (BSCS), a nonprofit organization, was among those receiving grants for curriculum study and revision. Working with scientists and teachers, BSCS developed a series of biology texts which, although emphasizing different aspects of biology, incorporated the theory of evolution as a major theme. The success of the BSCS effort is shown by the fact that 50 percent of American schoolchildren currently use BSCS books directly and the curriculum is incorporated indirectly in virtually all biology texts (testimony of Nelkin Mayer; Plaintiffs' Exhibit 1).[6]

In the early 1960s, there was again a resurgence of concern among fundamentalists about the loss of traditional values and a fear of growing secularism in society. The fundamentalist movement became more active and has steadily grown in numbers and political influence. There is an emphasis among current fundamentalists on the literal interpretation of the Bible and the book of Genesis as the sole source of knowledge about origins.

The term "scientific creationism" first gained currency around 1965 following publication of *The Genesis Flood* in 1961 by Whitcomb and Morris. There is undoubtedly some connection between the appearance of the BSCS texts emphasizing evolutionary thought and efforts of fundamentalist to attach the theory (Mayer).

In the 1960s and early 1970s, several fundamentalist organizations were formed to promote the idea that the book of Genesis was supported by scientific data. The terms "creation science" and "scientific creationism" have been adopted by these fundamentalists as descriptive of their study of creation and the origins of man. Perhaps the leading creationist organization is the Institute for Creation Research (ICR), which is affiliated with the Christian Heritage College and supported by the Scott Memorial Baptist Church in San Diego, California. The ICR, through the Creation-Life Publishing Company, is the leading publisher of creation science material. Other creation science organizations include the Creation Science Research Center (CSRC) of San Diego and the Bible Science Association of Minneapolis, Minnesota. In 1963, the Creation Research Society (CRS) was formed from a schism in the American Scientific Affiliation (ASA). It is an organization of literal fundamentalists[7] who have the equivalent of a master's degree in some recognized area of science. A purpose of the organization is "to reach all people with the vital message of the scientific and historical truth about creation" (Nelkin, *The Science Textbook Controversies and the Politics of Equal Time*, 66). Similarly, the CSRC was formed in 1970 from a split in the CRS. Its aim has been "to reach the 63 million children of the United States with the scientific teaching of Biblical creationism" (ibid., 69).

Among creationist writers who are recognized as authorities in the field by other creationists are Henry M. Morris, Duane Gish, G. E. Parker, Harold S. Slusher, Richard B. Bliss, John W. Moore, Martin E. Clark, W. L. Wysong,

Robert E. Kofahl, and Kelly L. Segraves. Morris is director of ICR, Gish is associate director, and Segraves is associated with CSRC.

Creationists view evolution as a source of society's ills, and the writings of Morris and Clark are typical expressions of that view.

> Evolution is thus not only anti-Biblical and anti-Christian, but it is utterly unscientific and impossible as well. But it has served effectively as the pseudo-scientific basis of atheism, agnosticism, socialism, fascism, and numerous other false and dangerous philosophies over the past century. [Morris and Clark, *The Bible Has the Answer* (Plaintiffs' Exhibit 31 and Plaintiffs' Pretrial Exhibit 89)][8]

Creationists have adopted the view of fundamentalists generally that there are only two positions with respect to the origins of the earth and life: belief in the inerrancy of the Genesis story of creation and of a worldwide flood as fact, or a belief in what they call evolution.

Henry Morris has stated, "It is impossible to devise a legitimate means of harmonizing the Bible with evolution" [Morris, "Evolution and the Bible," *ICR Impact Series*, Number 5 (undated, unpaged), quoted in Mayer, Plaintiffs' Exhibit 8, 3]. This dualistic approach to the subject of origins permeates the creationist literature.

The creationist organizations consider the introduction of creation science into the public schools part of their ministry. The ICR has published at least two pamphlets[9] containing suggested methods for convincing school boards, administrators, and teachers that creationism should be taught in public schools. The ICR has urged its proponents to encourage school officials to voluntarily add creationism to the curriculum.[10]

Citizens For Fairness In Education is an organization based in Anderson, South Carolina, formed by Paul Ellwanger, a respiratory therapist who is trained in neither law nor science. Mr. Ellwanger is of the opinion that evolution is the forerunner of many social ills, including Nazism, racism, and abortion (Ellwanger Deposition, 32–34). About 1977, Ellwanger collected several proposed legislative acts with the idea of preparing a model state act requiring the teaching of creationism as science in opposition to evolution. One of the proposals he collected was prepared by Wendell Bird, who is now a staff attorney for ICR.[11] From these various proposals, Ellwanger prepared a "model act" which calls for "balanced treatment" of "scientific creationism" and "evolution" in public schools. He circulated the proposed act to various people and organizations around the country.

Mr. Ellwanger's views on the nature of creation science are entitled to some weight since he personally drafted the model act which became Act 590. His evidentiary deposition with exhibits and unnumbered attachments (produced in response to a subpoena *duces tecum* speaks to both the intent of the act and the scientific merits of creation science. Mr. Ellwanger does not believe creation science is a science. In a letter to Pastor Robert E. Hays he states, "While neither evolution nor creation can qualify as a scientific theory, and since it is virtually impossible at this point to educate the whole world that evolution is not a true scientific theory, we have freely used these terms—the evolution theory and the theory of scientific creationism—in the bill's text" (unnumbered attachment to Ellwanger Deposition, 2). He further states in a letter to Mr. Tom Bethell, "As we examine evolution (remember, we're not making any scientific claims for creation, but we are challenging evolution's claim to be scientific, . . ." (unnumbered attachment to Ellwanger Deposition, 1).

Ellwanger's correspondence on the subject shows an awareness that Act 590 is a religious crusade, coupled with a desire to conceal this fact. In a letter to State Senator Bill Keith of Louisiana, he says, "I view this whole battle as one between God and anti-God forces, though I know there are a large number of evolutionists who believe in God." And further, ". . . it behooves Satan to do all he can to thwart our efforts and confuse the issue at every turn." Yet Ellwanger suggest to Senator Keith, "If you have a clear choice between having grassroots leaders of this statewide bill promotion effort to be ministerial or non-ministerial, be sure to opt for the non-ministerial. It does the bill effort no good to have ministers out there in the public forum and the adversary will surely pick at this point. . . . Ministerial persons can accomplish a tremendous amount of work from behind the scenes, encouraging their congregations to take the organizational and P.R. initiatives. And they can lead their churches in storming Heaven with prayers for help against so tenacious an adversary" (unnumbered attachment to Ellwanger Deposition, 1).

Ellwanger shows a remarkable degree of political candor, if not finesse, in a letter to State Senator Joseph Carlucci of Florida:

> It would be very wise, if not actually essential, that all of us who are engaged in this legislative effort be careful not to present our position and our work in a religious framework. For example, in written communications that might somehow be shared with those other persons whom we may be trying to convince, it would be well to exclude our own personal testimony and/or witness for Christ, but rather, if we are so moved, to give that testimony on a separate attached note. (Unnumbered attachment to Ellwanger Deposition, 1)

The same tenor is reflected in a letter by Ellwanger to Mary Ann Miller, a member of FLAG (Family, Life, America under God) who lobbied the Arkansas legislature in favor of Act 590:

> . . . we'd like to suggest that you and your co-workers be very cautious about mixing creation science with creation-religion. . . . Please urge your co-workers not to allow themselves to get sucked into the "religion" trap of mixing the two together, for such mixing does incalculable harm to the legislative thrust. It could even bring public opinion to bear adversely upon the higher courts that will eventually have to pass judgment on the constitutionality of this new law. (Exhibit 1 to Miller Deposition)

Perhaps most interesting, however, is Mr. Ellwanger's testimony in his deposition as to his strategy for having the model act implemented:

> Q. You're trying to play on other people's religious motives.
>
> A. I'm trying to play on their emotions, love, hate, their likes, dislikes, because I don't know any other way to involve, to get humans to become involved in human endeavors. I see emotions as being a healthy and legitimate means of getting people's feelings into action, and . . . I believe that the predominance of population in America that represents the greatest potential for taking some kind of action in this area is a Christian community. I see the Jewish community as far less potential in taking action . . . but I've seen a lot of interest among Christians and I feel, why not exploit that to get the bill going if that's what it takes. (Ellwanger Deposition, 146–47)

Mr. Ellwanger's ultimate purpose is revealed in the closing of his letter to Mr. Tom Bethell: "Perhaps all this is old hat to you, Tom, and if so, I'd appreciate your telling me so and perhaps where you've heard it before—the idea of killing evolution instead of playing these debating games that we've been playing for nigh over a decade already" (unnumbered attachment to Ellwanger Deposition, 3).

It was out of this milieu that Act 590 emerged. The Reverend W. A. Blount, a biblical literalist who is a pastor of a church in the Little Rock area and was, in February 1981, chairman of the Greater Little Rock Evangelical Fellowship, was among those who received a copy of the model act from Ellwanger.[12]

At Reverend Blount's request, the Evangelical Fellowship unanimously adopted a resolution to seek an introduction of Ellwanger's act in the Arkansas legislature. A committee composed of two ministers, Curtis Thomas and W. A. Young, was appointed to implement the resolution. Thomas obtained from Ellwanger a revised copy of the model act which he transmitted to Carl Hunt, a business associate of Senator James L. Holsted, with the request that Hunt prevail upon Holsted to introduce the act.

Holsted, a self-described "born again" Christian fundamentalist, introduced the act in the Arkansas Senate. He did not consult the State Department of Education, scientists, science educators, or the Arkansas attorney general.[13] The act was not referred to any Senate committee for hearing and was passed after only a few minutes' discussion on the Senate floor. In the House of Representatives, the bill was referred to the Education Committee, which conducted a perfunctory fifteen-minute hearing. No scientist testified at the hearing, nor was any representative form the State Department of Education called to testify.

Ellwanger's model act was enacted into law in Arkansas as Act 590 without amendment or modification other than minor typographical changes. The legislative "finding of fact" in Ellwanger's act and Act 590 are identical, although no meaningful fact-finding process was employed by the General Assembly.

Ellwanger's efforts in preparation of the model act and campaign for its adoption in the states were motivated by his opposition to the theory of evolution and his desire to see the biblical version of creation taught in the public schools. There is no evidence that the pastors, Blount, Thomas, Young, or the Greater Little Rock Evangelical Fellowship were motivated by anything other than their religious convictions when proposing its adoption or during their lobbying efforts in its behalf. Senator Holsted's sponsorship and lobbying efforts in behalf of the act were motivated solely by his religious beliefs and desire to see the biblical version of creation taught in the public schools.[14]

The state of Arkansas, like a number of states whose citizens have relatively homogeneous religious beliefs, has a long history of official opposition to evolution which is motivated by adherence to fundamentalist beliefs in the inerrancy of the book of Genesis. This history is documented in Justice Fortas's opinion in *Epperson v. Arkansas*, 393 U.S. 97 (1968), which struck down Initiated Act 1 of 1929, Ark. Stat. Ann. §80–1627–1628, prohibiting the teaching of the theory of evolution. To this same tradition may be attributed Initiated Act 1 of 1930, Ark. Stat. Ann. §80–1606 (repealed 1980), requiring "the reverent daily reading of a portion of the English Bible" in every public school classroom in the state.[15]

It is true, as defendants argue, that courts should look to legislative statements of a statute's purpose in Establishment Clause cases and accord such pronouncements great deference [see, e.g., *Committee for Public Education & Religious Liberty v. Nyquist*, 413 U.S. 756, 773 (1973) and *McGowan v. Maryland*, 366 U.S. 420, 445 (1961)]. Defendants also correctly state the principle that remarks by the sponsor or author of a bill are not considered controlling in analyzing legislative intent [see, e.g., *United States v. Emmons*, 410 U.S. 396 (1973) and *Chrysler Corp v. Brown*, 441 U.S. 281 (1979)].

Courts are not bound, however, by legislative statements of purpose or legislative disclaimers [*Stone v. Graham*, 449 U.S. 39 (1980); *Abbington School District v. Schempp*, 374 U.S. 203 (1963)]. In determining the legislative purpose of a statute, courts may consider evidence of the historical context of the act [*Epperson v. Arkansas*, 393 U.S. 97 (1968)], the specific sequence of events leading up to passage of the act, departures from normal procedural sequences, substantive departures from the normal [*Village of Arlington Heights v. Metropolitan Housing Corp.*, 429 U.S. 252 (1977)], and contemporaneous statements of the legislative sponsor [*Federal Energy Administration v. Algonquin SNG Inc.*, 426 U.S. 548, 564 (1976)].

The unusual circumstances surrounding the passage of Act 590, as well as the substantive law of the First Amendment, warrant an inquiry into the stated legislative purposes. The author of the act has publicly proclaimed the sectarian purpose of the proposal. The Arkansas residents who sought legislative sponsorship of the bill did so for a purely sectarian purpose. These circumstances alone may not be particularly persuasive, but when considered with the publicly announced motives of the legislative sponsor made contemporaneously with the legislative process; the lack of any legislative investigation, debate, or consultation with any educators or scientists; the unprecedented intrusion in school curriculum;[16] and official history of the state of Arkansas on the subject, it is obvious that the statement of purpose has little, if any, support in fact. The state failed to produce any evidence which would warrant an inference or conclusion that at any point in the process anyone considered the legitimate educational value of the act. It was simply and purely an effort to introduce the biblical version of creation into the public school curricula. The only inference which can be drawn from these circumstances is that the act was passed with the specific purpose by the General Assembly of advancing religion. The act therefore fails the first prong of the three-pronged test, that of secular legislative purpose, as articulated in *Lemon v. Kurtzman* and *Stone v. Graham.*

III

If the defendants are correct and the court is limited to an examination of the language of the act, the evidence is overwhelming that both the purpose and the effect of Act 590 is the advancement of religion in the public schools.

Section 4 of the act provides:

> Definitions, as used in this Act:
>
> (a) "Creation science" means the scientific evidences for creation and inferences from those scientific evidences. Creation science includes the scientific evidences and related inferences that indicate: (1) Sudden creation of the universe, energy, and life from nothing; (2) The insufficiency of mutation and natural selection in bringing about development of all living kinds from a single organism; (3) Changes only within fixed limits of originally created kinds of plants and animals; (4) Separate ancestry for man and apes; (5) Explanation of the earth's geology by catastrophism, including the occur-

rence of a worldwide flood; and (6) A relatively recent inception of the earth and living kinds.

(b) "Evolution-science" means the scientific evidences for evolution and inferences from those scientific evidences. Evolution-science includes the scientific evidences and related inferences that indicate: (1) Emergence by naturalistic processes of the universe from disordered matter and emergence of life from nonlife; (2) The sufficiency of mutation and natural selection in bringing about development of present living kinds from simple earlier kinds; (3) Emergence by mutation and natural selection of present living kinds from simple earlier kinds; (4) Emergence of man from a common ancestor with apes; (5) Explanation of the earth's geology and the evolutionary sequence by uniformitarianism; and (6) An inception several billion years ago of the earth and somewhat later of life.

(c) "Public schools" means public secondary and elementary schools.

The evidence establishes that the definition of "creation science" contained in 4(a) has as its unmentioned reference the first eleven chapters of the book of Genesis. Among the many creation epics in human history, the account of sudden creation from nothing, or *creatio ex nihilo*, and subsequent destruction of the world by flood is unique to Genesis. The concepts of 4(a) are the literal fundamentalists' view of Genesis. Section 4(a) is unquestionably a statement of religion, with the exception of 4(a)(2) which is a negative thrust aimed at what the creationists understand to be the theory of evolution.[17]

Both the concepts and the wording of Section 4(a) convey an inescapable religiosity. Section 4(a)(1) describes "sudden creation of the universe, energy and life from nothing." Every theologian who testified, including defense witnesses, expressed the opinion that the statement referred to a supernatural creation which was performed by God.

Defendants argue that: (1) the fact that 4(a) conveys ideas similar to the literal interpretation of Genesis does not make it conclusively a statement of religion; (2) that reference to a creation from nothing is not necessarily a religious concept since the act only suggests a creator who has power, intelligence, and a sense of design and not necessarily the attributes of love, compassion, and justice;[18] and (3) that simply teaching about the concept of a creator is not a religious exercise unless the student is required to make a commitment to the concept of a creator.

The evidence fully answers these arguments. The ideas of 4(a)(1) are not merely similar to the literal interpretation of Genesis; they are identical and parallel to no other story of creation.[19]

The argument that creation from nothing in 4(a)(1) does not involve a supernatural deity has no evidentiary or rational support. To the contrary, "creation out of nothing" is a concept unique to Western religions. In traditional Western religious thought, the conception of a creator of the world is a conception of God. Indeed, creation of the world "out of nothing" is the ultimate religious statement because God is the only actor. As Dr. Langdon Gilkey noted, the act refers to one who has the power to bring all the universe into existence from nothing. The only "one" who has this power is God.[20]

The leading creationist writers, Morris and Gish, acknowledge that the idea of creation described in 4(a)(1) is the concept of creation by God and make no pretense to the contrary.[21] The idea of sudden creation from nothing, or *creatio ex nihilo*, is an inherently religious concept (testimony of Vawter, Gilkey, Geisler, Ayala, Blount, Hicks).

The argument advanced by defendants' witness Dr. Norman Geisler, that teaching the existence of God is not religious unless the teaching seeks a commitment, is contrary to common understanding and contradicts settled case law [*Stone v. Graham*, 449 U.S. 39 (1980), *Abbington School District v. Schempp*, 374 U.S. 203, 222 (1963)].

The facts that creation science is inspired by the book of Genesis and that Section 4(a) is consistent with a literal interpretation of Genesis leave no doubt that a major effect of the act is the advancement of particular religious beliefs. The legal impact of this conclusion will be discussed further at the conclusion of the court's evaluation of the scientific merit of creation science.

IV

(A)

The approach to teaching "creation science" and "evolution-science" found in Act 590 is identical to the two-model approach espoused by the Institute for

Creation Research and is taken almost verbatim from ICR writings. It is an extension of fundamentalists' view that one must either accept the literal interpretation of Genesis or else believe in the godless system of evolution.

The two-model approach of the creationists is simply a contrived dualism[22] which has not scientific factual basis or legitimate educational purpose. It assumes only two explanations for the origins of life and existence of man, plants, and animals: it was either the work of a creator or it was not. Application of these two models, according to creationists, and the defendants, dictates that all scientific evidence which fails to support the theory of evolution is necessarily scientific evidence in support of creationism and is, therefore, creation science "evidence" in support of Section 4(a).

(B)

The emphasis on origins as an aspect of the theory of evolution is peculiar to the creationist literature. Although the subject of origins of life is within the province of biology, the scientific community does not consider origins of life a part of evolutionary theory. The theory of evolution assumes the existence of life and is directed to an explanation of *how* life evolved. Evolution does not presuppose the absence of a creator or God and the plain inference conveyed by Section 4 is erroneous.[23]

As a statement of the theory of evolution, Section 4(b) is simply a hodgepodge of limited assertions, many of which are factually inaccurate.

For example, although 4(b)(2) asserts, as a tenet of evolutionary theory, "sufficiency of mutation and natural selection in bringing about development of present living kinds from simple earlier kinds," Drs. Ayala and Gould both stated that biologists know that these two processes do not account for all significant evolutionary change. They testified to such phenomena as recombination, the founder effect, genetic drift, and the theory of punctuated equilibrium, which are believed to play important evolutionary roles. Section 4(b) omits any reference to these. Moreover, 4(b) utilizes the term "kinds" which all scientists have said is not a word of science and has no fixed meaning. Additionally, the act presents both evolution and creation science as "package deals." Thus, evidence critical to some aspect of what the creationists define as evolution is taken as support for a theory which includes a worldwide flood and a relatively young earth.[24]

(C)

In addition to the fallacious pedagogy of the two-model approach, Section 4(a) lacks legitimate educational value because "creation science" as defined in that section is simply not science. Several witnesses suggested definitions of science. A descriptive definition was said to be that science is what is "accepted by the scientific community" and is "what scientists do." The obvious implication of this description is that, in a free society, knowledge does not require the imprimatur of legislation in order to become science.

More precisely, the essential characteristics of science are:

(1) It is guided by natural law;
(2) It has to be explanatory by reference to nature law;
(3) It is testable against the empirical world;
(4) Its conclusions are tentative, i.e., are not necessarily the final word; and
(5) It is falsifiable. (Ruse and other science witnesses)

Creation science as described in Section 4(a) fails to meet these essential characteristics. First, the section revolves around 4(a)(1) which asserts a sudden creation "from nothing." Such a concept is not science because it depends upon a supernatural intervention which is not guided by natural law. It is not explanatory by reference to natural law, is not testable, and is not falsifiable.[25]

If the unifying idea of supernatural creation by God is removed from Section 4, the remaining parts of the section explain nothing and are meaningless assertions.

Section 4(a)(2), relating to the "insufficiency of mutation and natural selection in bringing about development of all living kinds from a single organism," is an incomplete negative generalization directed at the theory of evolution.

Section 4(a)(3), which describes "changes only within fixed limits of originally created kinds of plants and animals," fails to conform to the essential characteristics of science for several reasons. First, there is no scientific definition of "kinds" and none of the witnesses was able to point to any scientific authority which recognized the term or knew how many "kinds" existed. One defense witness suggested there may be 100 to 10,000 different "kinds."

Another believes there were "about 10,000, give or take a few thousand." Second, the assertion appears to be an effort to establish outer limits of changes within species. There is no scientific explanation for these limits which is guided by natural law and the limitations, whatever they are, cannot be explained by natural law.

The statement in 4(a)(4) of "separate ancestry of man and apes" is a bald assertion. It explains nothing and refers to no scientific fact or theory.[26]

Section 4(a)(5) refers to "explanation of the earth's geology by catastrophism, including the occurrence of a worldwide flood." This assertion completely fails as science. The act is referring to the Noachian flood described in the book of Genesis.[27] The creationist writers concede that *any* kind of Genesis Flood depends upon supernatural intervention. A worldwide flood as an explanation of the world's geology is not the product of natural law, nor can its occurrence be explained by natural law.

Section 4(a)(6) equally fails to meet the standards of science. "Relatively recent inception" has no scientific meaning. It can only be given in reference to creationist writings which place the age at between 6,000 and 20,000 years because of the genealogy of the Old Testament [see, e.g., Plaintiffs' Exhibit 78, Gish (6,000 to 10,000); Plaintiffs' Exhibit 87, Segraves (6,000 to 20,000)]. Such a reasoning process is not the product of natural law; not explainable by natural law; nor is it tentative.

Creation science, as defined in Section 4(a), not only fails to follow the canons of dealing with scientific theory, it also fails to fit the more general descriptions of "what scientists think" and "what scientists do." The scientific community consists of individuals and groups, nationally and internationally, who work independently in such varied fields as biology, paleontology, geology, and astronomy. Their work is published and subject to review and testing by their peers. The journals for publication are both numerous and varied. There is, however, not one recognized scientific journal which has published an article espousing the creation science theory described in Section 4(a). Some of the state's witnesses suggested that the scientific community was "close-minded" on the subject of creationism and that explained the lack of acceptance of the creation science arguments. Yet no witness produced a scientific article for which publication has been refused. Perhaps some members of the scientific community are resistant to new ideas. It is, however, inconceivable that such a loose-knit group of independent thinkers in all the varied fields of science could, or would, so effectively censor new scientific thought.

The creationists have difficulty maintaining among their ranks consistency in the claim that creationism is science. The author of Act 590, Ellwanger, said that neither evolution nor creationism was science. He thinks that both are religious. Duane Gish recently responded to an article in *Discover* critical of creationism by stating:

> Stephen Jay Gould states that creationists claim creation is a scientific theory. This is a false accusation. Creationists have repeatedly stated that neither creation nor evolution is a scientific theory (and each is equally religious). (Gish, letter to editor of *Discover*, July 1981, Appendix 30 to Plaintiffs' Pretrial Brief)

The methodology employed by creationists is another factor which is indicative that their work is not science. A scientific theory must be tentative and always subject to revision or abandonment in light of facts that are inconsistent with, or falsify, the theory. A theory that is by its own terms dogmatic, absolutist, and never subject to revision is not a scientific theory.

The creationists' methods do not take data, weigh it against the opposing scientific data, and thereafter reach the conclusions stated in Section 4(a). Instead, they take the literal wording of the book of Genesis and attempt to find scientific support for it. The method is best explained in the language of Morris in his book (Plaintiffs' Exhibit 31) *Studies in the Bible and Science*, page 114:

> ...it is...quite impossible to determine anything about Creation through a study of present processes, because present processes are not creative in character. If man wishes to know anything about Creation (the time of Creation, the duration of Creation, the order of Creation, the methods of Creation, or anything else) his sole source of true information is that of divine revelation. God was there when it happened. We were not there.... Therefore, we are completely limited to what God has seen fit to tell us, and this information is in His written Word. This is our textbook on the science of Creation!

The Creation Research Society employs the same unscientific approach to the issue of creationism. Its applicants for membership must subscribe to the belief that the book of Genesis is "historically and scientifically true in all of the original autographs."[28] The court would never criticize or discredit any person's testimony based on his or her religious beliefs. While anybody is free to approach a scientific inquiry in any fashion they choose, they cannot properly describe the methodology as scientific, if they start with the conclusion and refuse to change it regardless of the evidence developed during the course of the investigation.

(D)

In efforts to establish "evidence" in support of creation science, the defendants relied upon the same false premise as the two-model approach contained in Section 4, that is, all evidence which criticized evolutionary theory was proof in support of creation science. For example, the defendants established that the mathematical probability of a chance chemical combination resulting in life from nonlife is as remote that such an occurrence is almost beyond imagination. Those mathematical facts, the defendants argue, are scientific evidences that life was the product of a creator. While the statistical figures may be impressive evidence against the theory of chance chemical combinations as an explanation of origins, it requires a leap of faith to interpret those figures so as to support a complex doctrine which includes a sudden creation from nothing, a worldwide flood, separate ancestry of man and apes, and a young earth.

The defendants' argument would be more persuasive if, in fact, there were only two theories or ideas about the origins of life and the world. That there are a number of theories was acknowledged by the State's witnesses, Dr. Wickramasinghe and Dr. Geisler. Dr. Wickramasinghe testified at length in support of a theory that life on earth was "seeded" by comets which delivered genetic material and perhaps organisms to the earth's surface from interstellar dust far outside the solar system. The "seeding" theory further hypothesizes that the earth remains under the continuing influence of genetic material from space which continues to affect life. While Wickramasinghe's theory[29] about the origins of life on earth has not received general acceptance within the scientific community, he has, at least, used scientific methodology to produce a theory of origins which meets the essential characteristics of science.

The court is at a loss to understand why Dr. Wickramasinghe was called in behalf of the defendants. Perhaps it was because he was generally critical of the theory of evolution and the scientific community, a tactic consistent with the strategy of the defense. Unfortunately for the defense, he demonstrated that the simplistic approach of the two-model analysis of the origins of life is false. Furthermore, he corroborated the plaintiffs' witnesses by concluding that "no rational scientist" would believe the earth's geology could be explained by reference to a worldwide flood or that the earth was less than one million years old.

The proof in support of creation science consisted almost entirely of efforts to discredit the theory of evolution through a rehash of data and theories which have been before the scientific community for decades. The arguments asserted by the creationists are not based upon new scientific evidence or laboratory data which has been ignored by the scientific community.

Robert Gentry's discovery of radioactive polonium haloes in granite and coalified woods is, perhaps, the most recent scientific work which the creationists use as argument for a "relatively recent inception" of the earth and a "worldwide flood." The existence of polonium haloes in granite and coalified wood is thought to be inconsistent with radiometric dating methods based upon constant radioactive decay rates. Mr. Gentry's findings were published almost ten years ago and have been the subject of some discussion in the scientific community. The discoveries have not, however, led to the formulation of any scientific hypothesis or theory which would explain a relatively recent inception of the earth or a worldwide flood. Gentry's discovery has been treated as a minor mystery which will eventually be explained. It may deserve further investigation, but the National Science Foundation has not deemed it to be of sufficient import to support further funding.

The testimony of Marianne Wilson was persuasive evidence that creation science is not science. Ms. Wilson is in charge of the science curriculum for Pulaski County Special School District, the largest school district in the state of Arkansas. Prior to the passage of Act 590, Larry Fisher, a science teacher in the district, using materials from the ICR, convinced the school board that it should voluntarily adopt creation science as part of its science curriculum.

The district superintendent assigned Ms. Wilson the job of producing a creation science curriculum guide. Ms. Wilson's testimony about the project was particularly convincing because she obviously approached the assignment with an open mind and no preconceived notions about the subject. She had not heard of creation science until about a year ago and did not know its meaning before she began her research.

Ms. Wilson worked with a committee of science teachers appointed from the district. They reviewed practically all of the creationist literature. Ms. Wilson and the committee members reached the unanimous conclusion that creationism is not science; it is religion. They so reported to the board. The board ignored the recommendation and insisted that a curriculum guide be prepared.

In researching the subject, Ms. Wilson sought the assistance of Mr. Fisher, who initiated the board action and asked professors in the science departments of the University of Arkansas at Little Rock and the University of Central Arkansas[30] for reference material and assistance, and attended a workshop conducted at Central Baptist College by Dr. Richard Bliss of the ICR staff. Act 590 became law during the course of her work so she used Section 4(a) as a format for her curriculum guide.

Ms. Wilson found all available creationists' materials unacceptable because they were permeated with religious references and reliance upon religious beliefs.

It is easy to understand why Ms. Wilson and other educators find the creationists' textbook material and teaching guides unacceptable. The materials misstate the theory of evolution in the same fashion as Section 4(b) of the act, with emphasis on the alternative mutually exclusive nature of creationism and evolution. Students are constantly encouraged to compare and make a choice between the two models, and the material is not presented in an accurate manner.

A typical example is *Origins* (Plaintiffs' Exhibit 76) by Richard B. Bliss, director of curriculum development of the ICR. The presentation begins with a chart describing "preconceived ideas about origins" which suggests that some people believe that evolution is atheistic. Concepts of evolution, such as "adaptative radiation," are erroneously presented. At page 11, figure 1.6 of the text, a chart purports to illustrate this "very important" part of the evolution model. The chart conveys the idea that such diverse mammals as a whale, bear, bat, and monkey all evolved from a shrew through the process of adaptive radiation. Such a suggestion is, of course, a totally erroneous and misleading application of the theory. Even more objectionable, especially when viewed in light of the emphasis on asking the student to elect one of the models, is the chart presentation at page 17. That chart purports to illustrate the evolutionists' belief that man evolved from bacteria to fish to reptile to mammals and, thereafter, into man. The illustration indicates, however, that the mammal which evolved was a rat.

Biology, A Search for Order in Complexity[31] is a high school biology text typical of creationists' materials. The following quotations are illustrative:

> Flowers and roots do not have a mind to have purpose of their own: therefore, this planning must have been done for them by the Creator. (12)

> The exquisite beauty of color and shape in flowers exceeds the skill of poet, artist, and king. Jesus said (from Matthew's gospel), "Consider the lilies in the field, how they grow; they toil not, neither do they spin..." (Plaintiffs' Exhibit 129, 363)

The "public school edition" texts written by creationists simply omit biblical references but the content and message remain the same. For example, *Evolution—The Fossils Say No!*[32] contains the following:

> Creation. By creation we mean the bringing into being by a supernatural Creator of the basic kinds of plants and animals by the process of sudden, or fiat, creation.
>
> We do not know how the Creator created, what processes He used, for he used processes which are not now operating anywhere in the natural universe. This is why we refer to creation as Special Creation. We cannot discover by scientific investigation anything about the creative processes used by the Creator. (40)

Gish's book also portrays the large majority of evolutionists as "materialistic atheists or agnostics."

Scientific Creationism (Public School Edition) by Morris is another text reviewed by Ms. Wilson's committee and rejected as unacceptable. The following quotes illustrate the purpose and theme of the text:

> *Forward*
>
> Parents and youth leaders today, and even many scientists and educators, have become concerned about the prevalence and influence of evolutionary philosophy in modern curriculum. Not only is the system inimical to orthodox Christianity and Judaism, but also, as many are convinced, to a healthy society and true science as well. (iii)
>
> . . .
>
> The rationalist of course finds the concept of special creation insufferably naive, even "incredible." Such a judgment, however, is warranted only if one categorically dismisses the existence of an omnipotent God. (17)

Without using creationist literature, Ms. Wilson was unable to locate one genuinely scientific article or work which supported Section 4(a). In order to comply with the mandate of the board she used such materials as an article from *Reader's Digest* about "atomic clocks" which inferentially suggested that the earth was less than 4½ billion years old. She was unable to locate any substantive teaching material for some parts of Section 4 such as the worldwide flood. The curriculum guide which she prepared cannot be taught and has no educational value as science. The defendants did not produce any text or writing in response to this evidence which they claimed was usable in the public school classroom.[33]

The conclusion that creation science has no scientific merit or educational value as science has legal significance in light of the court's previous conclusion that creation science has, as one major effect, the advancement of religion. The second part of the three-pronged test for establishment reaches only those statutes as having their *primary* effect the advancement of religion. Secondary effects which advance religion are not constitutionally fatal. Since creation science is not science, the conclusion is inescapable that the only real effect of Act 590 is the advancement of religion. The act therefore fails both the first and second portions of the test in *Lemon v. Kurtzman*, 403 U.S. 602 (1971).

(E)

Act 590 mandates "balanced treatment" for creation science and evolution science. The act prohibits instruction in any religious doctrine or references to religious writings. The act is self-contradictory and compliance is impossible unless the public schools elect to forgo significant potions of subjects such as biology, world history, geology, zoology, botany, psychology, anthropology, sociology, philosophy, physics, and chemistry. Presently, the concepts of evolutionary theory as described in 4(b) permeate the public textbooks. There is no way teachers can teach the Genesis account of creation in a secular manner.

The State Department of Education, through its textbook selection committee, school boards, and school administrators, will be required to constantly monitor materials to avoid using religious references. The school boards, administrators, and teachers face an impossible task. How is the teacher to respond to questions about a creation suddenly and out of nothing? How will a teacher explain the occurrence of a worldwide flood? How will a teacher explain the concept of a relatively recent age of the earth? The answer is obvious because the only source of this information is ultimately contained in the book of Genesis.

References to the pervasive nature of religious concepts in creation science texts amply demonstrate why state entanglement with religion is inevitable under Act 590. Involvement of the state in screening texts for impermissible religious references will require state officials to make delicate religious judgments. The need to monitor classroom discussion in order to uphold the act's prohibition against religious instruction will necessarily involve administrators in questions concerning religion. These continuing involvements of state officials in questions and issues of religion create an excessive and prohibited entanglement with religion (*Brandon v. Board of Education*).

V

These conclusions are dispositive of the case and there is no need to reach legal conclusions with respect to the remaining issues. The plaintiffs raised two other issues questioning the constitutionality of the act and, insofar as the factual findings relevant to these issues are not covered in the preceding discussion, the court will address these issues. Additionally, the defendants raise two other issues which warrant discussion.

(A)

First, plaintiff teachers argue the act is unconstitutionally vague to the extent that they cannot comply with its mandate of "balanced" treatment without jeopardizing their employment. The argument centers around the lack of a precise definition in the act for the word "balanced." Several witnesses expressed opinions that the word has such meanings as equal time, equal weight, or equal legitimacy. Although the act could have been more explicit, "balanced" is a word subject to ordinary understanding. The proof is not convincing that a teacher using a reasonably acceptable understanding of the word and making a good faith effort to comply with the act will be in jeopardy of termination. Other portions of the act are arguably vague, such as the "relatively recent" inception of the earth and life. The evidence establishes, however, that relatively recent means from 6,000 to 20,000 years, as commonly understood in creation science literature. The meaning of this phrase, like Section 4(a) generally, is, for purposes of the Establishment Clause, all too clear.

(B)

The plaintiffs' other argument revolves around the alleged infringement by the defendants upon the academic freedom of teachers and students. It is contended this unprecedented intrusion in the curriculum by the state prohibits teachers from teaching what they believe should be taught or requires them to teach that which they do not believe is proper. The evidence reflects that traditionally the State Department of Education, local school boards, and administration officials exercise little, if any, influence upon the subject matter taught by classroom teachers. Teachers have been given freedom to teach and emphasize those portions of subjects the individual teacher considered important. The limits to this discretion have generally been derived from the approval of textbooks by the State Department and preparation of curriculum guides by the school districts.

Several witnesses testified that academic freedom for the teacher means, in substance, that the individual teacher should be permitted unlimited discretion subject only to the bounds of professional ethics. The court is not prepared to adopt such a broad view of academic freedom in the public schools.

In any event, if Act 590 is implemented, many teachers will be required to teach materials in support of creation science which they do not consider academically sound. Many teachers will simply forgo teaching subjects which might trigger the "balanced treatment" aspects of Act 590 even though they think the subjects are important to a proper presentation of a course.

Implementation of Act 590 will have serious and untoward consequences for students, particularly those planning to attend college. Evolution is the cornerstone of modern biology, and many courses in public schools contain subject matter relating to such varied topics as the age of the earth, geology, and relationships among living things. Any student who is deprived of instruction as to the prevailing scientific thought on these topics will be denied a significant part of science education. Such a deprivation through the high school level would undoubtedly have an impact upon the quality of education in the state's colleges and universities, especially including the pre-professional and professional programs in the health sciences.

(C)

The defendants argue in their brief that evolution is, in effect, a religion, and that by teaching a religion which is contrary to some students' religious views, the state is infringing upon the student's free exercise rights under the First Amendment. Mr. Ellwanger's legislative findings, which were adopted as a finding of fact by the Arkansas legislature in Act 590, provides:

> Evolution-science is contrary to the religious convictions or moral values or philosophical beliefs of many students and parents, including individuals of many different religious faiths and with diverse moral and philosophical beliefs. [Act 590, §7(d)]

The defendants argue that the teaching of evolution alone presents both a free exercise problem and an establishment problem which can only be redressed by giving balanced treatment to creation science, which is admittedly consistent with some religious beliefs. This argument appears to have its genesis in a student note written by Mr. Wendell Bird, "Freedom of Religion and Science Instruction in Public Schools," *Yale Law Journal* 87 (1978): 515. The argument has no legal merit.

If creation science is, in fact, science and not religion, as the defendants claim, it is difficult to see how the teaching of such a science could "neutralize" the religious nature of evolution.

Assuming for the purposes of argument, however, that evolution is a religion or religious tenet, the remedy is to stop the teaching of evolution, not establish another religion in opposition to it. Yet it is clearly established in the case law, and perhaps also in common sense, that evolution is not a religion and that teaching evolution does not violate the Establishment Clause (*Epperson v. Arkansas*, *Willoughby v. Stever*, *Wright v. Houston Independant School District*).

(D)

The defendants presented Dr. Larry Parker, a specialist in devising curricula for public schools. He testified that the public school's curriculum should reflect the subjects the public wants in schools. The witness said that polls indicated a significant majority of the American public thought creation science should be taught if evolution was taught. The point of this testimony was never placed in a legal context. No doubt a sizable majority of Americans believe in the concept of a Creator or, at least, are not opposed to the concept and see nothing wrong with teaching schoolchildren the idea.

The application and content of First Amendment principles are not determined by public opinion polls or by a majority vote. Whether the proponents of Act 590 constitute the majority or the minority is quite irrelevant under a constitutional system of government. No group, no matter how large or small, may use the organs of government, of which the public schools are the most conspicuous and influential, to foist its religious beliefs on others.

The court closes this opinion with a thought expressed eloquently by the great Justice Frankfurter:

> We renew our conviction that "we have at stake the very existence of our country on the faith that complete separation between the state and religion

> is best for the state and best for religion" [*Everson v. Board of Education*, 330 U.S. 59]. If nowhere else, in the relation between Church and State, "good fences make good neighbors." [*McCollum v. Board of Education*, 333 U.S. 203, 232 (1948)]

An injunction will be entered permanently prohibiting enforcement of Act 590.

It is ordered this January 5, 1982.

NOTES

1. The complaint is based on 42 U.S.C. §1983, which provides a remedy against any person who, acting under color of state law, deprives another of any right, privilege, or immunity guaranteed by the United States Constitution or federal law. This court's jurisdiction arises under 28 U.S.C. §1331, 1343 (3), and 1343 (4). The power to issue declaratory judgments is expressed in 28 U.S.C. §2201 and 2202.

2. The facts necessary to establish the plaintiff's standing to sue are contained in the joint stipulation of facts, which is hereby adopted and incorporated herein by reference. There is no doubt that the case is ripe for adjudication.

3. The state of Arkansas was dismissed as a defendant because of its immunity from suit under the Eleventh Amendment. *Hans v. Louisiana*, 134 U.S. 1 (1890).

4. The authorities differ as to generalizations which may be made about fundamentalism. For example, Dr. Geisler testified to the widely held view that there are five beliefs characteristic of all fundamentalist movements, in addition, of course, to the inerrancy of scripture: (1) belief in the virgin birth of Christ, (2) belief in the deity of Christ, (3) belief in the substitutional atonement of Christ, (4) belief in the second coming of Christ, and (5) belief in the physical resurrection of all departed souls. Dr. Marsden, however, testified that this generalization, which has been common in religious scholarship, is now thought to be historical error. There is no doubt, however, that all fundamentalists take the scriptures as inerrent and probably most take them as literally true.

5. Initiated Act 1 of 1929, Ark. Stat. Ann. §80–1627 *et seq.*, which prohibited the teaching of evolution in Arkansas schools, is discussed *infra* at text accompanying note 26.

6. Subsequent references to the testimony will be made by the last name of the witness only. References to documentary exhibits will be by the name of the author and the exhibit number.

7. Applicants for membership in the CRS must subscribe to the following statement of belief: "(1) The Bible is the written Word of God, and because we believe it to be inspired thruout [*sic*], all of its assertions are historically and scientifically true in all of the original autographs. To the student of nature, this means that the account of origins in Genesis is a factual presentation of simple historical truths. (2) All basic types of living things, including man, were made by direct creative acts of God during Creation Week as described in Genesis. Whatever biological changes have occurred since Creation have accomplished only changes within the original created kinds. (3) The great Flood described in Genesis, commonly referred to as the Noachian Deluge, was an historical event, worldwide in its extent and effect. (4) Finally, we are an organization of Christian men of science, who accept Jesus Christ as our Lord and Savior. The account of the special creation of Adam and Eve as one man and one woman, and their subsequent Fall into sin, is the basis for our belief in the necessity of a Savior for all mankind. Therefore, salvation can come only thru [*sic*] accepting Jesus Christ as our Savior" (Plaintiffs' Exhibit 115).

8. Because of the voluminous nature of the documentary exhibits, the parties were directed by pretrial order to submit their proposed exhibits for the court's convenience prior to trial. The numbers assigned to the pretrial submissions do not correspond with those assigned to the same documents at trial and, in some instances, the pretrial submissions are more complete.

9. Plaintiffs' Exhibit 130, Morris, *Introducing Scientific Creationism into the Public Schools* (1975), and Bird, "Resolution for Balanced Presentation of Evolution and Scientific Creationism," *ICR Impact Series*, no. 71, Appendix 14 to Plaintiff's Pretrial Brief.

10. The creationists often show candor in their proselytization. Henry Morris has stated, "Even if a favorable statute or court decision is obtained, it will probably be declared unconstitutional, especially if the legislation or injunction refers to the Bible account of creation." In the same vein he notes, "The only effective way to get creationism taught properly is to have it taught by teachers who are both willing and able to do it. Since most teachers now are neither willing nor able, they must first be both persuaded and instructed themselves." Plaintiffs' Exhibition 130, Morris, *Introducing Scientific Creationism into the Public Schools* (1975)(unpaged).

11. Mr. Bird sought to participate in this litigation by representing a number of individuals who wanted to intervene as defendants. The application for intervention was denied by this court....

12. The model act had been revised to insert "creation science" in lieu of creationism because Ellwanger had the impression people thought that creationism was too religious a term (Ellwanger Deposition, 79).

13. The original model act had been introduced in the South Carolina legislature, but had died without action after the South Carolina attorney general had opined that the act was unconstitutional.

14. Specifically, Senator Holsted testified that he holds to a literal interpretation of the Bible; that the bill was compatible with his religious beliefs; that the bill does favor the position of literalists; that his religious convictions were a factor in his sponsorship of the bill; and that he stated publicly to the *Arkansas Gazette* (although not on the floor of the Senate) contemporaneously with the legislative debate that the bill does presuppose the existence of a divine creator. There is no doubt that Senator Holsted knew he was sponsoring the teaching of a religious doctrine. His view was that the bill did not violate the First Amendment because, as he saw it, it did not favor one denomination over another.

15. This statute is, of course, clearly unconstitutional under the Supreme Court's decision in *Abbington School District v. Schempp*, 374 U.S. 203 (1963).

16. The joint stipulation of facts establishes that the following areas are the only information specifically required by statute to be taught in all Arkansas schools: (1) the effects of alcohol and narcotics on the human body, (2) conservation of national resources, (3) Bird Week, (4) Fire Prevention, and (5) Flag etiquette. Additionally, certain specific courses, such as American history and Arkansas history, must be completed by each student before graduation from high school.

17. Paul Ellwanger stated in his deposition that he did not know why Section 4(a)(2) (insufficiency of mutation and natural selection) was included as an evidence supporting creation science. He indicated that he was not a scientist, "but these are postulates that have been laid down by creation scientists" (Ellwanger Deposition, 136).

18. Although defendants must make some effort to cast the concept of creation in nonreligious terms, this effort surely causes discomfort to some of the act's more theologically sophisticated supporters. The concept of a creator God distinct from the God of love and mercy is closely similar to the Marcion and Gnostic heresies, among the deadliest to threaten the early Christian church. These heresies had much to do with development and adoption of the Apostle's Creed as the official creedal statement of the Roman Catholic Church in the West (testimony of Gilkey).

19. The parallels between Section 4(a) and Genesis are quite specific: (1) "sudden creation from nothing" is taken from Genesis 1:1–10 (testimony of Vawter and Gilkey); (2) destruction of the world by a flood of divine origin is a notion peculiar to Judeo-Christian tradition and is based on chapters 7 and 8 of Genesis (testimony of Vawter); (3) the term "kinds" has no fixed scientific meaning, but appears repeatedly in Genesis (all scientific witnesses); (4) "relatively recent inception" means

an age of the earth from 6,000 to 10,000 years and is based on the genealogy of the Old Testament using the rather astronomical ages assigned to the patriarchs (testimony of Gilkey and several of the defendants' scientific witnesses); (5) separate ancestry of man and ape focuses on the portion of the theory of evolution which fundamentalists find most offensive [*Epperson v. Arkansas*, 393 U.S. 97 (1968)].

20. "[C]oncepts concerning... a supreme being of some sort are manifestly religious.... These concepts do not shed that religiosity merely because they are presented as philosophy or as a science..." (*Malnak v. Yogi*, 1322).

21. See, e.g., Plaintiffs' Exhibit 76, Morris et al., *Scientific Creationism* (1980), p. 203. ("If creation really is a fact, this means there is a *Creator*, and the universe is His creation.") Numerous other examples of such admissions can be found in the many exhibits which represent creationist literature, but no useful purpose would be served here by a potentially endless listing.

22. Morris, the Director of ICR and one who first advocated the two-model approach, insists that a true Christian cannot compromise with the theory of evolution and that the Genesis version of creation and the theory of evolution are mutually exclusive. Plaintiffs' Exhibition 31, Morris, *Studies in the Bible & Science*, pp. 102–103. The two-model approach was the subject of Dr. Richard Bliss's doctoral dissertation (Defense Exhibit 35). It is presented in Bliss, *Origins: Two Models—Evolution, Creation* (1978). Moreover, the two-model approach merely casts in educationalist language the dualism which appears in all creationist literature—creation (i.e., God) and evolution are presented as two alternative and mutually exclusive theories. [See, e.g., Plaintiffs' Exhibition 75, Morris, *Scientific Creationism* (1974) (public school edition); Plaintiffs' Exhibition 59, Fox, *Fossils: Hard Facts from the Earth.*] Particularly illustrative is Plaintiffs' Exhibition 61, Boardman et al., *Worlds without End* (1975), a CSRC publication: One group of scientists, known as creationists, believe that God, in a miraculous manner, created all matter and energy...

"Scientists who insist that the universe just grew, by accident, from a mass of hot gases without the direction or help of a Creator are known as evolutionists."

23. The idea that belief in a creator and acceptance of the scientific theory of evolution are mutually exclusive is a false premise and offensive to the religious views of many (Hicks). Dr. Francisco Ayala, a geneticist of considerable reknown and a former Catholic priest who has the equivalent of a PhD in theology, pointed out that many working scientists who subscribe to the theory of evolution are devoutly religious.

24. This is so despite the fact that some of the defense witnesses do not subscribe to the young earth or flood hypotheses. Dr. Geisler stated his belief that the earth is several billion years old. Dr. Wickramasinghe stated that no rational scientist would

believe the earth is less than one million years old or that all the world's geology could be explained by a worldwide flood.

25. "We do not know how the Creator created, what processes He used, for he used processes which are not now operating anywhere in the natural universe. This is why we refer to creation as Special Creation. We cannot discover by scientific investigation anything about the creative processes used by God" [Plaintiffs' Exhibition 78, Gish, *Evolution—The Fossils Say No!*, 3rd ed. (1979), p. 42 (emphasis in original)].

26. The evolutionary notion that man and some modern apes have a common ancestor somewhere in the distant past has consistently been distorted by antievolutionists to say that man descended from modern monkeys. As such, this idea has long been more offensive to fundamentalists. See *Epperson v. Arkansas*, 393 U.S. 97 (1968).

27. Not only was this point acknowledged by virtually all the defense witnesses, it is patent in the creationist literature. See, e.g., Plaintiffs' Exhibit 89, Kofahl & Segraves, *The Creation Explanation*, p. 40: "The Flood of Noah brought about vast changes in the earth's surface, including vulcanism, mountain building, and the deposition of the major part of sedimentary strata. This principle is called 'Biblical catastrophism.'"

28. See note 7 for the full test of the CRS creed.

29. The theory is detailed in Wickramasinghe's book with Sir Fred Hoyle, *Evolution from Space* (1981), which is Defense Exhibit 79.

30. Ms. Wilson stated that some professors she spoke with sympathized with her plight and tried to help her find scientific materials to support Section 4(a). Others simply asked her to leave.

31. Plaintiffs' Exhibition 129, published by Zonderman Publishing House (1974), states that it was "prepared by the Textbook Committee of the Creation Research Society." It has a disclaimer pasted inside the front cover stating that it is not suitable for use in public schools.

32. Plaintiffs' Exhibit 77, by Duane Gish.

33. The passage of Act 590 apparently caught a number of its supporters off guard as much as it did the school district. The act's author, Paul Ellwanger, stated in a letter to "Dick" (apparently Dr. Richard Bliss at ICR): "And finally, if you know of any textbooks at any level and for any subjects that you think are acceptable to you and are also constitutionally admissible, these are things that would be of *enormous* to these bewildered folks who may be cause, as Arkansas now has been, by the sudden need to implement a whole new ball game with which they are quite unfamiliar" [*sic*] (unnumbered attachment to Ellwanger Deposition).

CASES

Abbingion School District v. Schempp, 374 U.S. 203 (1963).

Brandon v. Board of Education, 487 F. Supp. 1219 (N.D.N.Y.), *affirmed*, 635 F.2d 971 (2nd Circuit 1980).

Chrysler Corporation v. Brown, 441 U.S. 281(1979).

Committee for Public Education & Religious Liberty v. Nyquist, 413 U.S. 756 (1973).

Engel v. Vitale, 370 U.S. 421 (1962).

Epperson v. Arkansas, 393 U.S. 97 (1968).

Everson v. Board of Education, 330 U.S. 1 (1947).

Federal Energy Administration v. Algonquin SNG, Inc., 426 U.S. 548 (1976).

Hans v. Louisiana, 134 U.S. 1 (1890).

Lemon v. Kurtzman, 403 U.S. 602 (1971).

Malnak v. Yogi, 440 F. Supp. 1284 (D.N.J. 1977); *affirmed per curiuam*, 592 F.2d 197 (3rd Circuit 1979).

McCollum v. Board of Education, 333 U.S. 203 (1948).

McGowan v. Maryland, 366 U.S. 420 (1961).

Stone v. Graham, 449 U.S. 39 (1980).

United States v. Emmons, 410 U.S. 396 (1979).

Village of Arlington Heights v. Metropolitan Housing Corp., 429 U.S. 252 (1977).

Willoughby v. Stever, no. 15574–75 (Denver District Court, May 18, 1973), *affirmed* 504 F.2d 271 (D.C. Circuit 1974), *certiorari denied*, 420 U.S. 927 (1975).

Wright v. Houston Independent School District, 366 F. Supp. 1208 (Southern District of Texas, 1978), *affirmed* 486 F.2d 137 (5th Circuit 1973), *certiorari denied* 417 U.S. 969 (1974).

CHAPTER 15

THE DEMISE OF THE DEMARCATION PROBLEM

LARRY LAUDAN

1. INTRODUCTION

We live in a society that sets great store by science. Scientific "experts" play a privileged role in many of our institutions, ranging from the courts of law to the corridors of power. At a more fundamental level, most of us strive to shape our beliefs about the natural world in the "scientific" image. If scientists say that continents move or that the universe is billions of years old, we generally believe them, however counterintuitive and implausible their claims might appear to be. Equally, we tend to acquiesce in what scientists tell us not to believe. If, for instance, scientists say that Velikovsky was a crank, that the biblical creation story is hokum, that UFOs do not exist, or that acupuncture is ineffective, then we generally make the scientist's contempt for these things our own, reserving for them those social sanctions and disapprobations that are the just deserts of quacks, charlatans, and conmen. In sum, much of our intellectual life, and increasingly large portions of our

From Larry Laudan, "The Demise of the Demarcation Problem," in R. S. Cohen and L. Laudan, eds., *Physics, Philosophy, and Psychoanalysis* (Dordrecht: Reidel, 1983), pp. 111–27.

social and political life, rests on the assumption that we (or, if not we ourselves, then someone whom we trust in these matters) can tell the difference between science and its counterfeit.

For a variety of historical and logical reasons, some going back more than two millennia, that "someone" to whom we turn to find out the difference usually happens to be the philosopher. Indeed, it would not be going too far to say that, for a very long time, philosophers have been regarded as the gatekeepers to the scientific estate. They are the ones who are supposed to be able to tell the difference between real science and pseudoscience. In the familiar academic scheme of things, it is specifically the theorists of knowledge and the philosophers of science who are charged with arbitrating and legitimating the claims of any sect to "scientific" status. It is small wonder, under the circumstances, that the question of the nature of science has loomed so large in Western philosophy. From Plato to Popper, philosophers have sought to identify those epistemic features which mark off science from other sorts of belief and activity.

Nonetheless, it seems pretty clear that philosophy has largely failed to deliver the relevant goods. Whatever the specific strengths and deficiencies of the numerous well-known efforts at demarcation (several of which will be discussed below), it is probably fair to say that there is no demarcation line between science and nonscience, or between science and pseudoscience, which would win assent from a majority of philosophers. Nor is there one which *should* win acceptance from philosophers or anyone else; but more of that below.

What lessons are we to draw from the recurrent failure of philosophy to detect the epistemic traits that mark science off from other systems of belief? That failure might conceivably be due simply to our impoverished philosophical imagination; it is conceivable, after all, that science really is *sui generis*, and that we philosophers have just not yet hit on its characteristic features. Alternatively, it may just be that there are no epistemic features which all and only the disciplines we accept as "scientific" share in common. My aim in this [chapter] is to make a brief excursion into the history of the science/nonscience demarcation in order to see what light it might shed on the contemporary viability of the quest for a demarcation device.

2. THE OLD DEMARCATIONIST TRADITION

As far back as the time of Parmenides, Western philosophers thought it important to distinguish knowledge (*episteme*) from mere opinion (*doxa*), reality from appearance, truth from error. By the time of Aristotle, these epistemic concerns came to be focused on the question of the nature of scientific knowledge. In his highly influential *Posterior Analytics*, Aristotle described at length what was involved in having scientific knowledge of something. To be scientific, he said, one must deal with causes, one must use logical demonstrations, and one must identify the universals that "inhere" in the particulars of sense. But above all, to have science one must have *apodictic certainty*. It is this last feature that, for Aristotle, most clearly distinguished the scientific way of knowing. What separates the sciences from other kinds of beliefs is the infallibility of their foundations and, thanks to that infallibility, the incorrigibility of their constituent theories. The first principles of nature are directly intuited from sense; everything else worthy of the name of science follows demonstrably from these first principles. What characterizes the whole enterprise is a degree of certainty that distinguishes it most crucially from mere opinion.

But Aristotle sometimes offered a second demarcation criterion, orthogonal to this one between science and opinion. Specifically, he distinguished between know-how (the sort of knowledge which the craftsman and the engineer possess) and what we might call "know-why" or demonstrative understanding (which the scientist alone possesses). A shipbuilder, for instance, knows how to form pieces of wood together so as to make a seaworthy vessel; but he does not have, and has no need for, a syllogistic, causal demonstration based on the primary principles or first causes of things. Thus, he needs to know that wood, when properly sealed, floats; but he need not be able to show by virtue of what principles and causes wood has this property of buoyancy. By contrast, the scientist is concerned with what Aristotle calls the "reasoned fact"; until he can show why a thing behaves as its does by tracing its causes back to first principles, he has no scientific knowledge of the thing.

Coming out of Aristotle's work, then, is a pair of demarcation criteria. Science is distinguished from opinion and superstition by the certainty of its principles; it is marked off from the crafts by its comprehension of first causes. This set of contrasts comes to dominate discussions of the nature of science throughout the later Middle Ages and the Renaissance, and thus to provide a crucial backdrop to the reexamination of these issues in the seventeenth century.

It is instructive to see how this approach worked in practice. One of the most revealing examples is provided by premodern astronomy. By the time of Ptolemy, mathematical astronomers had largely abandoned the (Aristotelian) tradition of seeking to derive an account of planetary motion from the causes or essences of the planetary material. As Duhem and others have shown in great detail,[1] many astronomers sought simply to correlate planetary motions, independently of any causal assumptions about the essence or first principles of the heavens. Straightaway, this turned them from scientists into craftsmen.[2] To make matters worse, astronomers used a technique of post hoc testing of their theories. Rather than deriving their models from directly intuited first principles, they offered hypothetical constructions of planetary motions and positions and then compared the predictions drawn from their models with the observed positions of the heavenly bodies. This mode of theory testing is, of course, highly fallible and nondemonstrative; and it was known at the time to be so. The central point for our purposes is that, by abandoning a demonstrative method based on necessary first principles, the astronomers were indulging in mere opinion rather than knowledge, putting themselves well beyond the scientific pale. Through virtually the whole of the Middle Ages, and indeed up until the beginning of the seventeenth century, the predominant view of mathematical astronomy was that, for the reasons indicated, it did not qualify as a "science." (It is worth noting in passing that much of the furor caused by the astronomical work of Copernicus and Kepler was a result of the fact that they were claiming to make astronomy "scientific" again.)

More generally, the seventeenth century brought a very deep shift in demarcationist sensibilities. To make a long and fascinating story unconscionably brief, we can say that most seventeenth-century thinkers accepted Aristotle's first demarcation criterion (viz., between infallible science and fallible opinion), but rejected his second (between know-how and understanding). For instance, if we look to the work of Galileo, Huygens, or Newton, we see a refusal to prefer know-why to know-how; indeed, all three were prepared to regard as entirely scientific, systems of belief which laid no claim to an understanding grounded in primary causes or essences. Thus Galileo claimed to know little or nothing about the underlying causes responsible for the free fall of bodies, and in his own science of kinematics he steadfastly refused to speculate about such matters. But Galileo believed that he could still sustain his claim to be developing a "science of motion" because the results he reached were, so he claimed, infallible and demonstrative. Similarly, Newton in *Principia* was not indifferent to causal explanation, and freely admitted that he would like to know the causes of gravitational phenomena; but he was emphatic that, even without a knowledge of the causes of gravity, one can engage in a sophisticated and *scientific* account of the gravitational behavior of the heavenly bodies. As with Galileo, Newton regarded his noncausal account as "scientifical" because of the (avowed) certainty of its conclusions. As Newton told his readers over and again, he did not engage in hypotheses and speculations: he purported to be deriving his theories directly from the phenomena. Here again, the infallibility of results, rather than their derivability from first causes, comes to be the single touchstone of scientific status.

Despite the divergence of approach among thinkers of the seventeenth and eighteenth centuries, there is widespread agreement that scientific knowledge is apodictically certain. And this consensus cuts across most of the usual epistemological divides of the period. For instance, Bacon, Locke, Leibniz, Descartes, Newton, and Kant are in accord about this way of characterizing science.[3] They may disagree about how precisely to certify the certainty of knowledge, but none quarrels with the claim that science and infallible knowledge are coterminous.

As I have shown elsewhere,[4] this influential account finally and decisively came unraveled in the nineteenth century with the emergence and eventual triumph of a *fallibilistic* perspective in epistemology. Once one accepts, as most thinkers had by the mid-nineteenth century, that science offers no apodictic certainty, that all scientific theories are corrigible and may be subject to serious emendation, then it is no longer viable to attempt to distinguish science from nonscience by assimilating that distinction to the difference between knowledge and opinion. Indeed, the unambiguous implication of fallibilism is that there is no difference between knowledge and opinion: within a fallibilism framework, scientific belief turns out to be just a species of the genus opinion. Several nineteenth-century philosophers of science tried to take some of the sting out of this *volte-face* by suggesting that scientific opinions were more probable or more reliable than nonscientific ones; but even they conceded that it was no longer possible to make infallibility the hallmark of scientific knowledge.

With certainty no longer available as the demarcation tool, nineteenth-century philosophers and scientists quickly forged other tools to do the job. Thinkers as diverse as Comte, Bain, Jevons, Helmholtz, and Mach (to name only a few) began to insist that what really marks science off from everything else is its *methodology*. There was, they maintained, something called "the scientific method"; even if that method was not foolproof (the acceptance of fallibilism demanded that concession), it was at least a better technique for testing empirical claims than any other. And if it did make mistakes, it was sufficiently self-corrective that it would soon discover them and put them right. As one writer remarked a few years later: "if science lead us astray, more science will set us straight."[5] One need hardly add that the nineteenth century did not invent the idea of a logic of scientific inquiry; that dates back at least to Aristotle. But the new insistence in this period is on a fallible method which, for all its fallibility, is nonetheless superior to its nonscientific rivals.

This effort to mark science off from other things required one to show two things. First, that the various activities regarded as science utilized essentially the same repertoire of methods (hence the importance in the period of the so-called thesis of the "unity-of method"); second, the epistemic credentials of this method had to be established. At first blush, this program of identifying science with a certain technique of inquiry is not a silly one; indeed, it still persists in some respectable circles even in our time. But the nineteenth century could not begin to deliver on the two requirements just mentioned because there was no agreement about what the scientific method was. Some took it to be the canons of inductive reasoning sketched out by Herschel and Mill. Others insisted that the basic methodological principle of science was that its theories must be restricted to observable entities (the nineteenth-century requirement of "*vera causa*").[6] Still others, like Whewell and Peirce, rejected the search for *verae causae* altogether and argued that the only decisive methodological test of a theory involved its ability successfully to make surprising predictions.[7] Absent agreement on what "the scientific method" amounted to, demarcationists were scarcely in a position to argue persuasively that what individuated science was its method.

This approach was further embarrassed by a notorious set of ambiguities surrounding several of its key components. Specifically, many of the methodological rules proposed were much too ambiguous for one to tell when they were being followed and when breached. Thus, such common methodological rules as "avoid ad hoc hypotheses," "postulate simple theories," "feign no hypotheses," and "eschew theoretical entities" involved complex conceptions which neither scientists nor philosophers of the period were willing to explicate. To exacerbate matters still further, what most philosophers of science of the period offered up as an account of "the scientific method" bore little resemblance to the methods actually used by working scientists, a point made with devastating clarity by Pierre Duhem in 1908.[8]

As one can see, the situation by the late nineteenth century was more than a little ironic. At precisely that juncture when science was beginning to have a decisive impact on the lives and institutions of Western man, at precisely that time when "scientism" (i.e., the belief that science and science alone has the answers to all our answerable questions) was gaining ground, in exactly that quarter century when scientists were doing battle in earnest with all manner of "pseudoscientists" (e.g., homeopathic physicians, spiritualists, phrenologists, biblical geologists), scientists and philosophers found themselves empty-handed. Except at the rhetorical level, there was no longer any consensus about what separated science from anything else.

Surprisingly (or, if one is cynically inclined, quite expectedly), the absence of a plausible demarcation criterion did not stop fin de siècle scientists and philosophers from haranguing against what they regarded as pseudoscientific nonsense (any more than their present-day counterparts are hampered by a similar lack of consensus); but it did make their protestations less compelling than their confident denunciations of "quackery" might otherwise suggest. It is true, of course, that there was still much talk about "the scientific method"; and doubtless many hoped that the methods of science could play the demarcationist role formerly assigned to certainty. But, leaving aside the fact that agreement was lacking about precisely what the scientific method was, there was no very good reason as yet to prefer any one of the proposed "scientific methods" to any purportedly "nonscientific" ones, since no one had managed to show either that any of the candidate "scientific methods" qualified them as "knowledge" (in the traditional sense of the term) or, even more minimally, that those methods were epistemically superior to their rivals.

3. A METAPHILOSOPHICAL INTERLUDE

Before we move to consider and to assess some familiar demarcationist proposals from our own epoch, we need to engage briefly in certain metaphilosophical preliminaries. Specifically, we should ask three central questions: (1) What conditions of adequacy should a proposed demarcation criterion satisfy? (2) Is the criterion under consideration offering necessary or sufficient conditions, or both, for scientific status? (3) What actions or judgments are implied by the claim that a certain belief or activity is "scientific" or "unscientific"?

(1) Early in the history of thought it was inevitable that characterizations of "science" and "knowledge" would be largely stipulative and a priori. After all, until as late as the seventeenth century, there were few developed examples of empirical sciences that one could point to or whose properties one could study; under such circumstances, where one is working largely ab initio, one can be uncompromisingly legislative about how a term like "science" or "knowledge" will be used. But as the sciences developed and prospered, philosophers began to see the task of formulating a demarcation criterion as no longer a purely stipulative undertaking. Any proposed dividing line between science and nonscience would have to be (at least in part) explicative and thus sensitive to existing patterns of usage. Accordingly, if one were today to offer a definition of "science" that classified (say) the major theories of physics and chemistry as nonscientific, one would thereby have failed to reconstruct some paradigmatic cases of the use of the term. Where Plato or Aristotle need not have worried if some or even most of the intellectual activities of their time failed to satisfy their respective definitions of "science," it is inconceivable that we would find a demarcation criterion satisfactory that relegated to unscientific status a large number of the activities we consider scientific or that admitted as sciences activities which seem to us decidedly unscientific. In other words, the quest for a latter-day demarcation criterion involves an attempt to render explicit those shared but largely implicit sorting mechanisms whereby most of us can agree about paradigmatic cases of the

scientific and the nonscientific. (And it seems to me that there is a large measure of agreement at this paradigmatic level, even allowing for the existence of plenty of controversial problem cases.) A failure to do justice to these implicit sortings would be a grave drawback for any demarcation criterion.

But we expect more than this of a *philosophically* significant demarcation criterion between science and nonscience. Minimally, we expect a demarcation criterion to identify the *epistemic* or *methodological* features that mark off scientific beliefs from unscientific ones. We want to know what, if anything, is special about the knowledge claims and the modes of inquiry of the sciences. Because there are doubtless many respects in which science differs from nonscience (e.g., scientists may make larger salaries, or know more mathematics than nonscientists), we must insist that any philosophically interesting demarcative device must distinguish scientific and nonscientific matters in a way that exhibits a surer epistemic warrant or evidential ground for science than for nonscience, if it should happen that there is no such warrant, then the demarcation between science and nonscience would turn out to be of little or no philosophic significance.

Minimally, then, a philosophical demarcation criterion must be an adequate explication of our ordinary ways of partitioning science from nonscience and it must exhibit epistemically significant differences between science and nonscience. Additionally, as we have noted before, the criterion must have sufficient precision that we can tell whether various activities and beliefs whose status we are investigating do or do not satisfy it; otherwise it is no better than no criterion at all.

(2) What will the formal structure of a demarcation criterion have to look like if it is to accomplish the tasks for which it is designed? Ideally, it would specify a set of individually necessary and jointly sufficient conditions for deciding whether an activity or set of statements is scientific or unscientific. As is well known, it has not proved easy to produce a set of necessary and sufficient conditions for science. Would something less ambitious do the job? It seems unlikely. Suppose, for instance, that someone offers us a characterization that purports to be a necessary (but not sufficient) condition for scientific status. Such a condition, if acceptable, would allow us to identify certain activities as decidedly unscientific, but it would not help "fix our beliefs," because it would not specify which systems actually were scientific. We would

have to say things like: "Well, physics *might* be a science (assuming it fulfills the stated necessary conditions), but then again it might not, since necessary but, not sufficient conditions for the application of a term do not warrant application of the term." If, like Popper, we want to be able to answer the question, "when should a theory be ranked as scientific?"[9] then merely necessary conditions will never permit us to answer it.

For different reasons, merely sufficient conditions are equally inadequate. If we are only told: "satisfy these conditions and you will be scientific," we are left with no machinery for determining that a certain activity or statement is *unscientific.* The fact that (say) astrology failed to satisfy a set of *merely sufficient* conditions for scientific status would leave it in a kind of epistemic twilight zone—possibly scientific, possibly not. Here again, we cannot construct the relevant partitioning. Hence, if (in the spirit of Popper) we "wish to distinguish between science and pseudoscience,"[10] sufficient conditions are inadequate. The importance of these seemingly abstract matters can be brought home by considering some real-life examples. Recent legislation in several American states mandates the teaching of "creation science" alongside evolutionary theory in high school science classes. Opponents of this legislation have argued that evolutionary theory is authentic science, while creation science is not science at all. Such a judgment, and we are apt to make parallel ones all the time, would *not* be warranted by any demarcation criterion which gave only necessary or only sufficient conditions for scientific status. Without conditions which are both necessary and sufficient, we are never in a position to say "*this* is scientific: but *that* is unscientific." A demarcation criterion which fails to provide both sorts of conditions simply will not perform the tasks expected of it.

(3) Closely related to this point is a broader question of the purposes behind the formulation of a demarcation criterion. No one can look at the history of debates between scientists and "pseudoscientists" without realizing that demarcation criteria are typically used as *machines de guerre* in a polemical battle between rival camps. Indeed, many of those most closely associated with the demarcation issue have evidently had hidden (and sometimes not so hidden) agendas of various sorts. It is well known, for instance, that Aristotle was concerned to embarrass the practitioners of Hippocratic medicine; and it is notorious that the logical positivists wanted to repudiate metaphysics and

that Popper was out to "get" Marx and Freud. In every case, they used a demarcation criterion of their own devising as the discrediting device.

Precisely because a demarcation criterion will typically assert the epistemic superiority of science over nonscience, the formulation of such a criterion will result in the sorting of beliefs into such categories as "sound" and "unsound," "respectable" and "cranky," or "reasonable" and "unreasonable." Philosophers should not shirk from the formulation of a demarcation criterion merely because it has these judgmental implications associated with it. Quite the reverse, philosophy at its best should tell us what is reasonable to believe and what is not. But the value-loaded character of the term "science" (and its cognates) in our culture should make us realize that the labeling of a certain activity as "scientific" or "unscientific" has social and political ramifications that go well beyond the taxonomic task of sorting beliefs into two piles. Although the cleaver that makes the cut may be largely epistemic in character, it has consequences that are decidedly nonepistemic. Precisely because a demarcation criterion will serve as a rationale for taking a number of *practical* actions that may well have far-reaching moral, social, and economic consequences, it would be wise to insist that the arguments in favor of any demarcation criterion we intend to take seriously should be especially compelling.

With these preliminaries out of the way, we can turn to an examination of the recent history of demarcation.

4. THE NEW DEMARCATIONIST TRADITION

As we have seen, there was ample reason by 1900 to conclude that neither certainty nor generation according to a privileged set of methodological rules was adequate to denominate science. It should thus come as no surprise that philosophers of the 1920s and 1930s added a couple of new wrinkles to the problem. As is well known, several prominent members of the *Wiener Kreis* took a syntactic or logical approach to the matter. If, the logical positivists apparently reasoned, epistemology and methodology are incapable of distinguishing the scientific from the nonscientific, then perhaps the theory of meaning will do the job. A statement, they suggested, was scientific just in case it had a determinate meaning; and meaningful statements were those that could be exhaustively verified. As Popper once observed, the positivists thought that "verifiability, meaningfulness, and scientific character all coincide."[11]

Despite its many reformulations during the late 1920s and 1930s verificationism enjoyed mixed fortunes as a theory of meaning.[12] But as a would-be demarcation between the scientific and the nonscientific, it was a disaster. Not only are many statements in the sciences not open to exhaustive verification (e.g., all universal laws), but the vast majority of nonscientific and pseudoscientific systems of belief have verifiable constituents. Consider, for instance, the thesis that the Earth is flat. To subscribe to such a belief in the twentieth century would be the height of folly. Yet such a statement is verifiable in the sense that we can specify a class of possible observations which would verify it. Indeed, every belief which has ever been rejected as a part of science because it was "falsified" is (at least partially) verifiable. Because verifiable, it is thus (according to the "mature positivists" criterion) both meaningful and scientific.

A second familiar approach from the same period is Karl Popper's "falsificationism" criterion, which fares no better. Apart from the fact that it leaves ambiguous the scientific status of virtually every singular existential statement, however well supported (e.g., the claim that there are atoms, that there is a planet closer to the sun than the Earth, that there is a missing link), it has the untoward consequence of countenancing as "scientific" every crank claim that makes ascertainably false assertions. Thus flat Earthen, biblical creationists, proponents of laetrile or orgone boxes, Uri Geller devotees, Bermuda Triangulators, circle squarers, Lysenkoists, charioteers of the gods, perpetuum mobile builders, Big Foot searchers, Loch Nessians, faith healers, polywater dabblers, Rosicrucians, the-world-is-about-to-enders, primal screamers, water diviners, magicians, and astrologers all turn out to be scientific on Popper's criterion—just so long as they are prepared to indicate some observation, however improbable, which (if it came to pass) would cause them to change their minds.

One might respond to such criticisms by saying that scientific status is a matter of degree rather than kind. Sciences such as physics and chemistry have a high degree of testability, it might be said, while the systems we regard as pseudoscientific are far less open to empirical scrutiny. Acute technical difficulties confront this suggestion, for the only articulated theory of degrees of testability (Popper's) makes it impossible to compare the degrees of testability of two distinct theories *except when one entails the other*. Since (one hopes!) no "scientific" theory entails any "pseudoscientific" one, the relevant comparisons cannot be made. But even if this problem could be overcome, and if it were possible for us to conclude (say) that the general theory of relativity was more testable (and thus by definition more scientific) than astrology, it would not follow that astrology was any less worthy of belief than relativity—for testability is a semantic rather than an epistemic notion, which entails nothing whatever about belief-worthiness.

It is worth pausing for a moment to ponder the importance of this difference. I said before that the shift from the older to the newer demarcationist orientation could be described as a move from epistemic to syntactic and semantic strategies. In fact, the shift is even more significant than that way of describing the transition suggests. The central concern of the older tradition had been to identify those ideas or theories which were worthy of belief. To judge a statement to be scientific was to make a *retrospective* judgment about how that statement had stood up to empirical scrutiny. With the positivists and Popper, however, this retrospective element drops out altogether. Scientific status, on their analysis, is not a matter of evidential support or belief-worthiness, for all sorts of ill-founded claims are testable and thus scientific on the new view.

The failure of the newer demarcationist tradition to insist on the necessity of retrospective evidential assessments for determining scientific status goes some considerable way to undermining the practical utility of the demarcationist enterprise, precisely because most of the "cranky" beliefs about which one might incline to be dismissive turn out to be "scientific" according to falsificationist or (partial) verificationist criteria. The older demarcationist tradition, concerned with actual epistemic warrant rather than potential epistemic scrutability, would never have countenanced such an undemanding sense of the "scientific." More to the point, the new tradition has had to pay a hefty price for its scaled-down expectations. Unwilling to link scientific status to any evidential warrant, twentieth-century demarcationists have been forced into characterizing the ideologies they oppose (whether Marxism, psychoanalysis, or creationism) as unstable in principle. Very occasionally, that label is appropriate. But more often than not, the views in question can be tested, have been tested, and have failed those tests. But such failures cannot impugn their (new) scientific status: quite the reverse, *by virtue of failing the epistemic tests to which they are subjected, these views guarantee that they satisfy the relevant semantic criteria for scientific status!* The new demarcationism thus reveals itself as a largely toothless wonder, which serves neither to explicate the paradigmatic usages of "scientific" (and its cognates) nor to perform the critical stable-cleaning chores for which it was originally intended.

For these, and a host of other reasons familiar in the philosophical literature, neither verificationism nor falsificationism offers much promise of drawing a useful distinction between the scientific and the nonscientific.

Are there other plausible candidates for explicating the distinction? Several seem to be waiting in the wings. One might suggest, for instance, that scientific claims are well tested, whereas nonscientific ones are not. Alternatively (an approach taken by Thagard),[13] one might maintain that scientific knowledge is unique in exhibiting progress or growth. Some have suggested that scientific theories alone make surprising predictions which turn out to be true. One might even go in the pragmatic direction and maintain that science is the sole repository of useful and reliable knowledge. Or, finally, one might propose that science is the only form of intellectual system-building, which proceeds cumulatively, with later views embracing earlier ones, or at least retaining those earlier views as limiting cases.[14]

It can readily be shown that none of these suggestions can be a necessary and sufficient condition for something to count as "science," at least not as that term is customarily used. And in most cases, these are not even plausible as necessary conditions. Let me sketch out some of the reasons why these proposals are so unpromising. Take the requirement of well-testedness. Unfortunately, we have no viable overarching account of the circumstances under which a claim may be regarded as well tested. But even if we did, is it plausible to suggest that all the assertions in science texts (let alone science journals) have been well tested and that none of the assertions in such conventionally nonscientific fields as literary theory, carpentry, or football strategy are well tested? When a scientist presents a conjecture that has not yet been tested and is such that we are not yet sure what would count as a robust test of it, has that scientist ceased doing science when he discusses his conjecture? On the other side of the divide, is anyone prepared to say that we have no convincing evidence for such "nonscientific" claims as that "Bacon did not write the plays attributed to Shakespeare," that "a miter joint is stronger than a flush joint," or that "offside kicks are not usually fumbled"? Indeed, are we not entitled to say that all these claims are much better supported by the evidence than many of the "scientific" assumptions of (say) cosmology or psychology?

The reason for this divergence is simple to see. Many, perhaps most, parts of science are highly speculative compared with many nonscientific disciplines. There seems good reason, given from the historical record, to suppose that most scientific theories are false; under the circumstances, how plausible can be the claim that science is the repository of all and only reliable or well-confirmed theories?

Similarly, cognitive progress is not unique to the "sciences." Many disciplines (e.g., literary criticism, military strategy, and perhaps even philosophy) can claim to know more about their respective domains than they did fifty or one hundred years ago. By contrast, we can point to several "sciences" that, during certain periods of their history, exhibited little or no progress.[15] Continuous, or even sporadic, cognitive growth seems neither a necessary nor a sufficient condition for the activities we regard as scientific. Finally, consider the requirement of cumulative theory transitions as a demarcation criterion. As several authors[16] have shown, this will not do even as a necessary condition for marking off scientific knowledge, since many scientific theories—even those in the so-called mature science—do not contain their predecessors, not even as limiting cases.

I will not pretend to be able to prove that there is no conceivable philosophical reconstruction of our intuitive distinction between the scientific and the nonscientific. I do believe, though, that we are warranted in saying that none of the criteria which have been offered thus far promises to explicate the distinction.

But we can go further than this, for we have learned enough about what passes for science in our culture to be able to say quite confidently that it is not all cut from the same epistemic cloth. Some scientific theories are well tested; some are not. Some branches of science are presently showing high rates of growth; others are not. Some scientific theories have made a host of successful predictions of surprising phenomena some have made few if any such predictions. Some scientific hypotheses are ad hoc; others are not. Some have achieved a "consilience of inductions"; others have not. (Similar remarks could be made about several nonscientific theories and disciplines.) *The evident epistemic heterogeneity of the activities and beliefs customarily regarded as scientific should alert us to the probable futility of seeking an epistemic version of a demarcation criterion.* Where, even after detailed analysis, there appear to be no epistemic invariants, one is well advised not to take their existence for granted. But to say as much is in effect to say that the problem of demarcation—the very problem which Popper labeled "the central problem of epistemology"—is spurious, for that problem presupposes the existence of just such invariants.

In asserting that the problem of demarcation between science and nonscience is a pseudoproblem (at least as far as philosophy is concerned), I am manifestly not denying that there are crucial epistemic and methodological questions to be raised about knowledge claims, whether we classify them as scientific or not. Nor, to belabor the obvious, am I saying that we are never entitled to argue that a certain piece of science is epistemically warranted and that a certain piece of pseudoscience is not. It remains as important as it ever was to ask questions like: When is a claim well confirmed? When can we regard a theory as well tested? What characterizes cognitive progress? But once we have answers to such questions (and we are still a long way from that happy state!), there will be little left to inquire into which is epistemically significant.

One final point needs to be stressed. In arguing that it remains important to retain a distinction between reliable and unreliable knowledge, I am not trying to resurrect the science/nonscience demarcation under a new guise.[17] However we eventually settle the question of reliable knowledge, the class of statements falling under that rubric will include much that is not commonly regarded as "scientific" and it will exclude much that is generally considered "scientific." This, too, follows from the epistemic heterogeneity of the sciences.

5. CONCLUSION

Through certain vagaries of history, some of which I have alluded to here, we have managed to conflate two quite distinct questions: What makes a belief well founded (or heuristically fertile)? And what makes a belief scientific? The first set of questions is philosophically interesting and possibly even tractable;

the second question is both uninteresting and, judging by its checkered past, intractable. If we would stand up and be counted on the side of reason, we ought to drop terms like "pseudoscience" and "unscientific" from our vocabulary; they are just hollow phrases that do only emotive work for us. As such, they are more suited to the rhetoric of politicians and Scottish sociologists of knowledge than to that of empirical researchers.[18] Insofar as our concern is to protect ourselves and our fellows from the cardinal sin of believing what we wish were so rather than what there is substantial evidence for (and surely that is what most forms of "quackery" come down to), then our focus should be squarely on the empirical and conceptual credentials for claims about the world. The "scientific" status of those claims is altogether irrelevant.

NOTES

*I am grateful to NSF and NEH for support of this research. I have profited enormously from the comments of Adolf Grünbaum, Ken Alpern, and Andrew Lugg on an earlier version of this paper.

1. See especially his *To Save the Phenomena* (Chicago: University of Chicago Press, 1969).

2. This shifting in orientation is often credited to the emerging emphasis on the continuity of the crafts and the sciences and to Baconian-like efforts to make science "useful." But such an analysis surely confuses agnosticism about first causes—which is what really lay behind the instrumentalism of medieval and Renaissance astronomy—with a utilitarian desire to be practical.

3. For much of the supporting evidence for this claim, see the early chapters of Laudan, *Science and Hypothesis* (Dordrecht: D. Reidel, 1981).

4. See especially chapter 8 of *Science and Hypothesis.*

5. E. V. Davis, writing in 1914.

6. See the discussions of this concept by Kavaloski, Hodge, and R. Laudan.

7. For an account of the history of the concept of surprising predictions, see Laudan, *Science and Hypothesis*, chapters 8 and 10.

8. See Duhem's classic *Aim and Structure of Physical Theory* (New York: Atheneum, 1962).

9. Karl Popper, *Conjectures and Refutations* (London: Routledge and Kegan Paul, 1963), p. 33.

10. Ibid.

11. Ibid., p. 40.

12. For a very brief historical account, see C. G. Hempel's classic, "Problems and Changes in the Empiricist Criterion of Meaning," *Revue Inrernavionale tie Philosophie* 11 (1950): 41–63.

13. See, for instance, Paul Thagard, "Resemblance, Correlation and Pseudo-Science," in M. Hanen et al., *Science, Pseudo-Science and Society* (Waterloo, ON: W. Laurier University Press, 1980), pp. 17–28.

14. For proponents of this cumulative view, see Popper, *Conjectures and Refutations*; Hilary Putnam, *Meaning and the Moral Sciences* (London: Routledge and Kegan Paul, 1978); Wladyslaw Krajewski, *Correspondence Principle and Growth of Science* (Dordrecht, Boston: D. Reidel, 1977); Heinz Post, "Correspondence, Invariance and Heuristics," *Studies in History and Philosophy of Science* 2 (1971): 213–55; and L. Szumilewicz, "Incommensurability and the Rationality of Science," *British Journal for the Philosophy of Science* 28 (1977): 348ff.

15. Likely tentative candidates: acoustics from 1750 to 1780; human anatomy from 1900 to 1920; kinematic astronomy from 1200 to 1500; rational mechanics from 1910 to 1940.

16. See, among others: T. S. Kuhn, *Structure of Scientific Revolutions* (Chicago: University of Chicago Press, 1962); A. Grünbaum, "Can a Theory Answer More Questions than One of Its Rivals?" *British Journal for the Philosophy of Science* 27 (1976): 1–23; L. Laudan, "Two Dogmas of Methodology," *Philosophy of Science* 43 (1976): 467–72; L. Laudan, "A Confutation of Convergent Realism," *Philosophy of Science* 48 (1981): 19–49.

17. In an excellent study ["Theories of Demarcation between Science and Metaphysics," in *Problems in the Philosophy of Science* (Amsterdam: North-Holland, 1968), p. 4010], William Bartley has similarly argued that the (Popperian) demarcation problem is not a central problem of the philosophy of science. Bartley's chief reason for devaluing the importance of a demarcation criterion is his conviction that it is less important whether a system is empirical or testable than whether a system is "criticizable." Because he thinks many nonempirical systems are nonetheless open to criticism, he argues that the demarcation between science and nonscience is less important than the distinction between the revisable and the nonrevisable. I applaud Bartley's insistence that the empirical/nonempirical (or, what is for a Popperian the same thing, the scientific/nonscientific) distinction is not central; but I am not convinced, as Bartley is, that we should assign pride of place to the revisable/nonrevisable dichotomy. Being willing to change one's mind is a commendable trait, but

it is not clear to me that such revisability addresses the central epistemic question of the well-foundedness of our beliefs.

18. I cannot resist this swipe at the efforts of the so-called Edinburgh school to recast the sociology of knowledge in what they imagine to be the "scientific image." For a typical example of the failure of that group to realize the fuzziness of the notion of the "scientific," see David Bloor's *Knowledge and Social Imagery* (London: Routledge and Kegan Paul, 1976), and my criticism of it, "The Pseudo-Science of Science?" *Philosophy of the Social Sciences* 11 (1981): 173–98.

CHAPTER 16

SCIENCE AT THE BAR—CAUSES FOR CONCERN

LARRY LAUDAN

In the wake of the decision in the Arkansas creationism trial (*Mclean v. Arkansas*), the friends of science are apt to be relishing the outcome. The creationists quite clearly made a botch of their case and there can be little doubt that the Arkansas decision may, at least for a time, blunt legislative pressure to enact similar laws in other states. Once the dust has settled, however, the trial in general and Judge William R. Overton's ruling in particular may come back to haunt us; for, although the verdict itself is probably to be commended, it was reached for all the wrong reasons and by a chain of argument which is hopelessly suspect. Indeed, the ruling rests on a host of misrepresentations of what science is and how it works.

The heart of Judge Overton's Opinion is a formulation of "the essential characteristics of science." These characteristics serve as touchstones for contrasting evolutionary theory with creationism; they lead Judge Overton ultimately to the claim, specious in its own right, that since creationism is not "science," it must be religion. The Opinion offers five essential properties that demarcate scientific knowledge from other things: "(1) It is guided by natural

---

From Larry Laudan, "Science at the Bar—Causes for Concern," *Science, Technology & Human Values* 7, no. 41 (1982): 16–19.

law; (2) it has to be explanatory by reference to natural law; (3) it is testable against the empirical world; (4) its conclusions are tentative, that is, are not necessarily the final word; and (5) it is falsifiable."

These fall naturally into two families: properties (1) and (2) have to do with lawlikeness and explanatory ability; the other three properties have to do with the fallibility and testability of scientific claims. I shall deal with the second set of issues first, because it is there that the most egregious errors of fact and judgment are to be found.

At various key points in the Opinion, creationism is charged with being untestable, dogmatic (and thus nontentative), and unfalsifiable. All three charges are of dubious merit. For instance, to make the interlinked claims that creationism is neither falsifiable nor testable is to assert that creationism makes no empirical assertions whatever. That is surely false. Creationists make a wide range of testable assertions about empirical matters of fact. Thus, as Judge Overton himself grants (apparently without seeing its implications), the creationists say that the earth is of very recent origin (say 6,000 to 20,000 years old); they argue that most of the geological features of the earth's surface are diluvial in character (i.e., products of the postulated worldwide Noachian deluge); they are committed to a large number of factual historical claims with which the Old Testament is replete; they assert the limited variability of species. They are committed to the view that, since animals and man were created at the same time, the human fossil record must be paleontologically coextensive with the record of lower animals. It is fair to say that no one has shown how to reconcile such claims with the available evidence—evidence that speaks persuasively to a long earth history, among other things.

In brief, these claims are testable, they have been tested, and they have failed those tests. Unfortunately, the logic of the Opinion's analysis precludes saying any of the above. By arguing that the tenets of creationism are neither testable nor falsifiable, Judge Overton (like those scientists who similarly charge creationism with being untestable) deprives science of its strongest argument against creationism. Indeed, if any doctrine in the history of science has ever been falsified, it is the set of claims associated with "creation science." Asserting that creationism makes no empirical claims plays directly, if inadvertently, into the hands of the creationists by immunizing their ideology from empirical confrontation. The correct way to combat creationism is to confute the empirical claims it does make, not to pretend that it makes no such claims at all.

It is true, of course, that some tenets of creationism are not testable in isolation (e.g., the claim that man emerged by a direct supernatural act of creation). But that scarcely makes creationism "unscientific." It is now widely acknowledged that many scientific claims are not testable in isolation, but only when embedded in a larger system of statements, some of whose consequences can be submitted to test.

Judge Overton's third worry about creationism centers on the issue of revisability. Over and over again, he finds creationism and its advocates "unscientific" because they have "refuse[d] to change it regardless of the evidence developed during the course of the[ir] investigation." In point of fact, the charge is mistaken. If the claims of modern-day creationists are compared with those of their nineteenth-century counterparts, significant shifts in orientation and assertion are evident. One of the most visible opponents of creationism, Stephen Gould, concedes that creationists have modified their views about the amount of variability allowed at the level of species change. Creationists do, in short, change their minds from time to time. Doubtless they would create these shifts to their efforts to adjust their views to newly emerging evidence, in what they imagine to be a scientifically respectable way.

Perhaps what Judge Overton had in mind was the fact that some of creationism's core assumptions (e.g., that there was a Noachian flood, that man did not evolve from lower animals, or that God created the world) seem closed off from any serious modification. But historical and sociological researches on science strongly suggest that the scientists of any epoch likewise regard some of their beliefs as so fundamental as not to be open to repudiation or negotiation. Would Newton, for instance, have been tentative about the claim that there were forces in the world? Are quantum mechanicians willing to contemplate giving up the uncertainty relation? Are physicists willing to specify circumstances under which they would give up energy conservation? Numerous historians and philosophers of science (e.g., Kuhn, Mitroff, Feyerabend, Lakatos) have documented the existence of a certain degree of dogmatism about core commitments in scientific research and have argued that such dogmatism plays a constructive role in promoting the aims of science. I am not denying that there may be subtle but important differences between the dogmatism of scientists and that exhibited by many creationists; but one does not even begin to get at those differences by pretending that science is characterized by an uncompromising open-mindedness.

Even worse, the ad hominem charge of dogmatism against creationism egregiously confuses doctrines with the proponents of those doctrines. Since no law mandates that creationists should be invited into the classroom, it is quite irrelevant whether they themselves are close-minded. The Arkansas statute proposed that creationism be taught, not that creationists should teach it. What counts is the epistemic status of creationism, not the cognitive idiosyncrasies of the creationists. Because many of the theses of creationism are testable, the mind-set of creationists has no bearing in law or in the fact on the merits of creationism.

What about the other pair of essential characteristics which the *McLean* Opinion cites, namely, that science is a matter of natural law and explainable by natural law? I find the formulation in the Opinion to be rather fuzzy; but the general idea appears to be that it is inappropriate and unscientific to postulate the existence of any process or fact which cannot be explained in terms of some known scientific laws—for instance, the creationists' assertion that there are outer limits to the change of species "cannot be explained by natural law." Earlier in the Opinion, Judge Overton also writes "there is no scientific explanation for these limits which is guided by natural law," and thus concludes that such limits are unscientific. Still later, remarking on the hypothesis of the Noachian flood, he says, "A worldwide flood as an explanation of the world's geology is not the product of natural law, nor can its occurrence be explained by natural law." Quite how Judge Overton knows that a worldwide flood "cannot" be explained by the laws of science is left opaque; and even if we did not know how to reduce a universal flood to the familiar laws of physics, this requirement is an altogether inappropriate standard for ascertaining whether a claim is scientific. For centuries scientists have recognized a difference between establishing the existence of a phenomenon and explaining that phenomenon in a lawlike way. Our ultimate goal, no doubt, is to do both. But to suggest, as the *McLean* Opinion does repeatedly, that an existence claim (e.g., there was a worldwide flood) is unscientific until we have found the laws on which the alleged phenomenon depends is simply outrageous. Galileo and Newton took themselves to have established the existence of gravitational phenomena, long before anyone was able to give a causal or explanatory account of gravitation. Darwin took himself to have established the existence of natural selection almost a half century before geneticists were able to lay out the laws of heredity on which natural selection depended. If we took the *McLean* Opinion criterion seriously, we should have to say that Newton and Darwin were unscientific; and, to take an example from our own time, it would follow that plate tectonics is unscientific because we have not yet identified the laws of physics and chemistry that account for the dynamics of crustal motion.

The real objection to such creationist claims as that of the (relative) invariability of species is not that such invariability has not been explained by scientific laws, but rather that the evidence for invariability is less robust than the evidence for its contrary, variability. But to say as much requires renunciation of the Opinion's order charge—to wit, that creationism is not testable.

I could continue with this tale of woeful fallacies in the Arkansas ruling, but that is hardly necessary. What is worrisome is that the Opinion's line of reasoning—which neatly coincides with the predominant tactic among scientists who have entered the public fray on this issue—leaves many loopholes for the creationists to exploit. As numerous authors have shown, the requirements of testability, revisability, and falsifiability are exceedingly *weak* requirements. Leaving aside the fact that (as I pointed out above) it can be argued that creationism already satisfies these requirements, it would be easy for a creationist to say the following: "I will abandon my views if we find a living specimen of a species intermediate between man and apes." It is, of course, extremely unlikely that such an individual will be discovered. But, in that statement the creationist would satisfy, in one fell swoop, all the formal requirements of testability, falsifiability, and revisability. If we set very weak standards for scientific status—and, let there be no mistake, I believe that all of the Opinion's last three criteria fall in this category—then it will be quite simple for creationism to qualify as "scientific."

Rather than taking on the creationists obliquely and in wholesome fashion by suggesting that what they are doing is "unscientific" *tout court* (which is doubly silly because few authors can even agree on what makes an activity scientific), we should confront their claims directly and in piecemeal fashion by asking what evidence and arguments can be marshaled for and against each of them. The core issue is not whether creationism satisfies some undemanding and highly controversial definitions of what is scientific; the real question is whether the existing evidence provides stronger arguments for evolutionary theory than for creationism. Once that question is settled, we will know what belongs in the classroom and what does not. Debating the scientific status of creationism (especially when "science" is construed in such an unfortunate manner) is a red herring that diverts attention away from the issues that should concern us.

Some defenders of the scientific orthodoxy will probably say that my reservations are just nit-picking ones, and that—at least to a first order of approximation—Judge Overton has correctly identified what is fishy about creationism. The apologists for science, such as the editor of the *Skeptical Inquirer*, have already objected to those who criticize this whitewash of science "on arcane, semantic grounds . . . [drawn] from the most remote reaches of the academic philosophy of science."[1] But let us be clear about what is at stake. In setting out in the *McLean* Opinion to characterize the "essential" nature of science, Judge Overton was explicitly venturing into philosophical terrain. His *obiter dicta* are about as remote from well-founded opinion in the philosophy of science as creationism is from respectable geology. It simply will not do for the defenders of science to invoke philosophy of science when it suits them (e.g., their much-loved principle of falsifiability comes directly from the philosopher Karl Popper) and to dismiss it as "arcane" and "remote" when it does not. However noble the motivation, bad philosophy makes for bad law.

The victory in the Arkansas case was hollow, for it was achieved only at the expense of perpetuating and canonizing a false stereotype of what science is and how it works. If it goes unchallenged by the scientific community, it will raise grave doubts about that community's intellectual integrity. No one familiar with the issues can really believe that anything important was settled through anachronistic efforts to revive a variety of discredited criteria for distinguishing between the scientific and the nonscientific. Fifty years ago, Clarence Darrow asked, a propos the *Scopes* trial, "Isn't it difficult to realize that a trial of this kind is possible in the twentieth century in the United States of America?" We can raise that question anew, with the added irony that, this time, the pro-science forces are defending a philosophy of science which is, in its way, every bit as outmoded as the "science" of the creationists.

CHAPTER 17
PRO JUDICE
MICHAEL RUSE

As always, my friend Larry Laudan writes in an entertaining and provocative manner, but, in his complaint against Judge William Overton's ruling in *McLean v. Arkansas*, Laudan is hopelessly wide of the mark. Laudan's outrage centers on the criteria for the demarcation of science which Judge Overton adopted, and the judge's conclusion that, evaluated by these criteria, creation science fails as science. I shall respond directly to this concern—after making three preliminary remarks.

First, although Judge Overton does not need defense from me or anyone else, as one who participated in the *Arkansas* trial, I must go on record as saying that I was enormously impressed by his handling of the case. His written judgment is a first-class piece of reasoning. With cause, many have criticized the state of Arkansas for passing the "Creation Science Act," but we should not ignore that, to the state's credit, Judge Overton was born, raised, and educated in Arkansas.

Second, Judge Overton, like everyone else, was fully aware that proof that something is not science is not the same as proof that it is religion. The issue of what constitutes science arose because the creationists claim that their

---

From Michael Ruse, "Pro Judice," *Science, Technology & Human Values* 7, no. 41 (1982): 19–23.

ideas qualify as genuine science rather than as fundamentalist religion. The attorneys developing the American Civil Liberties Union (ACLU) case believed it important to show that creation science is not genuine science. Of course, this demonstration does raise the question of what creation science really is. The plaintiffs claimed that creation science always was (and still is) religion. The plaintiffs' lawyers went beyond the negative argument (against science) to make the positive case (for religion). They provided considerable evidence for the religious nature of creation science, including such things as the creationists' explicit reliance on the Bible in their various writings. Such arguments seem about as strong as one could wish, and they were duly noted by Judge Overton and used in support of his ruling. It seems a little unfair, in the context, therefore, to accuse him of "specious" argumentation. He did not adopt the naïve dichotomy of "science or religion but nothing else."

Third, whatever the merits of the plaintiffs' case, the kinds of conclusions and strategies apparently favored by Laudan are simply not strong enough for legal purposes. His strategy would require arguing that creation science is weak science and therefore ought not to be taught:

> The core issue is not whether creationism satisfies some undemanding and highly controversial definitions of what is scientific; the real question is whether the existing evidence provides stronger arguments for evolutionary theory than for creationism. Once that question is settled, we will know what belongs in the classroom and what does not.[2]

Unfortunately, the US Constitution does not bar the teaching of weak science. What it bars (through the Establishment Clause of the First Amendment) is the teaching of religion. The plaintiffs' tactic was to show that creation science is less than weak or bad science. It is not science at all.

Turning now to the main issue, I see three questions that must be addressed. Using the five criteria listed by Judge Overton, can one distinguish science from nonscience? Assuming a positive answer to the first question, does creation science fail as genuine science when it is judged by these criteria? And, assuming a positive answer to the second, does the Opinion in *McLean* make this case?

The first question has certainly tied philosophers of science in knots in recent years. Simple criteria that supposedly give a clear answer to every case—for example, Karl Popper's single stipulation of falsifiability[3]—will not do. Nevertheless, although there may be many gray areas, white does seem to be white and black does seem to be black. Less metaphorically, something like psychoanalytic theory may or may not be science, but there do appear to be clear-cut cases of real science and of real nonscience. For instance, an explanation of the fact that my son has blue eyes, given that both parents have blue eyes, done in terms of dominant and recessive genes and with an appeal to Mendel's first law, is scientific. The Catholic doctrine of transubstantiation (i.e., that in the Mass the bread and wine turn into the body and blood of Christ) is not scientific.

Furthermore, the five cited criteria of demarcation do a good job of distinguishing the Mendelian example from the Catholic example. Law and explanation through law come into the first example. They do not enter the second. We can test the first example, rejecting it if necessary. In this case, it is tentative, in that something empirical might change our minds. The case of transubstantiation is different. God may have His own laws, but neither scientist nor priest can tell us about those that turn bread and wine into flesh and blood. There is no explanation through law. No empirical evidence is pertinent to the miracle. Nor would the believer be swayed by any empirical facts. Microscopic examination of the Host is considered irrelevant. In this sense, the doctrine is certainly not tentative.

One pair of examples certainly do not make for a definitive case, but at least they do suggest that Judge Overton's criteria are not quite as irrelevant as Laudan's critique implies. What about the types of objections (to the criteria) that Laudan does or could make? As far as the use of law is concerned, he might complain that scientists themselves have certainly not always been that particular about reference to law. For instance, consider the following claim by Charles Lyell in his *Principles of Geology* (1830): "We are not, however, contending that a real departure from the antecedent course of physical events cannot be traced in the introduction of man."[4] All scholars agree that in this statement Lyell was going beyond law. The coming of man required special divine intervention. Yet, surely the *Principles* as a whole qualify as a contribution to science.

Two replies are open: either one agrees that the case of Lyell shows that science has sometimes mingled law with nonlaw; or one argues that Lyell (and others) mingled science and nonscience (specifically, religion at this point). My inclination is to argue the latter. Insofar as Lyell acted as scientist, he appealed only to law. A century and a half ago, people were not as conscientious as today about separating science and religion. However, even if one argues the former alternative—that some science has allowed place for non-lawbound events—this hardly makes Laudan's case. Science, like most human cultural phenomena, has evolved. What was allowable in the early nineteenth century is not necessarily allowable in the late twentieth century. Specifically, science today does not break with law. And this is what counts for us. We want criteria of science for today, not for yesterday. (Before I am accused of making my case by fiat, let me challenge Laudan to find one point within the modern geological theory of plate tectonics where appeal is made to miracles, that is, to breaks with law. Of course, saying that science appeals to law is not asserting that we know all of the laws. But, who said that we did? Not Judge Overton in his Opinion.)

What about the criterion of tentativeness, which involves a willingness to test and reject if necessary? Laudan objects that real science is hardly all that tentative: "[H]istorical and sociological researches on science strongly suggest that the scientists of any epoch likewise regard some of their beliefs as so fundamental as not to be open to repudiation or negotiation."[5]

It cannot be denied that scientists do sometimes—frequently—hang on to their views, even if not everything meshes precisely with the real world. Nevertheless, such tenacity can be exaggerated. Scientists, even Newtonians, have been known to change their minds. Although I would not want to say that the empirical evidence is all-decisive, it plays a major role in such mind changes. As an example, consider a major revolution of our own time, namely, that which occurred in geology. When I was an undergraduate in 1960, students were taught that continents do not move. Ten years later, they were told that they do move. Where is the dogmatism here? Furthermore, it was the new empirical evidence—for example, about the nature of the seabed—which persuaded geologists. In short, although science may not be as open-minded as Karl Popper thinks it is, it is not as close-minded as, say, Thomas Kuhn[6] thinks it is.

Let me move on to the second and third questions, the status of creation science and Judge Overton's treatment of the problem. The slightest acquaintance with the creation science literature and creationism movement shows that creation science fails abysmally as science. Consider the following passage, written by one of the leading creationists, Duane T. Gish, in *Evolution: The Fossils Say No!*:

> CREATION. By creation we mean the bringing into being by a supernatural Creator of the basic kinds of plants and animals by the process of sudden, or fiat, creation.
>
> We do not know how the Creator created, what processes He used, for He used processes which are not operating anywhere in the natural universe. This is why we refer to creation as Special Creation. We cannot discover by scientific investigations anything about the creative processes used by the Creator.[7]

The following similar passage was written by Henry M. Morris, who is considered to be the founder of the creation science movement:

> [I]t is . . . quite impossible to determine anything about Creation through a study of present processes, because present processes are not created in character. If man wishes to know anything about Creation (the time of Creation, the duration of Creation, the order of Creation, the methods of Creation, or anything else) his sole source of true information is that of divine revelation. God was there when it happened. We were not there . . . therefore, we are completely limited to what God has seen fit to tell us, and this information is in His written Word. This is our textbook on the science of Creation![8]

By their own words, therefore, creation scientists admit that they appeal to phenomena not covered or explicable by any laws that humans can grasp as laws. It is not simply that the pertinent laws are not yet known. Creative processes stand outside law as humans know it (or could know it) on Earth—at least there is no way that scientists can know Mendel's laws through observation and experiment. Even if God did use His own laws, they are necessarily veiled from us forever in this life, because Genesis says nothing of them.

Furthermore, there is nothing tentative or empirically checkable about the central claims of creation science. Creationists admit as much when they join the Creation Research Society (the leading organization of the movement). As a condition of membership applicants must sign a document specifying that they now believe and will continue to believe:

> (1) The Bible is the written Word of God, and because we believe it to be inspired throughout, all of its assertions are historically and scientifically true in all of the original autographs. To the student of nature, this means that the account of origins in Genesis is a factual presentation of simple historical truths. (2) All basic types of living things, including man, were made by direct creative acts of God during Creation Week as described in Genesis. Whatever biological changes have occurred since Creation have accomplished only changes within the original created kinds. (3) The great flood described in Genesis, commonly referred to as the Noachian Deluge, was an historical event, worldwide in its extent and effect. (4) Finally, we are an organization of Christian men of science, who accept Jesus Christ as our Lord and Savior. The account of the special creation of Adam and Eve as one man and one woman, and their subsequent fall into sin, is the basis for our belief in the necessity of a Savior for all mankind. Therefore, salvation can come only thru accepting Jesus Christ as our Savior.[9]

It is difficult to imagine evolutionists signing a comparable statement, that they will never deviate from the literal text of Charles Darwin's *On the Origin of Species*. The nonscientific nature of creation science is evident for all to see, as is also its religious nature. Moreover, the quotes I have used above were all used by Judge Overton, in the *McLean* Opinion, to make exactly the points I have just made. Creation science is not genuine science, and Judge Overton showed this.

Finally, what about Laudan's claim that some parts of creation science (e.g., claims about the flood) are falsifiable and that other parts (e.g., about the originally created "kinds") are revisable? Such parts are not falsifiable or revisable in a way indicative of genuine science. Creation science is not like physics, which exists as part of humanity's common cultural heritage and domain. It exists solely in the imaginations and writing of a relatively small group of people. Their publications (and stated intentions) show that, for example, there is no way they will relinquish belief in the Flood, whatever the evidence.[10] In this sense, their doctrines are truly unfalsifiable.

Furthermore, any revisions are not genuine revisions, but exploitations of the gross ambiguities in the creationists' own position. In the matter of origins, for example, some elasticity could be perceived in the creationist position, given the conflicting claims about the possibility of (degenerative) change within the originally created "kinds." Unfortunately, any open-mindedness soon proves illusory, for creationists have no real idea about what God is supposed to have created in the beginning, except that man was a separate species. They rely solely on the book of Genesis:

> And God said, Let the waters bring forth abundantly the moving creature that hath life, and the fowl that may fly above the earth in the open firmament of heaven.
>
> And God created great whales, and every living creature that moveth, which the waters brought forth abundantly, after their kind, and every winged fowl after his kind: and God saw that it was good.
>
> And God blessed them, saying Be fruitful, and multiply, and fill the waters in the seas, and let fowl multiply in the earth.
>
> And the evening and the morning were the fifth day.
>
> And God said, Let the earth bring forth the living creature after his kind, cattle, and creeping thing, and beast of the earth after his kind: and it was so.
>
> And God made the beast of the earth after his kind, and cattle after their kind, and everything that creepeth upon the earth after his kind: and God saw that it was good.[11]

But the definition of "kind," what it really is, leaves creationists as mystified as it does evolutionists. For example, creationist Duane Gish makes this statement on the subject:

> [W]e have defined a basic kind as including all of those variants which have been derived from a single stock.... We cannot always be sure, however, what constitutes a separate kind. The division into kinds is easier the more the divergence observed. It is obvious, for example, that among invertebrates the protozoa, sponges, jellyfish, worms, snails, trilobites, lobsters, and bees are all different kinds. Among the vertebrates, the fishes, amphibians, reptiles, birds, and mammals are obviously different basic kinds.

> Among the reptiles, the turtles, crocodiles, dinosaurs, pterosaurs (flying reptiles), and ichthyosaurs (aquatic reptiles) would be placed in different kinds. Each one of these major groups of reptiles could be further subdivided into the basic kinds within each.
>
> Within the mammalian class, duck-billed platypus, bats, hedgehogs, rats, rabbits, dogs, cats, lemurs, monkeys, apes, and men are easily assignable to different basic kinds. Among the apes, the gibbons, orangutans, chimpanzees, and gorillas would each be included in a different basic kind.[12]

Apparently, a "kind" can be anything from humans (one species) to trilobites (literally thousands of species). The term is flabby to the point of inconsistency. Because humans are mammals, if one claims (as creationists do) that evolution can occur within but not across kinds, then humans could have evolved from common mammalian stock—but because humans themselves are kinds such evolution is impossible.

In brief, there is no true resemblance between the creationists' treatment of their concept of "kind" and the openness expected of scientists. Nothing can be said in favor of creation science or its inventors. Overton's judgment emerges unscathed by Laudan's complaints.

NOTE

1. For the text of Judge Overton's Opinion, see *Science, Technology & Human Values* 40 (Summer 1982): 28–42; and Marcel LaFollette, *Creationism, Science, and the Law* (Cambridge, MA: MIT Press, 1983).

CHAPTER 18
MORE ON CREATIONISM
LARRY LAUDAN

Michael Ruse is distressed that I have taken exception to Judge William Overton's opinion in *McLean v. Arkansas*. Where I saw that ruling as full of sloppy arguments and non sequiturs, he hails it as "a first-class piece of reasoning." Since Ruse has claimed that my reservations are "hopelessly wide of the mark," I feel obliged to enter the fray once again, in the hope that reiteration will achieve what my initial argument has evidently failed to pull off, namely, to convince knee-jerk demarcationists like Ruse that things are more complicated than they have conceded.

In my short commentary, I sought to show: (1) that the criteria that Judge Overton offered as "essential conditions" of science are nothing of the sort, since many parts of what we all call "science" fail to satisfy those conditions; (2) that several of Overton's criteria constitute extremely weak demands from an epistemic point of view, so weak that if creationism does not already satisfy them (which I believe it manifestly does), it would be child's play for creationists to modify their position slightly—thus making their enterprise (by Overton's lights) "scientific"; and (3) that Overton's preoccupation with the

From Larry Laudan, "More on Creationism," *Science, Technology & Human Values* 8, no. 42 (1983): 36–38.

dogmatism and closemindedness of the advocates of creation science has led him into a chronic confusion of doctrines and their advocates.

Ruse makes no reply to the second point. Quite why is unclear since, standing entirely alone, it is more than sufficient to give one pause about the worrying precedents set in *McLean v. Arkansas*. But Ruse does deal, after a fashion, with points (1) and (3). Since we do not see eye-to-eye about these matters, I shall try to redirect his gaze.

THE LOGIC OF "ESSENTIAL CONDITIONS"

Consider the following parable: Suppose that some city dweller said that the "essential conditions" for something to be a sheep were that it be a medium-sized mammal and that it invariably butt into any human beings in its vicinity. A country fellow might try to suggest that his city cousin evidently did not understand what a sheep was. He might show, for instance, that there are plenty of things we call sheep that never butt into anything, let alone human beings. He might go further to say that what the city fellow is calling a sheep is what all the rest of us regard as a goat. Suppose, finally, that a second city fellow, on hearing his town friend abused by the bucolic bumpkin, entered the discussion, saying, "I once knew a sheep that butted into human beings without hesitation, and besides I once saw a goat which never bothered human beings. Accordingly, it is correct to say that the essential conditions of being a sheep are exactly what my friend said they were!"

Confronted by Michael Ruse's efforts to defend Overton's definition of science in the face of my counterexamples, I find myself as dumbfounded as the mythical farm boy. Overton offered five "essential characteristics of science." I have shown that there are respectable examples of science that violate *each* of Overton's desiderata, and moreover that there are many activities we do not regard as science which satisfy many of them. Stepping briskly to Overton's defense, Ruse points to *one* example of a scientific principle (Mendel's first law) that does fit Overton's definition and to one example of a nonscience (the thesis of transubstantiation) that does not. "You see," Ruse seems to conclude, "Laudan is simply wrongheaded and Overton got it basically right."

At the risk of having to tell Ruse that he does not know how to separate the sheep from the goats, I beg to differ. To make his confusion quite explicit, I shall drop the parable and resort to some symbols. Whenever someone lists a set of conditions, C_1, C_2,..., C_n, and characterizes them as "essential characteristics" of a concept (say, science), that is strictly equivalent to saying that each condition, C_1, taken individually is a necessary condition for scientific status. One criticizes a proposed set of necessary or essential conditions by showing that some things that clearly fall under the concept being explicated fail to satisfy the proposed necessary conditions. In my short essay, and elsewhere, I have offered plausible counterexamples to each of Overton's five characteristics.

Ruse mounts no challenge to those counterexamples. Instead, he replies by presenting instances (actually only one, but it would have been no better if he had given a hundred) of science that do satisfy Overton's demands. This is clearly to no avail because I was not saying that *all* scientific claims were (to take but one of Overton's criteria) untestable, only that *some* were. Indeed, so long as there is but one science that fails to exemplify Overton's features, then one would be ill advised to use his demarcation criterion as a device for separating the "scientific" from the "nonscientific." Ruse fails to see the absolute irrelevance to my argument of his rehearsing examples that "fit" Overton's analysis. Similarly, I did not say that *all* nonscientific claims were testable, only that *some* were. So once again, Ruse is dangling a red herring before us when he reminds us that the thesis of transubstantiation is untestable. Finding untestable bits of nonscience to buttress the claim that testability is the hallmark of science is rather like defending the claim that butting humans is essential to being a sheep by pointing out that there are many nonsheep (e.g., tomatoes) that fail to butt humans.

BELIEFS AND BELIEVERS

There is a more interesting—if equally significant—confusion running through much of Ruse's discussion, a confusion revealing a further failure to come to terms with the case I was propounding in "Science at the Bar." I refer to his (and Overton's) continual slide between assessing doctrines and assessing those who hold the doctrines. Ruse reminds us (and this loomed large in the *McLean* opinion as well) that many advocates of creation science tend to be dogmatic, slow to learn from experience, and willing to resort to all manner of ad hoc strategies so as to hold onto their beliefs in the face of counterevidence. For the sake of argument, let all that be granted; let us assume that the creationists exhibit precisely those traits of intellectual dishonesty that the friends of science scrupulously and unerringly avoid. Ruse believes (and Judge Overton appears to concur) that, if we once establish these traits to be true of creationists, then we can conclude that creationism is untestable and unfalsifiable (and "therefore unscientific").

This just will not do. Knowing something about the idiosyncratic mindset of various creationists may have a bearing on certain practical issues (such as "Would you want your daughter to marry one?"). But we learned a long time ago that there is a difference between *ad hominem* and *ad argumentum*. Creationists make assertions about the world. Once made, those assertions take on a life of their own. Because they do, we can assess the merits or demerits of creationist theory without having to speculate about the unsavoriness of the mental habits of creationists. What we do, of course, is to examine the empirical evidence relevant to the creationist claims about earth history. If those claims are discredited by the available evidence (and by "discredited" I mean impugned by the use of rules of reasoning that legal and philosophical experts on the nature of evidence have articulated), then creationism can safely be put on the scrap heap of unjustified theories.

But, intone Ruse and Overton, what if the creationists *still* do not change their minds, even when presented with what most people regard as thoroughly compelling refutations of their theories? Well, that tells us something interesting about the psychology of creationists, but it has no bearing whatever on an assessment of their doctrines. After all, when confronted by comparable problems in other walks of life, we proceed exactly as I am proposing, that is, by distinguishing beliefs from believers. When, for instance, several experiments turn out contrary to the predictions of a certain theory, we do not care whether the scientist who invented the theory is prepared to change his mind. We do not say that his theory cannot be tested, simply because he refuses to accept the results of the test. Similarly, a jury may reach the conclusion, in light of the appropriate rules of evidence, that a defendant who pleaded innocent is, in fact, guilty. Do we say that the defendant's assertion "I am innocent" can be tested only if the defendant himself is prepared to admit his guilt when finally confronted with the coup de grâce?

In just the same way, the soundness of creation science can and must be separated from all questions about the dogmatism of creationists. Once we make that rudimentary separation, we discover both (a) that creation science is testable and falsifiable, and (b) that creation science has been tested and falsified—insofar as any theory can be said to be falsified. But, as I pointed out in the earlier essay, that damning indictment cannot be drawn so long as we confuse creationism and creationists to such an extent that we take the creationists' mental intransigence to entail the immunity of creationist theory from empirical confrontation.

CHAPTER 19

COMMENTARY: PHILOSOPHERS AT THE BAR—SOME REASONS FOR RESTRAINT

BARRY R. GROSS

> ... [P]roblems of all sorts (including empirical ones) arise within a certain context of inquiry, and are partly defined by that context.
>
> —Larry Laudan[1]

> ... [T]oo many discussions of scientific rationality and progress have been both uninformed by, and inapplicable to, the actual course of the evolution of science. The various well-known philosophical models of rationality have been shown to be inapplicable to most of those cases in the history of science where, at least intuitively, we are convinced that sensible rational choices are being made.... [W]e should nonetheless be able to demand of any model of science that it substantially "fit" the actual course of scientific change.
>
> —Larry Laudan[2]

> Winning isn't the most important thing; it's everything.
>
> —attributed to Leo Durocher

From Barry R. Gross, "Commentary: Philosophers at the Bar—Some Reasons for Restraint," *Science, Technology & Human Values* 8, no. 4 (Fall 1983): 30–38.

Having apparently forgotten these admirable precepts, Larry Laudan presents in his jeremiad on *McLean v. Arkansas* a perfect example of a philosopher richly deserving exclusion from "the conversation of mankind." In commenting on an article by Michael Ruse, the philosopher who produced the excellent sentiments above not only missed the context of this inquiry and the essential features of the creationist position but has also shown lack of comprehension of the constitutional issues and standards of proof involved, of the nature of adversary trial, of the weight of legal decision, of the dynamics of preparation for trial undertaken by a large team of attorneys, and of the nature of state and local textbook decisions.

Laudan has confused the outlines of a constitutional conflict with a colloquium in philosophy. He thinks that Judge William R. Overton's powerful and clear opinion has reached a laudable decision, but for the wrong reasons and via a chain of hopelessly suspect reasoning that misrepresents what science is and how it works. The judge's tendentious arguments, according to Laudan, are probably traceable to the testimony of overzealous expert witnesses. Laudan further complains that, in particular, Judge Overton has claimed speciously that because creationism is not science it must be religion (committing, I suppose, in Laudan's view another version of the black/white fallacy committed by creationists when they argue that because evolution is false, creationism must be true).

Moreover, says Laudan, creationism really is empirical because creationists make a range of testable assertions, a fact he holds that the court granted without seeing its implications. Besides, claiming that the tenets of creationism are untestable deprives science of one of its strongest arguments against creationism, because "... if any doctrine in the history of science has ever been falsified ... it's 'creation science.'" Indeed, by immunizing the creationists' ideology from empirical confrontation, Judge Overton played into their hands. For, according to Laudan, "[t]he right way to combat creationism is by confuting the empirical claims it does make." Although

> some tenets of creationism are not testable in isolation (e.g., the claim that man emerged by a direct supernatural act of creation) ... that scarcely makes creationism "unscientific." It is now widely acknowledged that many scientific claims are not testable in isolation, but only when embedded in part of a larger system of statements, some of whose consequences can be submitted to test.[3]

Laudan admits that the core assumptions of creationism—the flood, nonevolution, literal creation—seem closed from serious modification, but so were many of the core beliefs of the sciences of any epoch. And Laudan believes that it is error to claim that creationism is not revisable because comparison with nineteenth-century versions shows many changes. The dangers are (a) that creationists will recognize and credit such shifts to efforts to adjust their views to newly emerging evidence in what they imagine to be a scientifically respectable way, and (b) that Judge Overton's weak criteria leave too many loopholes for creationists to slip through. Of course, these loopholes would require creationists to accept positions that are inimical to their views.

Because few authors can agree on the criteria of scientific activity, Laudan continues, we ought to confute the creationist claims head on and ask what evidence supports them: "The core issue is not whether creationism satisfies some undemanding and highly controversial definition of what is scientific, the real question is whether the existing evidence provides stronger arguments for evolutionary theory than for creationism." Presumably having decided this issue in favor of evolution, we will then know what does or does not belong in the classroom. Anyway, according to Arkansas Act 590 (which is what is at issue), it is not necessary that creationists teach, only that creationism be taught. Thus, Laudan concludes, the immovable mind-set of creationists is of no interest.

What are we to make of all this? Laudan appears to think that the plaintiffs were lucky to have won because the issues and the arguments were all wrong. If Laudan is correct, then there must be a better way to have conducted the case. What can that way have been? He thinks the rival claims should have been tested for truth and the stronger case should have prevailed. But what is this line of approach likely to have gained?

COURTS AND TRUTH

Testing the truth claims of evolution and creationism might be a good strategy outside a courtroom facing a constitutional issue. But Anglo-American law does not place the highest value on finding truth (even in criminal proceedings). Barriers to the introduction of certain kinds of evidence or of evidence obtained in certain ways, restriction upon the introduction of evidence to the two opposing parties, curbs on investigatory powers and conduct, bars against self-incrimination, spousal privilege, and absolute liability represent some among many impediments to arriving at truth in a courtroom. Even the adversary and jury trial systems themselves have been called, correctly, impediments both to truth and to justice. One is not convicted because one did the deed but because a jury has decided that one is guilty, even if the verdict is in the teeth of the evidence.[4]

Indeed, in many areas of legal concern—liability, responsibility, punishment, contracts, and other commercial transactions—there may be no truth independent of legal construction: From whom is the plaintiff done a mischief entitled to collect? (consider the doctrine of deep pockets); when is a malefactor not liable for punishment? (consider statutes of limitations); when is a nonmalefactor liable? (consider strict liability); who owns the land? (consider eminent domain and adverse possession). The law may parallel concerns for truth, but it often diverges from them because its primary aim is to preserve order. And to do so, the law must be framed for arriving at a decision after due time, not at truth after all the evidence is in. "But scientific reasoning and legal thinking are quite distinct," A. G. Guest has written. "The object of scientific inquiry is discovery; the object of a legal inquiry is decision."[5]

In fact, it has been cogently demonstrated that, in administrative and scientific matters, and sometimes even in judicial management, courts are not necessarily even minimally equipped to discover the truth. No court may pursue a question absent a complaint properly brought before it. No court has independent investigatory powers. So the only evidence a court may hear is what the parties choose to put before it. No court has independent experts to assess technical evidence.[6]

These observations are common in the jurisprudence of the last thirty years or so. Our courts have been compared quite unfavorably with those in civil jurisdictions where truth is a sought-after value. It is true that Americans appeal to the courts in numbers several orders of magnitude greater than even the English and certainly than other Europeans or the Japanese, and on a far wider range of issues. Tocqueville noticed this litigiousness and offered structural reasons for it. And courts, in responding to these appeals for judgment, have often rendered silly or incompetent ones.

For these and more reasons, courts are not well suited to scientific or scholarly inquiry designed to reveal truth—but suppose that, in spite of all I have said, they were. Three questions would still arise. First, do courts have the power to set the curricula for public schools? Second, is the falsity of any doctrine a legal or constitutional bar to its being required by state law to be taught? Third, if the truth of creationism were demonstrated, would that make it permissible to require its teaching in the public schools? The answer to all three questions is, plainly, no.

In general, the courts are reluctant to inject themselves into school affairs. Even in a period of heightened sensibilities about rights and equity, for example, they rarely overturn personnel decisions. Even when the court is sympathetic to a plaintiff's case, as in the recent suspension for plagiarism of a Princeton senior's degree, both the lower court and the New Jersey Court of Appeals refused to direct university authorities either to rescind their decision or to award damages.[7]

The US Constitution does not forbid the propagation of false doctrine in the public schools. Nor am I aware of any judicial decisions or interpretations that have inclined in that direction. Indeed, if there were, school boards would find it hard to set curriculum, because much of what is taught is not literally true. In some cases, the curriculum has not caught up with recent scholarship; in many other areas of history, philosophy, literary criticism, sociology, political science, and (as Laudan himself seems to hold) science, literal truth is either unobtainable, nonexistent, or irrelevant.

The Establishment Clause of the First Amendment reads: "The state shall establish no religion. . . ." Therefore, Judge Overton's opinion in *McLean* properly cites the full case law interpreting what is meant by "establish" and rightly concludes that the teaching of religions doctrine as doctrine is prohibited. This finding is not controversial and is admitted by creationists. Neither the truth nor the falsity of either creationism or evolution is constitutionally relevant. If creationism is false, it may still be taught as long as it is not a religion within the meaning of the Establishment Clause. If creationism is true, and if God himself so testified in Federal District Court, then it may nevertheless not be taught in public schools as long as it is religion within the meaning of the Establishment Clause. That is the issue: the religious character of creationism.

LAUDAN'S STRATEGY

The strategy Laudan recommends—either to discover the truth of the matter or where the preponderance of the evidence lies—is poor and would have resulted in the dismissal of the plaintiffs' case. Moreover, even a minimally competent opposing attorney would have studied Laudan's own book, *Progress and Its Problems*, and would have remarked on such passages as:

> *In appraising the merits of theories, it is more important to ask whether they constitute adequate solutions to significant problems than it is to ask whether they are "true," "corroborated," "well-confirmed" or otherwise justifiable within the framework of contemporary epistemology.*[8]
>
> In very rough form, we can say that an empirical problem is solved when, within a particular context of inquiry, scientists properly no longer regard it as an unanswered question, i.e., when they believe they understand why the situation propounded by the problem is the way it is.... [I]t is theories which are meant to provide such understanding.... [A] theory may solve a problem so long as it entails even an *approximate* statement of the problem, in determining if a theory solves a problem, *it is irrelevant* whether the theory is true, false, well or poorly confirmed....[9]
>
> But whatever role questions of truth have in the scientific enterprise..., one need not, and scientists generally do not, consider matters of truth or falsity when determining whether a theory does or does not solve a particular empirical problem....
>
> *Generally, any theory T can be regarded as having solved an empirical problem, if T functions (significantly) in any scheme of inference whose conclusion is a statement of the problem.*[10]

Imagine the delight of opposing counsel upon reading Laudan's pages 27–30, where he writes that truth or corroboration is irrelevant to the assessment of the worth of a scientific theory and that for every theory "a class of non-refuting anomalies exists."[11] Is not the opposition's case made by arguing that each unproven creationist empirical prediction is a member of this class?

Of course, Laudan will (rightly) object that this interpretation misreads his subtle thesis. But we are in court, not in a seminar room. And, in court, subtle theses are rare and rarely win the case. Thus does Laudan's strategy arm his opponents.

LEGAL ARGUMENT AND LEGAL DECISION

Although the function of a philosophy colloquium may be to seek truth (though cognoscenti may be excused for doubt), the function of court is different. For in science and in philosophy, the precision of concepts and arguments is the desired aim, in court it is not.

> The usefulness of logic in the law is inhibited by the fact that the concepts which are formulated from an examination are often so imprecisely expressed that by far the most important task of the judge is to discover, clarify and define the concepts involved. Some concepts in the law maintain their position solely by the fact that they are so vague and general as to allow the judge to subsume, or to refuse to subsume, individual cases within the rule virtually without any form of conceptual restraint.[12]
>
> In courts of law it sometimes happens that opposing counsel are agreed as to facts and are not trying to settle a question of further fact,... but are concerned with whether Mr. A... did or did not exercise reasonable care, whether a ledger is or is not a document, whether a certain body was or was not a public authority.
>
> In such cases we notice that the process of argument is not a chain of demonstrative reasoning. It is a presenting and representing of those features of the case which *severally cooperate* in favour of the conclusion, in favour of saying what the reasoner wishes said, in favour of calling the situation by the name which he wishes to call it. The reasons are like the legs of a chair, not the links of a chain.[13]

Admittedly, some may argue that every judgment ought to "follow logically on statues and precedents"[14] or "that legal decisions might be arrived at automatically by a computer into which all the conditions and precedents of the case would be fed and purely mechanical process of logical reduction would produce exactly the correct judgment."[15] But the most cursory introduction to legal method would show that, even in the statute-bound civil systems of continental Europe, this is an absurd suggestion. New types of cases arise, terms are elastic, precedents are conflicting, dicta become judgments, and conclusions do not follow necessarily from arguments used to support them.

A judicial opinion is a long legal argument. And legal argument is a species of rational persuasion that is not like a proof in logic or mathematics, or any form of rigorous deductive or inductive argument. In strict proof, only one of the opposing opinions, at most, may be right; others are mistaken and these mistakes can be demonstrated rigorously. In rational persuasion, people may reasonably differ; some views may be completely wrong but a spectrum of opinion exists, along which good cases may be made for differing points of view. Literary criticism, social theory, political argument, and philosophy share these traits. Even when legal argument may be thought to overlap scientific argument, their aims and standards of proof differ because legal argument seeks to make a court take a certain decision. External legal rules, principles, and precedents set the standard of proof. None of this is true in scientific argument.

> Judicial opinions are set out by way of justification of decisions. Decisions themselves as "acts of will" are not, of course, logical conclusions that follow necessarily from justificatory statements of judicial opinion.[16]

THE *RATIO DECIDENDI* OF JUDGE OVERTON'S OPINION AND ITS WEIGHT

Did, as Laudan asserts, Judge Overton argue that because creationism is not science it must therefore be religion? In the thirty-eight-page opinion, the judge devoted pages 5–19 to examinations of the history of attempts to inject religion into public schools in Arkansas, of the background and motives of the authors and sponsors of Act 590, of the legislative purpose of Act 590, and of the religious purposes and effects of the act, together with the relations of the act's language to that of the book of Genesis. Pages 28–32 showed that creationism has no scientific basis, only a religious one. Pages 10–12 and page 25 contained citations to the creationists' assertions of the nonscientific character of their theory, and pages 17–26 referred to defense testimony affirming the supernatural character of creation according to the theory.

Respectable opinion in science and in philosophy holds that argument be directed to issues and not ad hominem. However, a settled principle of legislative construction holds that, in addition to the plain meaning of the words of a statute, its purpose may be gleaned from the statements of its framers and supporters in legislative debate. Because there was less than one hour of debate on Act 590 and its framers were not members of the Arkansas legislature, Judge Overton could properly look beyond the statute and the debate to the expressed purposes of the framers and supporters, and to the nature and purpose of the organizations they used to advance the legislation.

What did Judge Overton find? (a) That Paul Ellwanger, the author of the model act that became Act 590, admitted his belief that creationism is no science (pages 10–11). (b) That Ellwanger advised creationists to hide their religious purposes as a tactic for getting the bill adopted (page 11). (c) That religious convictions alone motivated the bill (page 14). (d) That the state of Arkansas had a long history of religious interference in the public schools (page 14). (e) That the language of Act 590 is laden with religious concepts (pages 16–17). (f) That the language of creationism and of the act both parallel the first eleven chapters of the book of Genesis (pages 17–18, note 19). (g) That leading creationist writers admit these facts openly (page 19). And (h) that applicants for membership in the Creation Research Society must subscribe to the belief that the book of Genesis is historically and scientifically true in all of the original autographs.

Judge Overton's opinion establishes all this before he mentions either conceptual inadequacies in creationism or the characteristics of science. Thus, the proof of the religious character of creationism is already complete on its own and requires no argument of the sort Laudan supposes Judge Overton to have made (that because creationism is not science it must then be religion). The argument of the opinion should be read not as a deductive sequence but as a rational establishment of three independent conclusions, two of which can (but need not here be shown to) be linked logically, that is, as the three legs of a stool. These conclusions are that (1) creationism is religious doctrine, (2) creationism is conceptually confused, and (3) creationism is not science. Now, on any reasonable conceptions of religion and of science, (3) follows from (1). Religion, whatever else its characteristics, necessarily involves the supernatural; science, whatever else its characteristics, necessarily does not. In misunderstanding the nature of legal argument, Laudan forges links in a logical chain where the *McLean* opinion did not. Preoccupied with his own subject, Laudan takes the definition of science as paramount in the case—as did some of the consultants[17]—but the opinion did not. Perhaps Laudan was unduly struck by the sentence in the opinion which does say that, because creationism is not science, the conclusion is inescapable that Act 590 has only one real effect: The advancement of religion. Of course, if a thing is not *x*, it scarcely follows that it is another thing *y*, merely that, whatever it is, it is not *x*. Additional argument is needed to show that it is, in fact, *y*. But Judge Overton supplied this in his argument when he showed that creationism is religion, and he may properly have reached his decision on this basis alone. It is in the context of this finding, then, that the offending sentence—if it offends—must be read. But the judge goes further. He shows that creationism is conceptually confused, a conclusion that I suppose Laudan to grant, and also that it is not science, a conclusion that I hope he will grant despite his displeasure with the arguments given.

Laudan claims that Judge Overton's decision will return to haunt us. Far too many loopholes are left, he says, through which creationists may slip. But could holes in the decision actually affect future litigation? Only if the arguments were damagingly unsound and future litigants were bound by them. I have already challenged the first conjunct. Let me now turn to the second. Are future litigants bound by *McLean* and, if so, by which parts?

It seems clear to me that Laudan has overestimated the weight of *McLean*, mistaken who is bound by any binding decision, and failed to distinguish the three parts of a legal opinion and, thus, which part is binding. These three parts are (A) the ruling = holding, (b) the *ratio decidendi*, or principles by which the holding is reached, and (c) the arguments by which the principles compelling the holding are shown to apply to creationism and thus to Act 590.

Judge Overton held that creationism is a religious doctrine within the meaning of the Establishment Clause and therefore that the act mandating its teaching in the public schools must be struck down. The reasoning or principles that require this holding are that (1) any statute that fails to have a secular legislative purpose violates the Establishment Clause, and (2) any statute that advances or inhibits religion violates the clause.

When these settled and uncontroversial principles apply to a statute, it is struck down. Thus, the *McLean* case turned upon whether they applied to Act

590. There are six primary lines of argument so urging: (1) creationism invokes only religious concepts, (2) the language of creationism and of the Act closely parallel that of the first eleven chapters of Genesis, (3) creationists admit in print and as witnesses the religious character of creationism, (4) creationists admit supernatural causes as part of their doctrine, (5) in correspondence about strategy and tactics, creationists admit religious motivation in introducing Act 590, and (6) the state of Arkansas has a long history of religious interference in the public schools. Notice that none of these arguments requires any definition of science.

Who, then, is bound and by what? *McLean* arose in Federal District Court in Arkansas, a court of first instance. Its holdings bind Arkansas state officials, but no other court in any other jurisdiction. More important, only the holdings of an opinion bind, never the arguments used to arrive at a determination of fact, in this case, that creationism is religion within the ambit of the First Amendment. Future litigants need take no notice of *McLean* and if, per impossible, it were noticeable, only the court, not the litigants, would take notice and then only of the holding. Thus, even if Laudan's fear were real, even if Judge Overton's definition of science were wrong, and his arguments for the religious nature of creationism were wrong, plaintiffs and defendants and court in a separate case would be compelled to ignore them.

The arguments that make creationism a religious doctrine are sound. They do not depend upon showing that it is not science, although that follows, I think, from showing that it is religious. Even if Judge Overton got the criteria of science wrong, even if there were no criteria to get right, his arguments about the religious character of Act 590 are compelling. Even if the opinion is bad philosophy, that does not make it a bad decision:

> The caliber of a court's opinion, even in a Constitutional case, is not to be made dependent on its announcement of a principle that is fully satisfying to reason and that will indicate for us how future cases are to be decided.[18]

As this quotation from Levi emphasizes, it is legal reasoning we deal with, not some canon of philosophical reasoning or pure rationality—which are, in any case, things in which I should have thought Laudan disbelieved. And because the opinion can stand without the defining characteristics of science, we do not need to follow Laudan's strategy of weighing truth claims and arguing that those of creationism had been refuted.

TEXTBOOK ADOPTION

The effect of Laudan's strategy would be to dismiss the case from court. The battle would then shift to the processes of textbook adoption. To join battle here is for evolutionists to lose and creationists know it. In fact, this tactic now seems to be their preferred strategy.[19] In some Canadian school districts, it has already been a winning one because Canada's repatriated Constitution does not contain a clause prohibiting the establishment of religion. Indeed, many people predict that creationists will now pursue this method of attack vigorously. Before countless lay school boards susceptible to many sorts of pressures, creationists with ample resources will present their case and often win by default. Will evolutionists have sufficient resources and energy to debate year after year at every meeting where texts are chosen? Can they be expected to win by rational persuasion alone? Even in the face of *McLean*, creationists may succeed as school boards bend to their pressure;[20] moreover, publishers are influenced by decisions in those states where adoption is statewide and hence the financial stakes are great.[21]

ASSEMBLING THE CASE

One may wish for a latter-day Clarence Darrow to have argued the case in *McLean*, but such talent is in short supply and, anyway, Darrow lost. To win, plaintiffs will wish to use every resource at their command. In *McLean* the litigators for the plaintiffs were Skadden, Arps, Slate, Meagher, and Flom, a large and powerful New York firm. It was remarked that they litigated *McLean* like a corporate merger but, of course, that is what they normally litigate. Some seven or eight months were spent preparing the case. A large team of lawyers was involved in both the preparation and the battle. Many expert witnesses, and even more outside consultants, were used.

In presenting the *McLean* case, the plaintiffs' counsel had three objectives: to present arguments good enough to win, to present a case the court could grasp, and to avoid even the appearance of contradictory testimony. For, although philosophers or scientists may present considerations too complex for their professional contemporaries and yet may be judged more favorably by future generations, that course is not open to litigators. They must win at the appointed time and place or not at all. The door may not be left open to, say, clever insinuation in cross-examination. Witnesses need not testify in a powerful chain of argument but they also cannot appear to contradict each other's testimony. Lawyers must construct the testimony to make the case clear, simple, and direct, not subtle. If, as in *McLean*, some of the experts disagree on a sophisticated, narrow point, they lose. The skilled litigator must sketch with broad strokes; he will shape the testimony so that it has a clear outline and is graspable by a judge and, in a prominent case, by the press as well. Nicety of language and nuance of position will be shunned in favor of wide agreement. The philosophical testimony must agree with the scientific testimony and it must be understandable, must be "boiled down" for presentation at an adversary trial, and must avoid epicyclical complexity and neurotic maneuvering against *outré* counterexamples.

Finally, the sensibilities of the litigators must be faced. If a litigant is lucky, his or her attorney will be expert at litigation and the law, but it is unreasonable to require—because it is impossible to fulfill—that the attorney be expert in the subject of a case, whether politics, economics, marketing, medicine, philosophy, or science. The attorney must certainly understand the points being made so that they can be used to best effect in court and, to this end, lawyers need to know, simply and quickly, "Is it this?" or "Is it that?" or "How can we show it?"

In preparing the *McLean* case, many pages of qualifications about what counted as a science were boiled down to a brief statement of fact. When the lawyers did not like what their consultants wrote or said, they simply reformulated it. When the consultants criticized or refused to accept the rewrite, the lawyers became impatient and pointed out that philosophy should be kept to the seminar room. As a consultant, I objected strongly to the use of phrases like "natural law" to describe scientific laws, but with no success. Ultimately, and correctly, the counsels shaped the case, using the strategy and argumentation that they thought would win. Was this wrong? No. Given the boundary conditions and given the dynamics of impatient professional fighters aiming to win, what else could have been the outcome? And, they did win.

JUDGE OVERTON'S CHARACTERIZATION OF SCIENCE

In Section IV-C of the opinion, Judge Overton sets out what he takes to be the essential features of science. First, he offers a very loose definition: Science is what is accepted by the scientific community or whatever scientists do. He then adds five narrower criteria: (1) Science is guided by natural law; (2) Science is explanatory by reference to natural law; (3) Science is testable against the empirical world; (4) Its conclusions are tentative; and (5) It is falsifiable. Philosophically, these criteria may have been acceptable sixty or eighty years ago; but they are not rigorous, they are redundant, and they take no account of many distinctions nor of historical cases. The opinion does not state whether they are singly necessary or jointly sufficient. One would not recommend to graduate school a student who could do no better than this. Fortunately, Judge Overton and the litigators were not applying to graduate school. They were in a court of law. In this forum, confronted with creationism and religious "know-nothingism" as adversaries, are these criteria so far off the mark? Do they not represent what a thoughtful intelligent layman, looking at what scientists do and at how science has evolved, might conclude? Are they not sufficient to refute creationism? Consider, for example, the creationist doctrine of sudden creation from nothing. Judge Overton writes, "Such a concept is not science because it depends upon a supernatural intervention..." (p. 22). In footnote 25 he quotes from Duane Gish's book, *Evolution? The Fossils Say No!*: "We cannot by scientific investigation discover anything about the creative processes used by God." Now, at the very least, what is held to be science must not rule out of court, as creationism does, the scientific discovery and analysis of its main ideas. The claim that God exists and did what creationists claim he did is neither falsifiable nor abandonable for creationists. It is certainly true, as Laudan says, that throughout the history of science there have been concepts that individual scientists or groups of scientists have held immune from refutation or, at least, held not testable in isolation. Yet, many of these have been abandoned or refuted and the sciences in question have prospered. That people regard a tenet as immune from abandonment or falsification does not make it so. Although it is a semantic nicety to decide whether one ceases to do science when, in the face of overwhelming evidence, one refuses to give up a tenet or whether, in that case, one is simply doing bad science, or even outmoded science, it is not splitting hairs to say that one who holds a tenet to be beyond the reach of science, to be supernatural, is not a scientist with respect to that tenet. A person who locates the source of his supposed science and its principles in (to use Spinoza's felicitous phrase) that last refuge of ignorance, the mind of God, is not a scientist in his dealings with them, even if in other respects he may do irreproachable scientific work.

Laudan has claimed that because parts of a scientific theory are usually not testable in isolation, we cannot single out one creationist assertion and, showing that it is not testable, show that creationism is not science. But in isolation from what is the claim that God created things not testable? In conjunction with what is this claim to be tested? What ensemble does the creationist offer for testing? Of course, creationists do make some empirical claims, but so do astrologers, flat earthers, clairvoyants, and magicians. Does Laudan require that any ensemble of doctrine which attaches to itself a set of empirical assertions is, by that token, a science? False, of course, but nonetheless science?

Plato seems to have believed that if you could not define something you could not recognize it. We should not allow creationists to saddle us with that demonstrably false doctrine. Although there are sometimes troublesome borderline cases, who could not pick out a game or a mathematical theorem, a work of art, or a science?

Laudan also asserts that creationism does change its views, and comparison of its current views with those of the nineteenth-century version will show such changes. But Laudan has confused scientific creationism with christian fundamentalism, and other forms of literalism. Scientific creationism dates from the early 1960s. Bibliolatry, literalism, fundamentalism, and scientific creationism are related, but not the same. Regardless of which specific beliefs define fundamentalism, this does not argue that their beliefs change. All subscribe to the literal inerrancy of the Bible and do so because they believe it to be the word of God. This belief, at least, is immune to revision. Without it no one is a fundamentalist.

But there are different levels of immunity. A belief may be immune to revision in the sense that it is constitutive of some structure of thought (although that structure may be abandoned) or it may be immune in a "strong" sense because its proponents will doubt and reject all evidence rather than revise it. One could, for example, pick out a core set of beliefs constitutive of some version of evolutionary theory; but one could also point to evidence that, if found, would cause practitioners to abandon the theory. Creationists' beliefs, however, constitute a faith and are thus immune to revision in both senses. To admit revision is to destroy the belief structure; but they will not entertain the possibility of evidence that will either cause them to revise it or to abandon it. This is not science but dogmatism. In science, the outlooks of scientists change as one or another of their commitments arises or is eclipsed, whether because a paradigm is abandoned, or because problem-solving ability decreases, or for any other factor. The creationists are a monolithic community with a monolithic outlook that is immune to change because no member of the community will countenance any change and therefore no criterion for change exists.

McLean was a triumph, not a disaster; cause for rejoicing, not tears. The right side won for the right reasons; the necessary standard of proof was met in the case at bar. And there is no better reason for winning a legal case. Disaster is much more likely to occur when scholars eminent in one field venture to apply inappropriate standards to another. Mr. Laudan in proposing himself as the Socrates of the *Gorgias* has, instead, read us the lines of Euthyphro.

NOTES

1. Larry Laudan, *Progress and Its Problems* (Berkeley: University of California Press, 1977), p. 15.

2. Ibid., p. 7.

3. Larry Laudan, "Science at the Bar: Causes for Concern," *Science, Technology & Human Values* 7, no. 41 (Fall 1982): 17.

4. See, for example, Vincent Buglosi, *Helter Skelter: The True Story of the Manson Murders* (New York: Norton, 1974); Mirjan Damaska, "Evidentiary Barriers to Conviction and Two Models of Criminal Procedure," *University of Pennsylvania Law Review* 125 (1975), and "Structures of Authority and Comparative Criminal Procedure," *Yale Law Journal* 84 (1975); Marvin Frankael, "The Search for Truth: An Umpireal View," *University of Pennsylvania Law Review* 123 (1975); Barry Gross, "Adversaries, Juries, and Justice," *Loyola Law Review* 26 (1980); John Langbein, *Comparative Criminal Procedure: Germany* (St. Paul, MN: West Publishing Company, 1977); and Lloyd Weinreb, *Denial of Justice* (New York: Free Press, 1979).

5. A. G. Guest, "Logic in Law," *Oxford Essays in Jurisprudence* (Oxford: Oxford University Press, 1961), p. 188.

6. Donald L. Horowitz, *The Courts and Social Policy* (Washington, DC: Brookings Institution, 1977).

7. This was the *Napolitano* case: Docket A-4364-81-T1, Superior Court of New Jersey, Appellate Division.

8. Larry Laudan, *Progress and Its Problems*, p. 14; author's italics.

9. Ibid., p. 22; author's italics.

10. Ibid., pp. 24, 25; author's italics.

11. Ibid., pp. 27–30.

12. Guest, "Logic in Law," *Oxford Essays in Jurisprudence*, p. 193.

13. John Wisdom, "Gods," in *Logic and Language*, ed. Antony Flew (Oxford: Basil Blackwell, 1961), p. 195.

14. John Ziman, *Public Knowledge* (New York: Cambridge University Press, 1968), p. 13.

15. Ibid.

16. Neil MacCormick, "Legal Reasoning and Practical Reason," in *Social and Political Philosophy, Midwest Studies in Philosophy* 7, ed. Peter A. French, Theodore E. Uehling Jr., and Howard K. Wettstein (Minneapolis: University of Minnesota Press, 1982), p. 273.

17. Ibid.

18. Edward Levi, "Judicial Reasoning," in *Law and Philosophy*, ed. Sidney Hook (New York: New York University Press, 1964), p. 273. See also Levi's masterly *Introduction to Legal Reasoning* (Chicago: University of Chicago Press, 1949).

19. Gene Lyons, "Repealing the Enlightenment," *Harper's* (April 1982): 78.

20. See Frances Fitzgerald, *America Revised* (New York: Little, Brown, 1979).

21. For example, giants like IBM, ITT, Xerox, and CBS have entered the field. The tactics used for even small adoptions are fierce. On this point, I am indebted to Raymond English, especially for his references to his article, "The Politics of Textbook Adoption," *Phi Delta Kappan* (December 1980), and to that of Sherry Kieth, "Politics of Textbook Adoption," Project Report 81-A7, Institute For Research on Educational Finance and Governance, School of Education, Stanford University.

Nor should one underestimate the inanities of which state legislatures and school boards are capable. After World War I, Oregon required all students to attend only public schools, and Nebraska, Iowa, and Ohio banned the teaching of foreign languages. See Diane Ravitch, "The New Right and the Schools," *American Educator* (Fall 1982): 10.

PART III

INTELLIGENT DESIGN CREATIONISM AND THE *KITZMILLER* CASE

INTRODUCTION TO PART III

The first article in this section was written especially for this volume to document the early transition from the "creation science" of the 1981 *McLean* trial to the "intelligent design" (ID) of the 2005 *Kitzmiller* trial. Nicholas J. Matzke, then on the staff of the National Center for Science Education, was a consultant to the plaintiff attorneys in the latter case. The *Kitzmiller* case involved the use of the ID textbook *Of Pandas and People.* It was in the production of *Pandas* that the ID terminology, in its present use, was born. Matzke's historical detective work during the trial's discovery process unearthed previously unknown manuscript drafts of *Pandas* that were then subpoenaed for the trial, and these helped establish a direct paper trail between creation science and ID. The article functions as a continuation of the history of creationism presented in the excerpt from Ronald Numbers in chapter 11. In addition to reviewing some of the documentation he uncovered for the trial, Matzke introduces many of the key pioneering figures who formulated the ID arguments to try to overcome the legal defeats scientific creationism suffered in the *McLean* case and the *Edwards v. Aguillard* case. This is currently the only article available anywhere that covers this critical transitional period in the modern history of creationism.

The second piece is by law professor Phillip Johnson who is recognized by all as the pivotal leader who hammered out a new legal strategy for ID and who brokered a truce between young-earth and old-earth creationists, making ID a unified movement. Because of his importance to the movement, it is worth taking extra space here to provide some additional background information about his thinking. It was Johnson's 1991 book, *Darwin on Trial*, that put ID on the map. The article we include here was originally given as a talk at Hillsdale College in the year after his book was published and provides a clear overview of its main theses—ideas that became the ID movement's recurring themes. It was Johnson who organized many of the key players one now recognizes as the core of the ID movement. One of the most significant events was a 1992 symposium organized around Johnson's book that was held at Southern Methodist University. Organized with the help of *Pandas* publisher Jon Buell, the symposium included Michael Behe, William Dembski, and Steven Meyer, who would subsequently all be slated as expert witnesses to defend the Dover ID policy in the *Kitzmiller* trial. (By strange coincidence, both of us were also at that symposium—Ruse as the invited evolutionist to face Johnson on stage and Pennock in the audience doing research on the nascent movement for what would become a series of articles and his book *Tower of Babel: The Evidence against the New Creationism.*) For the next decade Johnson was the central leader of the movement and was known as "the cutting edge of the wedge," though he was forced to scale back his work following a stroke in 2001 and another in 2004. In 2003 Johnson was lauded for his efforts by being named *World Magazine*'s "Daniel of the Year." In an interview upon accepting the award he explained his thinking regarding how to search for an alternative to evolution:

> "I looked for the best place to start the search," Mr. Johnson says, "and I found it in the prologue to the Gospel of John: 'In the beginning was the Word.' And I asked this question: Does scientific evidence tend to support this conclusion, or the contrary conclusion of the materialists that 'in the beginning were the particles?'" (*World Magazine* 18, no. 48 [December 13, 2003])

He concluded:

> If we start with the Gospel's basic explanation of the meaning of creation, we see that it is far better supported by scientific investigation than the con-

> trary.... [A]ll I really want to do with the scientific evidence is to clear away the obstacle that it presents to a belief that the creator is the God of the Bible. (*World Magazine* 18, no. 48 [December 13, 2003])

Just as Duane Gish and the creation scientists did, Johnson claimed that his conclusions were made scientifically without reference to the Bible. One also sees the same dual-model argument, in which creation is purportedly proven necessarily if Darwinian evolution is false.

> "...I thought if Darwinism is not true, what is? If you can't do the creating without an Intelligent Designer, a creator, then there must be a creator." In subsequent books...Mr. Johnson argued persuasively that a supernatural power or Intelligent Designer had to have guided the creation and development of life. (*World Magazine* 18, no. 48 [December 13, 2003])

Johnson goes on to say how his biggest surprise and disappointment was the resistance his movement encountered from some Christians.

> The more frustrating thing has been the Christian leaders and pastors, especially Christian college and seminary professors. The problem is not just convincing them that the theory is wrong, but that it makes a difference. What's at stake isn't just the first chapter of Genesis, but the whole Bible from beginning to end, and whether or not nature really is all there is. (*World Magazine* 18, no. 48 [December 13, 2003])

One will notice echoes of Charles Hodge from chapter 6 in Johnson's piece; indeed, Johnson gives his article the same title as Hodge's book on Darwinism to emphasize the connections. Hodge's views about the relationship of evolution to Christianity were very influential to the ID movement. ID leader Jonathan Wells wrote his theology dissertation on Hodge and William Dembski founded a student group he called the Charles Hodge Society while a student at Princeton Theological Seminary. Like Hodge, ID theorists view Darwinism as equivalent to atheism. They explicitly reject what is known as theistic evolution, the mainstream Christian position that accepts evolution in its scientific form and does not take it to be incompatible with Christianity.

The next piece is an excerpt from an important law review article about the legal prospects for ID written by biologist Matthew Brauer, philosopher Barbara Forrest, and law professor Steven Gey. Written a couple of years prior to the *Kitzmiller* case, the article considers ID in light of the *McLean* case. This short excerpt from the long full article focuses on the question of whether ID could pass muster as science given the revised legal standards set by the Supreme Court in the 1993 *Daubert v. Merrell Pharmaceutical, Inc.* case, which some ID proponents had argued supported their claim to be recognized as scientific. The authors suggest that some theories are so inconsistent with current scientific understandings of the world that they cannot be reasonably construed as scientifically valid and that in the end it is irrelevant whether one calls such theories unscientific or merely "bad science." However, they also point out that *Daubert* speaks directly of what is to count as scientific knowledge and makes reference to testability, falsification, and general acceptance by the scientific community. This would rule out ID for much the same reasons that creation science was ruled out in *McLean*. Forrest later served as an expert witness in *Kitzmiller*, providing key testimony on the history of the ID movement and the ID textbook *Of Pandas and People.*

Starting with chapter 23, we turn to the *Kitzmiller* trial itself and how our philosophical question played out for ID when its proponents had their chance to present their case in court. We begin with excerpts from the testimony of the lead defense expert witness, Michael Behe. Next to Phillip Johnson, Behe is certainly the most important figure in the ID movement. His scientific credentials as a professor of biochemistry at Lehigh University made him the most credible public face for a movement that was dominated by law professors, philosophers, and others who did no scientific research. More substantively, his "irreducible complexity" argument against Darwinian or any natural mechanism of evolution (which the reader will recognize as a variation of Paley's watchmaker argument from part I) is the core ID argument and figured prominently in the case. The excerpts we include are the key parts of Behe's direct oral testimony in which he offers a definition of science that he argues should include ID as legitimately scientific. It is also interesting to see how Behe brings up Paley's design argument for the existence of God and argues that it, too, is scientific, and how he appeals to Richard Dawkins's views to try to support this claim. We also include a brief excerpt from Behe's testimony under cross-examination in which ACLU attorney Eric Rothschild questions him about the differences between Behe's definition of science and that of the National Academy of Sciences, and asks him about the implications of his more liberal definition, which would also license astrology as science. These excerpts cover the main portions of Behe's testimony that deal directly with the question of what science is, but readers are encouraged to examine his complete testimony, which is available online starting at http://www2.ncseweb.org/kvd/trans/2005_1017_day10_am.pdf and continuing through... 2005_1018_day12_pm.pdf to get the full context and for other indirect aspects of Behe's testimony that bear on the question. His written expert report is available at http://www2.ncseweb.org/kvd/experts/behe.pdf.

Next comes my (Pennock's) written testimony as an expert witness for the plaintiffs in which I lay out the argument that ID is not science but rather is a sectarian religious view. For my expert report I drew from the extended discussion of these topics in my book *Tower of Babel* and in various prior and subsequent articles, and supplemented this with additional evidence and arguments. The core of my written testimony dealt with methodological naturalism (MN), the scientific ground rule restricting science to natural explanations. Supernaturalism is not ruled out dogmatically, but rather for good reasons having to do with requirements of testability. I also presented evidence to document that ID is a form of creationism and does appeal substantively to the supernatural (though its advocates sometimes try to deny or obscure these facts). The element of supernatural design also independently established that ID is a religious view, and I also presented additional evidence to show how ID is not just a generic theistic view, but a sectarian one. Finally, I briefly described some of my experimental evolution research, which allows one to directly observe how evolution by natural selection can produce irreducible complexity. To see how I defended my written report from challenges by the defense attorney, one may read a transcription of my daylong deposition, which is available online at http://www2.ncseweb.org/kvd/depo/2005-06-14_deposition_Pennock.pdf. The transcript of my direct oral testimony and cross-examination in court may be found at http:// www2.ncseweb.org/kvd/trans/2005_0928_day3_am.pdf.

In chapter 25, University of Warwick sociologist of science Steve Fuller explains the argument he made as an expert witness for the defendants. Although he had not previously been much involved in the ID debate, he was known as a supporter of the movement in a general way, as described by ID proponent Tom Woodward (personal communication) and in comments in several articles. For instance, in a 2002 article he had written:

> [W]hile I support the teaching and research of intelligent design theory in mainstream universities, it is not in the spirit of the establishment of departments of Creation and Darwinian Science in "separate but equal" facilities on opposite sides of the campus. Rather, I would expect the cross-listing of courses and collaboration of faculty that one normally finds between any two intellectually overlapping disciplines. Among other things, I would hope that creationists come to see that some of their concerns better are addressed by Darwinists than if left to their own devices—and vice versa. (Fuller, "Demystifying Gnostic Scientism," *Rhetoric & Public Affairs* [2002]: 718–26)

In the article we reproduce here, one written after the trial, Fuller explains how his role was similar to that of Michael Ruse in the *McLean* trial, though he saw himself as providing the perspective of science studies to the question of whether ID counts as science. Fuller explains why he rejects methodological naturalism, arguing that "the history of science is full of hypotheses of 'supernatural' inspiration." The operative word here is "inspiration"; Fuller is making use of the distinction between what is known as the *context of discovery*, in which scientists generate hypotheses, and the *context of justification*, in which those hypotheses are tested. This defense turned out not to be especially helpful to ID, which makes a stronger claim that supernatural notions should be accepted as substantive explanations and not merely as a possible element of someone's inspiration. Again, the reader should seek out the online trial transcript of the trial for Fuller's complete testimony, which is available at http://www2.ncseweb.org/kvd/trans/2005_1024_day15_am.pdf and ... 2005_1024_day15_pm.pdf.

The next article, chapter 26, is by University of Wisconsin philosopher of biology Elliott Sober, who was a philosophy consultant during the preparation of the *Kitzmiller* case, standing ready to serve as a rebuttal witness had that been required. In this piece he steps back from the details of the ID position and considers only its most minimal thesis—what he calls "mini-ID"—namely, that complex adaptations that biological organisms display were crafted by an intelligent designer. Reviewing the notions of testability and falsifiability, he explains the role that auxiliary propositions play in any scientific test and how there are no independently attested auxiliary propositions available that allow mini-ID to make any differentiating observational prediction. Even in is most minimal form, he concludes, ID fails as a serious alternative theory.

Chapter 27 is an excerpt from the final ruling in *Kitzmiller*. We unfortunately did not have the space to include the entire text of the court's opinion issued by Judge John E. Jones III. The unusually detailed 139-page *Kitzmiller* opinion summarizes the wealth of evidence that both sides presented over weeks of testimony and a careful account of the legal reasoning behind the verdict that found the teaching of ID to be unconstitutional. The excerpted material includes a short section that lays out the background of the case and two longer sections that deal with the reasoning behind the court's findings that ID was a form of creationism and not science. The Dover board's introduction of ID in the schools as though it were science, the judge concluded in probably the most oft-quoted phrase from the ruling, was "breathtaking inanity." *Kitzmiller* has already been hailed as a legal landmark and we encourage readers to seek out and read the complete version of the ruling, which is available online at http://www2.ncseweb.org/kvd/all_legal/2005-12-20_Kitzmiller_decision.pdf.

This is a good place to point out that readers who wish to delve further into the details of the *Kitzmiller v. Dover* case have a wealth of online material they may examine. The most complete set of documents is hosted by the National Center for Science Education, which was a consultant to the plaintiff attorneys in the case. The center's *Kitzmiller* materials are indexed at http://www2.ncseweb.org/kvd/ and one can find additional background information including a timeline with news articles about the case at http://www2.ncseweb.org/wp/?page_id=5.

ID creationists were, of course, dismayed by the *Kitzmiller* verdict, which was a devastating defeat for their movement. In attempting to downplay its significance, they labeled Judge Jones an "activist judge" with "delusions of grandeur" who overstepped by inappropriately ruling on the question of whether ID was science. Taking us back to philosophical debate about demarcation from the previous section, they cited Larry Laudan to say that it was illegitimate for Judge Jones to have ruled on the question of whether ID is or is not science because Laudan had shown that demarcating science from non-science was a dead pseudoproblem. In the final chapter, I (Pennock) revisit the demarcation problem, arguing that Luadan and other anti-demarcationists were mistaken not only in many of their key pronouncements but also in how they framed the question. The problem of demarcation in the creation controversy is not the search for a set of necessary and sufficient defining conditions that draws a pinline border between science and anything else. Rather, the relevant notion is a more modest notion of demarcation that gives a ballpark ruling—creationism fails to be scientific because it violates a basic ground rule that science takes for granted. I conclude: "We do not need to precisely delimit the boundaries of science any more than we need the precise boundaries of a pin to conclude that it is not science to ask how many angels can dance on its head." If philosophers of science are not able to tell the difference between science and pseudoscience or religion even in cases such as creationism, then they have lost touch not only with their subject but also with the natural world itself.

CHAPTER 20

BUT ISN'T IT CREATIONISM? THE BEGINNINGS OF "INTELLIGENT DESIGN" IN THE MIDST OF THE *ARKANSAS* AND *LOUISIANA* LITIGATION

NICK MATZKE

In 1981, the state of Arkansas adopted Act 590, which mandated "balanced treatment" for creation science in the public schools. An ACLU-led coalition filed a constitutional challenge, producing the case *McLean v. Arkansas*. A two-week trial was held in December, and on January 5, 1982, the law was overturned in a forceful, detailed decision issued by Judge William Overton.[1] A similar law was passed in Louisiana, also in 1981, but it would not take effect until 1983. A complex series of litigations over the Louisiana law followed, eventually resulting in a summary judgment (decision without a trial) against its constitutionality in 1985, which creationists then appealed up to the Supreme Court.[2] On June 19, 1987, the Supreme Court issued a 7–2 decision upholding the rulings of the lower courts. Even without a trial, the Court said, the legislative record showed that the Louisiana law violated the Constitution by "advancing the religious belief that a supernatural being created humankind."[3]

These cases, their history, and their impact, have been discussed extensively, as have some of the issues that arose during their litigation.[4] Critics of "intelligent design" (ID) almost universally claim that ID is a form of creationism that was relabeled in an attempt to get around the *Edwards* decision.

Original essay. Reprinted by permission of the author.

The label "intelligent design creationism" (IDC) is often applied to make the connection explicit. Promoters of ID, on the other hand, go to great lengths to deny connections between ID and the creationism that the creationists were defending in the *McLean* and *Edwards* cases.

"Intelligent design," defined as the ID terminology plus the distinctive body of claims typically put forward by the ID movement, was not an invention of Phillip Johnson, Michael Behe, the Discovery Institute, or others who became prominent promoters of ID in the 1990s. The 1990s were important for ID, as the period in which its rhetoric solidified, the "Wedge Strategy" was constructed, and ID migrated from the outskirts of creationism to the headlines. But before the expansion of ID in the 1990s was its origin in the 1980s. The ID advocates of the 1990s were able to adopt with only minor variations a preexisting ID platform that had emerged fully formed in 1989 with the publication of the textbook *Of Pandas and People: The Central Question of Biological Origins.* The actual label "intelligent design" was adopted in a decision made by the editors of the *Pandas* books, in direct response to the Supreme Court's 1987 *Edwards* decision that ruled "creation science" unconstitutional. However, ID's distinctive body of claims was not assembled after 1987, but between 1982 and 1984: salvaged from the ruins of the creation scientists' spectacular collapse in the *McLean* trial, and retooled in preparation for the hoped-for and expected trial in the *Edwards* case, and a potential victory in that case. The ID claims are nothing more than a stripped-down version of the set of claims put forward under the "creation science" label in the *McLean* case, and are essentially identical to the set of claims put forward by the creationists in the *Edwards* case.

This picture is substantially different, at least in its details, than any history of ID currently available, either from ID critics, who have tended to focus on Phillip Johnson, the Discovery Institute, and the Wedge Strategy, all of which actually postdate the origin of ID; or from ID promoters, who indignantly deny any substantive connection between ID and creationism.[5] Numbers and Larson, the prominent historians of creationism, seem to have taken intermediate positions. However, neither has conducted a detailed study of the origins of ID in the 1980s. To help the reader see how I arrived at the present position, it is worth reviewing some recent discoveries about the origins of ID.

THE GREAT RELABELING EVENT OF 1987

Court cases have a remarkable way of focusing the mind. I say this from personal experience. As an employee of the National Center for Science Education (NCSE), I served as a researcher and advisor for the plaintiffs' legal team in the 2005 *Kitzmiller v. Dover* case on the constitutionality of "intelligent design" (ID). The Dover Area School District's ID policy recommended the supplementary textbook *Of Pandas and People* for "an understanding of what intelligent design actually involves."[6] And *Pandas*, first published in 1989, was the first book to systematically use terms like "intelligent design," "design proponents," and "design theory." It is true that occasional instances of the concatenation "intelligent design" or "intelligent Design" can be found in the creationist literature, and even (rarely) going back to Darwin and before, in discussions of teleology and the classical argument from design for the existence of God. But *Pandas* is the first work to enshrine the term "intelligent design" in a glossary, and the first work to disavow a connection between "intelligent design" and creationism. Before the *Kitzmiller* case made the fact embarrassing, Jon Buell,[7] in a 2004 preface to the third edition of *Pandas*,[8] noted proudly that *Pandas* was "the first place where the phrase 'intelligent design' appeared in its present use."[9]

For these reasons, it was apparent from the filing of the *Kitzmiller* case in December 2004 that it would be important to connect the "intelligent design" of *Pandas* to the creationism ruled unconstitutional in the 1987 Supreme Court case *Edwards v. Aguillard*. Initially, the plaintiffs' plan was to make this argument based only on the published text of *Pandas*. (A second edition of *Pandas* was published in 1993; this was the edition used in *Dover*.) This is an easy enough argument to make. The arguments in *Pandas* against biological evolution are mostly indistinguishable from claims previously made by the "creation science" movement of the 1980s—assertions about gaps in the vertebrate fossil record, the Cambrian explosion, the unacceptability of "macroevolution" as opposed to "microevolution," the improbability of the origin of life, and the evolution of biological complexity, etc.[10] Similary, *Pandas* regularly makes use of the old creationist tactic of "quote-mining" from evolutionary biologists; virtually every quote in *Pandas* had been previously exploited endlessly by the creation scientists.[11]

In addition, despite several explicit denials, various passages in *Pandas* make it unambiguously clear that the view the book advocates is not some intangible "guidance" of evolution, or intervention by space aliens, but the biblical doctrine of special creation. For example, *Pandas* describes the two longstanding creationist positions, young-earth and old-earth creationism, only mildly disguised in design terminology:

> An additional issue concerns the matter of the earth's age. While design proponents are in agreement on these significant observations about the fossil record, they are divided on the issue of the earth's age. Some take the view that the earth's history can be compressed into a framework of thousands of years, while others adhere to the standard old earth chronology.[12]

Another passage states that "[d]esign proponents point to the role of intelligence in shaping clay into living form."[13] This seems like an extremely odd passage, until it is realized that this is a not-so oblique reference to Genesis 2:7, which reads, "And the LORD God formed man of the dust of the ground, and breathed into his nostrils the breath of life; and man became a living soul."[14] Moreover, *Pandas* clearly refers to special creation of different "kinds" of life in various passages. A good example, regularly cited at the *Kitzmiller* trial, occurs on pages 99–100:

> Intelligent design means that various forms of life began abruptly through an intelligent agency, with their distinctive features already intact—fish with fins and scales, birds with feathers, beaks, and wings, etc.[15]

Faced with evidence of this sort, ever since 1989, critics of *Pandas* and its "intelligent design" proposal have been sure they smelled a creationist rat. Paleontologist Kevin Padian denounced the work as "straight fundamentalist creationism."[16] Eugenie C. Scott, executive director of the NCSE, said, "It is a classic 'equal time' tract."[17] Michael Ruse cited *Pandas* as the fulfillment of a prophecy that an ACLU attorney issued after the *McLean* case: "Don't think the creationists will go away. They won't! They'll just regroup and be smarter and sneakier next time."[18] In his *Tower of Babel: The Evidence against the New Creationism*, Robert Pennock even wrote that *Pandas* "substitutes the term 'designer' or 'master intellect' for 'Creator,'"[19] a prescient remark given subsequent discoveries. Other notable examples of the argument that ID is merely creationism relabeled include *Creationism's Trojan Horse: The Wedge of Intelligent Design*, by Barbara Forrest and Paul Gross,[20] and Leonard Krishtalka's pithy description of ID as "nothing more than creationism dressed in a cheap tuxedo."[21]

In the early stages of the *Kitzmiller* case, the general arguments connecting ID to creationism were incorporated into the expert witness reports and testimony of the *Kitzmiller* experts.[22] But one more argument was added to the plaintiffs' arsenal in *Kitzmiller*, and it turned out to be particularly devastating.

Here is where the mind-focusing aspect of court cases came into play. Could the argument that ID was creationism relabeled be made even more strongly? In early 2005, I began to look carefully into the origins of the *Pandas* book. The full story is told elsewhere,[23] but in short, an examination of creationist newspapers and other documents from the 1980s in the NCSE's archives revealed that the 1989 *Pandas* had been under development since 1981 or before—and, crucially, that the book had originally been on the topic of "creation" rather than "intelligent design." Furthermore, a draft version of *Pandas*, titled *Biology and Origins*, had been produced in substantial quantities around 1987 and distributed to reviewers and potential publishers.

This indicated that a physical draft of *Pandas* might still exist, using explicitly creationist terminology instead of "intelligent design" terminology. In April 2005, I summarized this information for the *Kitzmiller* lawyers, who promptly issued a subpoena to the Foundation for Thought and Ethics (FTE) for any drafts or any other material relating to the origins of the book. FTE filed a motion to quash the subpoena, but the motion was denied, and in July 2005 the requested documents were produced for the plaintiffs. Amid the documents were five distinct drafts of *Pandas*. They were eventually introduced into evidence in the *Kitzmiller* case as shown below. The two published versions of *Pandas* were also introduced into evidence, as was a chapter from the unpublished third edition, *The Design of Life*.[24]

Exhibit #	Title	Date
P-563	*Creation Biology* Textbook Supplement	1983 (based on file dates on the manuscript)
P-560	*Biology and Creation*	1986 (copyright page)
P-561, P-001, P-002	*Biology and Origins*	1987 (copyright page)
P-562	*Of Pandas and People*, version 1	1987 (copyright page)
P-652	*Of Pandas and People*, version 2	1987 (copyright page)
P-652	*Of Pandas and People*, published, first edition	1989 (copyright page)
P-652	*Of Pandas and People*, published, second edition	1993 (copyright page)
P-775	*The Design of Life* (draft of chapter 6)	2005 (unpublished)

As revealed during Barbara Forrest's testimony at the *Kitzmiller* trial, between the two 1987 drafts titled *Of Pandas and People*, the creationist terminology was expunged, and replaced with "intelligent design" terminology. For example, the previously quoted statement that "intelligent design means that various forms of life began abruptly..." reads as follows in a previous draft:

> Creation means that various forms of life began abruptly through the agency of an intelligent Creator with their distinctive features already intact—fish with fins and scales, birds with feathers, beaks and wings, etc.[25]

Similar changes occurred over one hundred times between the two 1987 drafts, as illustrated in charts presented in the *Kitzmiller* case.[26] As if this evidence wasn't remarkable enough, while examining the drafts, Forrest discovered a peculiar sentence in the second 1987 draft of *Pandas*: "Evolutionists think the former is correct, cdesign proponentsists [*sic*] accept the latter view."[27] Apparently, an editor was replacing the word "creationists" with the phrase "design proponents," but in the course of this tedious procedure the "c" and "ists" of "creationists" were accidentally left in the text. This has since been dubbed "the missing link" by critics of ID, and is now gleefully included in virtually every anti-ID talk and lecture.

For the foes of ID, the creationist terminology of *Pandas* seems to provide the final vindication of what they have always claimed: that ID is just warmed-over creationism. The argument was convincing in the *Kitzmiller* case—the judge called the evidence "astonishing" and drew the obvious conclusion: that the editors of *Pandas*, unwilling to abandon years of work on their creationist textbook, instead "re-branded" creationism after the 1987 *Edwards* decision. From the perspective of a federal judge in a district court, the legal implications are particularly compelling. Above all, the district court judge's job is to follow the precedents set down by the Supreme Court. The 1987 *Edwards* decision was clearly such a precedent. The *Pandas* drafts are the smoking gun that proves that the "intelligent design" label was adopted in response to the *Edwards* decision.

THE *PANDAS* DRAFTS IN CONTEXT

The *Pandas* drafts, even taken in isolation, are compelling evidence that something is deeply erroneous about any claim that ID is not creationism relabeled. But they should not be viewed in isolation. An examination of the events of the early 1980s will show that the *Pandas* drafts were not an anachronism, but instead fit neatly into the creationist legal struggles to defend the "balanced treatment" laws and their "two model" "equal time" approach.

Previous works have covered in depth the origins and development of "creation science" in the 1960s and 1970s, including the influence of Henry Morris, Duane Gish, and their associates at the Institute for Creation Research (ICR) and the Creation Research Society (CRS), as well as the origins of the creation science two-model/equal-time legal strategy in the work of young-earth creationist attorney Wendell Bird.[28] A convenient point at which to pick up the story of the origin of ID is the *McLean* trial in 1981. On March 19, 1981,

Arkansas governor Frank White signed into law the Balanced Treatment of Creation Science and Evolution-Science Act. The law provided that if "evolution science" were taught, equal time would have to be given to "creation science," which was defined as sudden creation of the universe, "kinds" of organisms, and humans; geological catastrophism including a worldwide flood and "a relatively recent inception of the earth and living kinds." This is obviously just a literal reading of Genesis, known as "young-earth creationism," given a pseudoscientific veneer; but Act 590 is full of assertions that creation science is a scientific view and not a religious doctrine, and the act explicitly forbade the use of "any religious doctrine or materials."

The ACLU filed suit on May 27 on behalf of a group of plaintiffs that included Arkansas teachers and clergy. A frantic six-month preparation for trial ensued, with the ACLU recruiting several prominent academics (including biologist Francisco Ayala, paleontologist Stephen Jay Gould, geologist Brent Dalrymple, chemist Harold Morowitz, philosopher Michael Ruse, historian George Marsden, and theologian Langdon Gilkey) as expert witnesses, and dozens more as advisors. Attorney General Steve Clark defended the law. The defense assembled its own roster of experts to counter the plaintiffs, but, somewhat surprisingly, left out the two biggest names in creation science, Henry Morris and Duane Gish, probably because of obvious religious baggage.[29] Only two of the defense experts, Donald Chittick and Robert Gentry, testified primarily about evidence for a young earth and a global flood. The main thrust of the defense case, however, was that scientific evidence, mostly biological, showed that evolution didn't work, and thus provided evidence for a scientific, nonreligious version of creation. It is now conventional wisdom on all sides that creation science is irrevocably derived from a literal reading of the Bible, but at the time, the whole point of creation science was to dodge this rap, at least in court. The defense presented creation science as being dramatically different from Bible-thumping fundamentalism.

The defense's lead expert was Norman Geisler, an old-earth creationist theologian at Dallas Theological Seminary. Geisler was to testify on the definition of religion and its inapplicability to creation science. However, he himself was unable to deny the obvious: "I think in all honesty that the people who devised this [act] probably got their model from the book of Genesis."[30] One might think this admission would sink the defense, but Geisler said, "the source does not matter," and argued at length that "belief in" God was religious, but "belie[f] that there is a God [has] no religious significance whatsoever."[31] He then made several arguments that will seem eerily familiar to twenty-first-century observers of the intelligent design movement. Geisler said that science has a narrow and a broad definition. The narrow definition relies on repeatability and natural law but only "deals with the present."[32] Creation is not science under this definition, but then, said Geisler, neither is evolution.[33] "Origins" topics were only science under Geisler's broad definition, where, as in "forensic medicine," "you can only make probable models," and "when we're talking about origins, we can't talk about the fact of evolution or the fact of creation, because it's really only an extrapolation . . . we can't repeat or observe."[34] Moreover, argued Geisler, "[Science about origins] has to take things we know to be true in the present and suppose that they were also true in the past[,] or argue from analogy."[35] and biological creation was well supported by arguments from analogy, "a dictionary is normally produced by intelligent activity, not [an] explosion in a printing shop."[36]

According to the description of the trial in *Science*, Geisler's argument that a creator was not necessarily religious "was the defense's principal thrust for being able to teach about the product of a creator without necessarily being religious. Judge Overton was clearly interested in this line of reasoning, until, under cross-examination, Geisler tarnished his credibility somewhat by declaring that UFOs "were agents of Satan."[37] The UFO remark, solicited during Geisler's cross-examination, made the headlines at the time and is now almost the only thing that anyone remembers about Geisler's testimony. But it is clear that even in 1981, Geisler was arguing for something essentially identical to what would now be called intelligent design, but in defense of an undeniably "creation science" law.

Several other defense experts made standard creationist arguments that later became key planks of ID.[38] Biologist Wayne Frair, a young-earth creationist and board member of the Creation Research Society, testified that "the various kinds of life" were "genetically unrelated," but that "kind" was a larger group than species, which could evolve "within limits." Frair asserted that Arkansas was "on the very cutting edge of an educational movement" with its two-model approach to origins.[39] Agnostic biochemist William Scot Morrow stated that, although he was an evolutionist, the plaintiffs' experts were "close-minded," that producing life or proteins "by chance" was "basically impossible," and that the fossil record supported creation.[40] In response to a skeptical question by the judge about the lack of references supporting Morrow's opinions, Morrow said that the scientific community practices "systematic censorship."[41] Retired botanist Margaret Helder, vice president of the Creation Science Association of Alberta,[42] made an argument similar to Frair's, but focused on plants. She claimed "[s]he never had to introduce any religious literature to discuss her creationist views."[43]

Ariel Roth, the director of the Geoscience Research Institute (a young-earth creationist, Seventh-Day Adventist group based at Loma Linda University) invoked standard creationist improbability arguments, which rely on the utterly mistaken idea that evolution is a purely random process. His examples included the "high improbability of random formation of life" and "the near impossibility of the random formation of chromosomes" and genes.[44] Roth also made an argument identical to the one Michael Behe would produce fifteen years later:

> Roth mentioned [the] difficulty of evolving complex integrated structures since each part of the integrated structure alone would be useless to the organism in which it first appeared and therefore would be weeded out by natural selection. ...As an example of the difficulty of evolving "complex integrated structures," Roth noted the relationship of "the ear, the brain, and the auricular nerve," and the respiratory system. Of the respiratory system, he said, "This system would not be functional until all of the parts were there.... How did these parts survive during evolution as useless parts under natural selection?"
>
> Asked if creation science could be taught only on scientific grounds without religious references, Roth said it could, since "origin by design" is a scientific idea....[45]

Harold Coffin, another employee of the Geoscience Research Institute, testified next. He argued that "sudden appearance of life in the 'Cambrian' rocks of the geological column indicates sudden creation instead of slow evolution." Evolution was called into question by gaps in the fossil record, said Coffin, who asserted an "absence of 'connecting links' between basic kinds of animals and plants in the fossil record." Coffin invoked the creationist "kinds" concept, asserting "the inability of scientists to cause or observe in modern life forms changes from one basic kind of life to another."[46] The concept of a created "kind" that can only evolve within strict limits is derived from the peculiar creationist exegesis of Genesis, where God commands various animals to reproduce "after their kind." The "kinds" concept was later adopted wholesale by the ID movement with only cosmetic changes.[47]

Perhaps the oddest creationist expert was N. Chandra Wickramasinghe. Wickramasinghe, an "agnostic Buddhist" physicist from the United Kingdom, had just coauthored a book with famed physicist Fred Hoyle. Hoyle, a maverick in many areas, was the last prominent holdout against the big bang model for the origin of the universe, advocating instead an infinitely old universe in a "steady state." In 1981 Hoyle and Wickramasinghe coauthored *Evolution from Space: A Theory of Cosmic Creationism*,[48] a book that proposed the natural origin of life on Earth and the major transitions in evolution were so improbable as to be effectively impossible. The book suggested instead that genetic material, apparently produced by some kind of cosmic intelligence embedded in the structure of the universe, rained down to Earth from space. This was supposed to have somehow produced the origin of life as well as the major evolutionary transitions.[49] In Arkansas, Wickramasinghe said

> [t]hat the probabilities of upward change by chance combination of the new bacteria with current life forms was so infinitely tiny that he and Hoyle had to postulate the idea that the "intelligent designer" arranged the times and places at which the interstellar bacteria would arrive on earth so that it would cause upward change.

But Wickramasinghe's testimony was a mixed blessing for the creation science advocates; apart from Geisler's UFO statement, the most-remembered statement from a witness for the creation science side seems to be Wickramasinghe's admission that "no rational scientist" would accept a young earth and a global flood.[50]

However, these "gotcha" moments were not key to Judge Overton's conclusions in his decision. When the trial testimony and Judge Overton's decision are reviewed, it becomes apparent that several points sunk the defense case. One was Overton's definition of science,[51] but numerous other points were significant. Overton noted that the unusual ideas of Wickramasinghe and others had not been published in peer-reviewed journals, and that the creationist tactic of using negative objections to evolution as positive evidence for creation constituted a "contrived dualism."

These findings rebutted the defense's positive argument that "creation science" was science and not religion. More important, though, was the plaintiffs' positive argument that creation science was a specific religious view. This was established through testimony that noted that views like a young earth, global flood, and sudden creation of humans and other "kinds" of organisms were derived from a particular literalist reading of the book of Genesis that, as a historical matter, was closely tied to Christian fundamentalism.

A particularly severe problem for the creation scientists was the lack of religiously neutral educational materials that Arkansas teachers could use to teach creation science. The topic of educational materials comes up again and again in the briefs and trial testimony, and occupies a major section of Overton's decision. There were only a few creation science texts even potentially usable in a classroom, and each had significant religious material and/or educational problems.[52] Overton concluded that educators were "unable to locate one genuinely scientific article or work which supported" creation science, that they rightly found "the creationists' textbook material and teaching guides unacceptable," and that "[t]he defendants did not produce any text or writing in response to this evidence which they claimed was usable in the public school classroom." Overton, thus dismissing the creationists' scientific/educational case, added this negative finding to the positive finding that creation science constituted a specific religious view, and concluded, "[s]ince creation science is not science, the conclusion is inescapable that the only real effect of Act 590 is the advancement of religion."[53]

THE BEGINNINGS OF ID

After the defeat in *McLean*, creationists took heart in the fact that they had a second chance in the courts: Louisiana's Balanced Treatment for Creation Science and Evolution-Science in Public School Instruction Act. Many of the apparent mistakes that doomed the Arkansas litigation were avoided in Louisiana. For starters, on May 28, 1981—the day after the ACLU filed suit in Arkansas—the Louisiana legislature stripped from the bill references to a young earth and global flood. Second, Wendell Bird was deputized by the Louisiana attorney general to be the attorney of record on the case, and was backed by the Creation Science Legal Defense Fund. Finally, Bird and the creationists "stole a march"[54] on the ACLU in Louisiana. On December 2, 1981—a day before the ACLU filed suit—Bird and the creationists filed their own lawsuit, as plaintiffs, to force the state to enforce Louisiana's law.[55] These moves meant that the creationist lawsuit would have to be resolved first, putting the creationists in the plaintiffs' driver's seat and guaranteeing years of litigation during which the creationists could perfect their case.

It was in this heated atmosphere that we can begin to trace the origin of ID in its proper context. The first public announcement of the *Pandas* project I have found appears in the Fall 1981 issue of the creationist student newspaper *Origins Research.* Beneath a prominent headline discussing the then-ongoing litigation—"Lawsuit prospects dim in Arkansas, bright in Louisiana"—appears a short announcement, titled "Unbiased biology textbook planned." It states, "[a] high school biology textbook is in the planning stages that will be sensitively written to 'present both evolution and creation while limiting discussion to scientific data.' Dr. Charles B. Thaxton is science advisor to the project.... The author selected is a teaching biologist with two McGraw-Hill books in print."[56] This is clearly an early announcement of what would eventually become the "intelligent design" textbook *Of Pandas and People.* A similar announcement was published in the August/September 1981 newsletter of the American Scientific Affiliation (ASA):

> Meanwhile, Charles B. Thaxton is beginning to work on the biology textbook situation. Charlie has just moved from Probe Ministries "down the road a piece" to the Foundation for Thought and Ethics, founded by Jon Buell, who had also been with Probe. The Foundation wants to produce "a sensitively written high school biology textbook that presents both evolution and creation while limiting discussion to scientific data." Charlie, as science adviser to the project, hopes to draw together an editorial board made up of both creationists and evolutionists. The author selected is "a teaching biologist who is a committed Christian, with two McGraw-Hill books in print."[57]

The players at this juncture will be introduced in turn.

Probe Ministries and Jon Buell

Probe Ministries is a Dallas-based ministry founded in the fall of 1973 by Jon Buell and Jim Williams, both previously employees of Campus Crusade for Christ.[58] According to Williams, Probe's purpose was to address "the seemingly irreconcilable gap in the minds of most college students and between biblical Christianity and their class studies." Early projects included campus presentations and forums (over 1000 presented by 1976), and summer camps for students about to enter college, "to prepare conferees for the secular biases they will encounter in college...and to provide training in witnessing and how to act on biblical concepts."[59] Another early project was the Christian Free University Curriculum, a series of short books presenting conservative evangelical positions on the Bible, science, morality, miracles, and other topics. Noteworthy products of the series include Buell's first book, a 1978 apologetics work titled *Jesus: God, Ghost or Guru?*; the 1977 *Fossils in Focus* by J. Kerby Anderson and Harold Coffin (a young-earth creationist and future *McLean* witness); the 1980 book *The Mysterious Matter of Mind* by old-earth creationist anthropologist Arthur C. Custance; the 1981 *The Necessity of Ethical Absolutes* by Erwin Lutzer; Norman Geisler's 1982 *Miracles and Modern Thought*; and the 1984 *Natural Limits to Biological Change*, by Lane Lester (a young-earther and CRS member) and Ray Bohlin (an old-earther).[60]

Charles Thaxton

In 1980, Charles Thaxton was on the staff of Probe Ministries, as well as on the faculty of the Center for Advanced Biblical Studies in Dallas.[61] His jobs at Probe included lecturing, as well as editing the above-mentioned books by Custance, Lutzer, and Geisler.[62] In August 1970, Thaxton earned a PhD in physical chemistry at Iowa State University for a dissertation on x-ray crystallography. According to Walter Hearn of the ASA, who was on Thaxton's committee, Thaxton "dumbfounded" some of his professors by immediately traveling to the L'Abri Fellowship in Switzerland to study with theologian Francis Schaeffer.[63] In 1981, *Newsweek* called Schaeffer "the guru of fundamentalists,"[64] and many evangelical thought leaders cite Schaeffer's works or visits to Switzerland as key steps in their own development: "It was Schaeffer who first made evangelicals aware of the culture war...no writer did more to activate and politicize evangelicals."[65] A prominent theme of Schaeffer's theology was that "evangelicals should apply a biblical worldview to every department of life."[66]

Thaxton met his future wife at L'Abri, and upon returning to the United States he spent two years at Harvard as a postdoctoral researcher in the history of science. While there, the only piece he published was a letter in *Pensee on Velikovsky*, an infamous catastrophist journal reviled by the scientific community. Thaxton complained of the scientific community's bias toward uniformitarianism and the "quite disturbing question of why a man eminently qualified should be held in the highest form of contempt by being ignored."[67] In the fall of 1973 Thaxton was back doing a postdoc in physical chemistry,[68] but in the fall of 1976 he joined the staff of Probe Ministries. In 1977 Thaxton began work on a book discussing the origin of life,[69] originally intended for the Christian Free University Curriculum series.[70] The project, originally titled *Life: The Crisis in Chemistry*, grew beyond these boundaries and was taken up by FTE in 1981. This eventually became the 1984 book *Mystery of Life's Origin*, coauthored by Thaxton, Walter Bradley, and Roger Olsen—three old-earth creationists.[71]

The Foundation for Thought and Ethics

The Foundation for Thought and Ethics (FTE), founded by Jon Buell, is the nonprofit think tank that gave birth to intelligent design. FTE was intended to be a Probe-like group that could produce materials that would be secular enough to get taken seriously in forums like universities and the public schools, whereas materials from a group with an obvious evangelism mission, like Probe Ministries, would experience difficulties.[72] However, Buell seems to have had trouble separating the secular from the sacred. According to FTE's "Articles of Incorporation," filed with the state of Texas on December 5, 1980,

> The primary purpose [of FTE] is both religious and educational, which includes, but is not limited to, proclaiming, publishing, preaching, teaching, promoting, broadcasting, disseminating, and otherwise making known the Christian gospel and understanding of the Bible and the light it sheds on the academic and social issues of our day.[73]

When questioned during the *Kitzmiller* case, Thaxton and Buell seemed surprised to discover that these statements were in FTE's "Articles of Incorporation." Thaxton was not a founder of FTE, and might well have been ignorant, but Buell's legally binding signature is on the documents, submitted to the government to obtain the tax exemptions that accrue to a registered nonprofit. Buell responded that the statements were "boilerplate" and that the attorney and accountant who prepared the documents got it wrong. Eric Rothschild, the exasperated attorney for the plaintiffs, asked, "So the accountant got it wrong and the attorney got it wrong?" Buell responded, "It's true."[74]

Even if Buell's explanation of FTE's "Articles of Incorporation" is given the benefit of the doubt, other documents cannot be explained away as attorney-derived boilerplate. A promotional statement in FTE's nonprofit application to the IRS, titled "What Is the Foundation for Thought and Ethics?" describes FTE's goals:

> Someone has rightly said, "If you wish to alter the destiny of a people, you have only to alter its ideas; actions are the blossoms of thought." The Foundation for Thought and Ethics has been established to introduce Biblical perspective into the mainstream of America's humanistic society, confronting the secular thought of modern man with the truth of God's Word.
>
> Nearing completion, our first project is a rigorous scientific critique of the theory of prebiotic evolution. Next, we will develop a two-model high school biology textbook that will fairly and impartially give scientific evidences for creation side by side with evolution. (In this case Scripture or even religious doctrine would violate the separation of church and state.) A credentialed author team and a consulting editorial board of scholars are being assembled for the project. The manuscript will be placed with a secular textbook publisher for publication.
>
> The Foundation's future projects will include publications on a wide range of topics, each vital to shaping the course of our nation's future. Operating primarily as a Christian think tank, the Foundation emphasizes first, publishing, and second, lectures and seminars. 95% of what the Christian press publishes is written to Christians. We've been talking to ourselves! Through the work of the Foundation, Christians are challenged to make their voice and view heard in the published arenas of discourse where the opinion leaders of society must give them genuine consideration.[75]

This document makes it clear that what was intended to be secular about FTE was not its mission, but its intended audience. With FTE, Buell wanted to escape the usual audience of evangelical ministries—evangelical Christians—and take the biblical evangelical worldview to secular audiences. Rather than "talking to ourselves," Buell wanted to talk to the "opinion leaders of society," secular academics and public school educators. *Mystery of Life's Origins* would do this for university audiences. The "two model high school biology textbook"—which became the 1989 *Of Pandas and People*—would do this through balanced treatment of creation and evolution.

FTE and the Creation Science "Two-Model" Approach

After the 2005 discovery that creationist terminology had been used instead of "intelligent design" in early *Pandas* manuscripts, Buell and Thaxton claimed that FTE's usage of "creation" meant something substantially different than creationism or creation science. In his *Kitzmiller* deposition, Thaxton said, in a passage now widely quoted in Discovery Institute propaganda,

> [W]e were trapped by the vocabulary that we were given, and the culture was talking about creation and evolution, God and the supernatural and natural.... We weren't comfortable—at least I wasn't comfortable with the typical vocabulary that for the most part creationists were using because it didn't express what I was trying to do. They were wanting to bring God into the discussion, and I was wanting to stay within the empirical domain and do what you can do legitimately there....[76]

Buell also argued that the "creation" originally advocated by FTE was something very different from the creationism of creation scientists. During his testimony in the *Kitzmiller* case, Buell claimed that the term "creation" in the drafts of *Pandas* was "just a place holder term until we came to grips with which of the plausible two or three terms that are in the scientific literature we would settle on. And that was the last thing we did before the book was... sent to the publisher."[77] Buell added, "[t]here was a new position that was being determined through dense extensive interaction between scientists and philosophy [of] science. We knew that it was fundamentally different from creation science." As evidence, Buell claimed that "we, on our own dime, flew to Little Rock, Arkansas, after *McClain* [*sic*] went down, and tried to appeal to the Attorney General not to appeal the verdict, because we felt that it was wrong—wrong minded."[78] Buell also claimed the same for the Louisiana bill: "[W]e flew to Atlanta, we met with the attorney, the lead attorney. We tried to persuade him to drop creation science."

Buell's claim about encouraging Arkansas attorney general Steve Clark not to appeal the *McLean* decision is corroborated by Geisler, who himself, though a major supporter of the creation science bills, thought that appealing *McLean* would just make the situation worse: "[f]ew creationists were ultimately disappointed that the decision was not appealed."[79] I have not been able to corroborate Buell's claim about FTE's lobbying against the *Edwards* case. The attorney in Atlanta is obviously Wendell Bird, who lived there at the time. It is possible that FTE lobbied against the Louisiana case at some point, perhaps when the legal situation became dire, but it seems extremely unlikely that it did so early on, or on the basis of a principled stand against creation science.[80] This conclusion is supported by Geisler's account:

> The [*McLean*] case was never appealed, since Jon Buell of the Dallas-based Foundation for Thought and Ethics, which eventually produced a textbook (*Of Pandas and People*) for teaching creation alongside evolution in public schools, requested that the Arkansas attorney general not appeal the case. The Foundation believed that a similar law that had been enacted in Louisiana was better worded, had less baggage, could be better argued, and, therefore, had a better chance of success when appealed to the Supreme Court.[81]

At least in 1982, FTE was not against creation science generally, but only against appealing the hopeless Arkansas case, in the hopes that the Louisiana case would turn out better. And there is plenty of evidence that Buell and Thaxton initially favored the "two model" and "balanced treatment" approach that had been widely promoted by the creation scientists in the 1970s and early 1980s. Rather than exhibiting substantial disagreement with the creation science approach, they thought they could do a better job at producing something that would be viable in the courts. In the June/July 1982 *ASA Newsletter*, which contained an extensive wrap-up on the *McLean* trial, comments from FTE were relayed:

> Jon Buell and Charlie Thaxton at Foundation for Thought and Ethics in Richardson, Texas, are trying to put the pieces back together in yet another way. They are producing "a sound biology textbook to banish all the misconceptions about teaching creation by demonstrating how it can be taught rationally and with scientific and constitutional integrity." They think that if the textbook they have in mind had been written before the Little Rock trial, "it would have disarmed the ACLU of virtually all of their allegations."[82]

There is no indication here that Buell and Thaxton thought they were working on something "fundamentally different" from creation science. Instead, they were announcing to the world that they were supporting the two-model creation science approach by producing the constitutional two-model textbook the *McLean* defense had lacked. In a February 1982 essay, Thaxton tells the same story. After a byline explaining that Thaxton "is an advisor to a project to publish a high school biology textbook sensitive to both evolution and creationism," Thaxton writes,

> I favor two-model creation/evolution teaching of origins, because I believe it is the best way to teach about events in the past which are unique and therefore, cannot now be observed, repeated or directly tested.[83]

There is no hint in the essay that Thaxton thinks that there is a problem with creation science—which certainly seems like it might be a pertinent topic if it was troubling him, given the litigation ongoing as the essay was being written. Instead, in the body of the essay, Thaxton offers a courtroom analogy for his two-model position. Essentially his argument is that origins events, like crimes, are unique and not replicable, and thus it is unfair for only one side to have a say. As in a courtroom, each side should be allowed to make its case, and the members of the jury (i.e., the students) can make up their own minds. This is the same sentiment expressed in Geisler's testimony about "forensic medicine." The courtroom analogy later reappeared in *The Mystery of Life's Origin* and the *Pandas* drafts. The essay concludes,

> One need not share the creationists' view to be in favor of introducing a two-model approach into the classroom. Davis and Solomon sum it up as follows:
>
> "We cannot imagine that the cause of truth is served by keeping unpopular or minority ideas under wraps.... Specious arguments can be exposed only by examining them. Nothing is so unscientific as the inquisition mentality that served, as it thought, the truth, by seeking to suppress or conceal dissent rather than by grappling with it." (P. William David [*sic*] and Eldra Pearl Solomon, *The World of Biology*, 2nd ed. [New York: McGraw-Hill, 1979], p. 610)[84]

Although Thaxton appears to be quoting just another standard biology textbook in his conclusion, this is not quite the case. The editions of *The World of Biology* coauthored by Percival William Davis (first edition 1974) were celebrated by creationists. They were unique in that the evolution section (pp. 409–17) is actually presented in a two-model style. The standard lines of evidence for evolution are put forward, but they are countered with standard creationist counterarguments, including a full-page picture of the bombardier beetle, a creationist favorite, complete with the irreducible complexity argument (from the caption: "Since neither the peroxide nor the catalyst is useful by itself, the creationist asks from what beginnings the beetle's present mechanism could have evolved"). Davis's textbook was celebrated as a rare success in creationist circles, cited widely in legal arguments for the two-model approach, and was exhibited by Wayne Frair at the 1974 meeting of the board of directors of the Creation Research Society.[85]

Percival William Davis

Sharp readers may recall that the initial 1981 announcement of the *Pandas* project mentioned that the author was to be a McGraw-Hill textbook author. This is, of course, a reference to Davis. In proto-ID and ID literature, Davis is universally described as a textbook author and college instructor. What is left tactically unsaid is that Davis was a member of the Creation Research Society and a published contributor to its journal, the *Creation Research Society Quarterly*,[86] and that he is a strict young-earth creationist who to this day teaches flood geology at a fundamentalist Bible college in Florida.[87] Davis also coauthored several editions of a creation science book, *The Case for Creation*, with *McLean* witness Wayne Frair.[88] Davis was an odd choice for FTE to make if it was deeply convinced that its textbook project was about something "fundamentally different" from creation science.

Dean Kenyon

Buell and Thaxton soon recruited a second author for the *Pandas* project: Dean Kenyon. Kenyon, too, is an odd choice if FTE had something "fundamentally different" from creation science in mind. Starting in the 1990s, Kenyon has been described in ID circles as a top origin-of-life researcher who saw the light and became an early leader in the intelligent design movement with his authorship of a foreword to *Mystery of Life's Origin*, his contribution to *Pandas*, and his victory in a 1993 dispute at San Francisco State over his right to devote a portion of his courses to intelligent design, over the objections of the biology department.[89]

Again, selective omission has distorted the picture. Kenyon was indeed an origin-of-life researcher and coauthored a well-regarded book on the topic, *Biochemical Predestination*, in 1969. But when he changed his views in the mid-1970s, he didn't exactly become an advocate of a sophisticated ID view; he became a strict Bible-based young-earth creationist. According to a December 1980 article in the *San Francisco Examiner*, Kenyon's position was that "[i]n the relatively recent past—10,000 to 20,000 years ago—the entire cosmos was brought into existence out of nothing at all by supernatural creation." Kenyon added that there are "no errors in the Bible." Kenyon's department limited him to spending only 5 percent of his class on creationism, but Kenyon told the paper that "[i]f I were to dream about it, I would [want] a 50–50 split." The article includes a large photo of Kenyon holding up Henry Morris's *Scientific Creationism* (general edition).[90]

Another inconvenient fact that is systematically obscured in ID-sympathetic histories is Kenyon's interesting activity in the legal arena. During the *McLean* trial, Kenyon was scheduled to testify for the defense and probably would have been the creationists' star expert witness in the trial. Kenyon actually flew to Arkansas to testify but "fled town after watching the demolition of four of the state's witnesses on day 1 of the second week."[91] According to *Christianity Today*, Kenyon left under the encouragement of Wendell Bird, who told several creationist experts, "I don't think you should jeopardize your reputation with the way [the trial] is being handled." Attorney General Clark said that he considered legal action against Bird for his interference.[92]

The Louisiana Litigation

Bird clearly wanted to "save" Kenyon for the Louisiana case. In preparation for his then-expected trial, Bird listed Kenyon as an expert, as well as future *Mystery* author Walter Bradley and *McLean* experts Wickramasinghe and Morrow.[93] Experts were deposed in preparation for the trial.[94] Unfortunately, in June 1982, Bird's lawsuit was dismissed,[95] giving the initiative to the ACLU's lawyers. A complex sideshow delayed consideration of constitutional matters until 1984. At that point, the ACLU filed for a summary judgment against the Louisiana bill, arguing that no possible trial testimony could save the law. Bird assembled five expert witness affidavits and a massive supplementary brief, hundreds of pages long, in support of the contention that a trial was necessary to show that creation science was science and therefore teaching it did not have a primarily religious purpose or effect. Kenyon was again recruited, this time to write the lead affidavit.

Kenyon's affidavit, signed on September 17, 1984, represented the essence of the "creation science" that Wendell Bird wanted to put before the court. Kenyon's definition of creation science is slightly unusual:

> Creation science means origin through abrupt appearance in complex form, and includes biological creation, biochemical creation (or chemical creation), and cosmic creation. [. . .] Creation science does not include as essential parts the concepts of catastrophism, a world-wide flood, a recent inception of the earth or life, from nothingness (ex nihilo), the concept of kinds, or any concepts from Genesis or other religious texts.[96]

The affidavit then argues for "abrupt appearance" along the lines of classic creation science objections to evolution—gaps in the fossil record, complexity and genetic information, the improbability of the origin of life, and the rest. Kenyon's definition was concocted purely in order to make it difficult for a court to link the Louisiana bill to the Bible. As Kenyon's 1980 interview showed, he actually did believe that the various items denied in his affidavit were part of creation science. Wendell Bird, closely associated with the ICR, certainly believed these things as well. Nevertheless, for legal purposes, Bird, Kenyon, and the Creation Science Legal Defense Fund were all willing to disavow the usual tenets of creation science—up to a point. A close reading of the definition shows that while a young-earth, global flood, and the rest are not necessarily part of creation science, they are not necessarily excluded from creation science, either. Nowhere does Kenyon say the only scientific respectable thing in this situation, namely, "the young-earth, global flood view is as false as the view that the Earth is flat." Instead, a studied agnosticism is maintained, at least in front of the courts.

Kenyon's affidavit also mentions that creation science now can meet the objection the ACLU raised in the *McLean* case—the lack of educational materials appropriate for public schools:

> [B]alanced presentation of creation science and evolution is educationally valuable, and in fact is more educationally valuable than indoctrination in just the viewpoint of evolution. Presentation of alternate scientific explanations has educational benefit, and balanced presentation of creation science and evolution does exactly that. Creation science can indeed be taught in the classroom in a strictly scientific sense, and a textbook can present creation science in a strictly scientific sense, either as a supplement or as part of a balanced presentation text.[97]

Conveniently enough, in 1984, Kenyon just happened to already be a coauthor on a supplemental textbook that allegedly offered just such a balanced presentation—the prototype of *Pandas*.

THE PROTOTYPE *PANDAS* DRAFT

The first known draft of *Pandas*, titled *Creation Biology Textbook Supplement*, was produced by Davis and Kenyon in 1983.[98] In Buell's response to *Kitzmiller*, he made some categorical statements about the *Pandas* drafts in support of his contention that ID was not creationism:

> Was there a systematic replacement in *Pandas* manuscripts of the words "creationism," "creationist," and "creation" with the term "intelligent design," as alleged? Let's look at the three terms one at a time. Neither "creationism" nor its synonym, "Creation Science," was ever used in any *Pandas* manuscript, as alleged. Although they differ by only one letter, "creationist," is not a variant of "creationism"; it is a variant of "creation," a modifier that means "of the viewpoint of creation." When we began work on the book, we agreed that if we couldn't make a convincing empirical case, we would not go forward. But during the roughly five years the manuscript was being written, we used the word "creation" and sometimes "creationist" as placeholder terminology. The complete absence of manuscripts or portions of manuscripts teaching the tenets of the six-part description of Creation Science is eloquent evidence that this and only this is why those two words are sprinkled throughout old drafts.[99]

The Discovery Institute's Casey Luskin has echoed these arguments in a post-*Kitzmiller* campaign to try to explain away the creationist language in the *Pandas* drafts. According to Luskin, "When certain pre-publication drafts of *Pandas* used terms such as 'creation' and 'creationist,' they used them in a way that rejected 'creationism' as defined by the courts and popular culture."[100]

Even if one has the saintlike innocence that would be required to believe that "creationists" advocating "creation" were talking about something "fundamentally different" from "creationism," the evidence from *Creation Biology* does not match the *Pandas* apologists' claims. Not only was *Creation Biology* written by two committed creation scientists, one of whom was at the time actively involved in defending creation science in court, but the draft textbook actually does refer to "creation science." Some passages will illustrate:

- What does creation science make of all this?[101]
- If the conventional timetable of billions of years is correct (many creationist scientists do not accept this timetable) then life may have been on Earth for about 4 billion years, when abundant fossil life presumably began about 600 million years ago (the approximate beginning of the Cambrian Period).[102]
- Now let us assume that instead of being billions of years old, the Earth is actually only 10,000–20,000 years old as many creationist scientists have concluded.[103]

- We do hope your main textbook contained a discussion of crossing-over. It's hard enough writing a creation science book without having to write an elementary biology book at the same time![104]

The claim that the *Pandas* drafts avoided the six-part description of creation science used in the *McLean* case, which ID advocates represent as a "strict" definition of creationism—a global flood, young Earth, and so on—is also dubious. As the reader can see in the above quotes, Creation Biology actually does treat the young-earth view as if it had scientific validity. For that matter, so does the published, "intelligent design" version of *Pandas* (cited near the beginning of this chapter)! The term "creationism" and the "strict" creation science concept of "special creation" are also found in *Creation Biology*. For instance:

> One of the criticisms advanced against creationism embodies this very point: which creation story shall we consider to be the true one? Yet it seems to us that if creation by an intelligent entity is taken to be the kernel of creationism, then the most fundamental postulate of creationism must be that species were originally created as species, which immediately raises the questions of just which ones, how many of them, and when?
>
> One might think that some higher taxonomic category might have been the original unit of creation—the genus, perhaps, or the family or even the phylum. Yet this begs the question. Categories of classification are for the most part artificial; indeed the surprising thing is that there is as much agreement among taxonomists as does exist. As it is, there are minor invertebrate phyla all of whose contained species number scarcely more than a dozen. If there were but a single species that could be considered to belong to that phylum, still, a species it would be. There can therefore be no doubt that if, indeed, a creator produced phyla initially, those phyla would contain specially created species. The original species were created, if creationism is true at all.[105]

This passage may seem familiar to some readers. Compare it to the published version of *Pandas*:

> With respect to the existence of a single, unified modern theory of intelligent design, holders of this view point out that, while there are difficulties that need to be worked out, all adhere to the same fundamental aspects. Most signifi-

> cantly, all design proponents hold that major groups of organisms had their own origins. While there is diversity among design proponents, it is not unlike the diversity among Darwinists with respect to modern evolutionary theory.
>
> What unit of classification was originally designed? Was it the species, or genus, or family, or even the phylum? Some Darwinists insist that those who hold true design must be able to answer this question. In point of fact, the question is irrelevant, because categories of classification are largely artificial, human groupings. [. . .] Indeed, it is surprising that taxonomists agree as much as they do. There are minor invertebrate phyla which contain not more than a dozen species. It is theoretically possible that a phylum and a species can be identical; a phylum could contain only a single species. [. . .] In any case, there can be no doubt that if a designer produced phyla initially, those phyla must contain specially fashioned representatives, or species.[106]

The creationists, and the intelligent design advocates, like to claim that evolution only occurs within strict limits. In biology, this is false; but in the evolution of creationism, it applies in spades.

CONCLUSION

This chapter has only explored the beginnings of the story of the emergence of *Pandas* and intelligent design. Even so, it is apparent that various key features of ID, usually attributed to Phillip Johnson or other figures, were in place quite early. For example, the "big tent" for young-earthers and old-earthers was already in place by 1984. Indeed, the *Pandas* book was being written by two young-earth creationists, under the editorship of two old-earth creationists.

In light of this study, a reinterpretation of other pieces of ID mythology is clearly in order. Topping the list is *The Mystery of Life's Origins*. Was it really as revolutionary as modern ID advocates say? Briefly, my opinion is that it is yet another work within the two-model genre, and its primary "innovation" was to reincarnate the creationists' hapless and much-ridiculed second law of thermodynamic argument in the form of an equally hapless technical argument about "information" and "specified complexity."[107] The intuitive "information" argument was of course already present in Geisler's *McLean* line about getting a dictionary from an explosion in a printshop.

The close intellectual relationship among Geisler, Thaxton, and proto-ID can be seen in many other places: another "new" concept in *Mystery*, the "origins science" versus "operation science" distinction, usually attributed to Thaxton, already existed in all but name in Geisler's *McLean* testimony. Geisler expanded on the concept of "origins science" in the 1987 book *Origin Science: A Proposal for the Creation-Evolution Controvesy*, after extensive discussions with Thaxton as well as Bradley, Kenyon, and Gish.[108] The "origins science" concept, relabeled as "historical science," soon became a major plank in Stephen Meyer's argumentation, for example, in his "Note to Teachers" in the second edition of *Pandas*.[109] From *McLean* through to *Pandas*, creationists/ID advocates have attempted to slice off a special "origins" category in order to create a kind of scientific ghetto where everything is speculative analogy, "theory not fact,"[110] and creation/design is just as good as evolution.

Appendix 1 of *Origin Science* is "Paley's Updated Argument," which closely paraphrases the famous watchmaker passage from *Natural Theology*, but substitutes Mount Rushmore for the watch. The appendix is essentially a reprint of Geisler's 1983 article in *Creation/Evolution*, which attributes the Mount Rushmore argument to Thaxton.[111] (The "Mount Rushmore" argument is now an endlessly repeated ID talking point, even forming a central theme in the ID-skeptical documentary *Flock of Dodos* [Prairie Starfish Productions, 2006].) However, the earliest publication of the Mount Rushmore argument is actually in Geisler's 1982 apologetics book on miracles.[112]

Continuing the theme of connection of proto-ID to the creation science litigation, appendix 4 of *Origin Science* reprints the dissenting opinion in the en banc appeal of the Louisiana Balanced Treatment Act, where the creationists lost by a narrow margin of 8 to 7 on December 12, 1985. Other connections between FTE and the *Edwards* litigation were investigated during *Kitzmiller*[113] and in other work, although much remains unexplored. Notably, the mysterious Ad Hoc Origins Committee, described in pro-ID histories as a academic group of "like-minded skeptics of evolution"[114] that formed the nucleus of the ID movement, actually originated because of the *Edwards* litigation. According to the *ASA Newsletter*, Thaxton's 1988 proto-ID conference in Tacoma, Washington, sometimes described as where the theory of ID was founded,[115] actually "had its origin in a 1986 gathering of a group whose advice on scientific and educational matters was sought by members of the

Christian Legal Society being drawn into public policy disputes in the courts," the "'Ad Hoc' meeting."[116] On June 19, 1986, the Christian Legal Society and National Association of Evangelicals submitted an amicus brief to the Supreme Court in support of the Louisiana Balanced Treatment Act.[117] The Counsel of Record was Michael Woodruff, who was a member of the Ad Hoc Committee, and who would later write an eleven-page "legal scrutiny" of *Pandas* and determine that it was constitutional under *Edwards*.[118]

The evolution of *Pandas* between 1984 and 1989, and 1989 and 1993, also needs further exploration. After *Kitzmiller*, attempts to distance *Pandas* from creationism and ID from *Pandas* have been on the increase. However, these attempts are hopeless. For example, another author of *Pandas*, Nancy Pearcey, was brought on to the project around 1988 to write the overview chapter. Pearcey was a longtime editor of the *Bible-Science Newsletter*, a young-earth creationist newspaper considered unsophisticated even by other creationists;[119] in 1989 she published creationist versions of most of the overview chapter in three articles in the *Newsletter*.[120] Many other "Critical Reviewers" on *Pandas* were young-earth creationists: John Baumgardner, Harold Coffin (of *McLean* fame), L. James Gibson, Paul Nelson, and Kurt Wise. Many others were old-earth creationists. Notables include Ray Bohlin, Walter Bradley, Norman Geisler (of *McLean* fame), Stephen Meyer, Gordon Mills, J. P. Moreland, Alvin Plantinga, and John Wiester. As for distancing ID from *Pandas*, Stephen Meyer's chapter of *Pandas* is posted in multiple locations on the Discovery Institute's Web sites and various other ID sites; and six *Pandas* coauthors are ID fellows, including Kenyon, Thaxton, Pearcey, Meyer, and Hartwig. The sixth is Michael Behe, who is strangely enough not listed as an author on *Pandas*, even though he wrote pages 141–46 of the second edition of *Pandas* in 1993. The section is on blood-clotting and presents Behe's irreducible complexity argument (but not the actual term). Behe's coauthorship was unknown until the *Kitzmiller* case, when a close reading of *Pandas* turned up the extremely close textual similarities to chapter 4 of Darwin's *Black Box*. All of the other major ID figures not already mentioned (including Phillip Johnson, William Dembski, and Jonathan Wells) have explicitly endorsed *Pandas* in print on FTE's Web site or elsewhere.

These topics, dealing with matters after 1989, while interesting, are not going to change the big picture of the origins of ID. The political/legal strategy of "creation science" collapsed suddenly in 1981–1982. The next legal strategy was to strip down creation science to make it appear even less sectarian than before. When the stripped-down creationism strategy failed in *Edwards*, a few creationists decided to try again with a new label, and that is why "intelligent design" is the term on our lips today.

ACKNOWLEDGMENTS

I would like to thank Robert T. Pennock, Glenn Branch, and Barbara Forrest for help improving this essay through many comments; and Ted Davis and Susan Spath for useful discussions.

NOTES

1. *Mclean v. Arkansas Board of Education*, 529 F.Supp 1255 (E.D.Ark. 1982).

2. E. J. Larson, *Trial and Error: The American Controversy over Creation and Evolution* (New York: Oxford University Press, 2003).

3. *Edwards v. Aguillard*, 482 US 578 (S.Ct. 1987).

4. M. Ruse, *But Is It Science?: The Philosophical Question in the Creation/Evolution Controversy* (Amherst, NY: Prometheus Books, 1996).

5. For examples, see the various pages returned as results from a Google search of the Discovery Institute Web site on the phrase "Intelligent design is not creationism" (19 hits returned in April 2007). The argument over this issue can also be found throughout the *Kitzmiller* briefs and expert witness reports, available online at http://www2.ncseweb.org/kvd/.

6. Dover Area School District, "Biology Curriculum Press Release," http://www2.ncseweb.org/kvd/exhibits/DASD/2004-11-19_DASD_press_release_Biology_Curriculum.pdf (accessed March 30, 2007).

7. The president of the Foundation for Thought and Ethics (FTE), the think tank that produced *Pandas*. Buell and FTE will be introduced later in the essay.

8. The preface, along with several other sections of the book, was publicly posted on the Web site of ID leader William Dembski in 2004. The book, to be titled *The Design of Life: Discovering Signs of Intelligence in Biological Systems*, has been promised for several years, but has not yet been published. FTE fund-raising letters indicate

that the necessary funds are still being gathered. The 2004 version of *Design of Life* lists Behe, Davis, Dembski, Kenyon, and Wells as authors, however in his 2005 deposition for the *Kitzmiller* case, Behe denied being an author on the revised edition. New descriptions of the book posted at Dembski's overwhelmingevidence.com Web site in 2006 indicate that only Dembski and Wells are authors. Dembski has been FTE's academic editor since at least 2002, according to an FTE fund-raising letter.

9. J. Buell, preface, in *The Design of Life: Discovering Signs of Intelligence in Biological Systems*, ed. M.J. Behe, Percival Davis, William A. Dembski, Dean H. Kenyon, and Jonathan Wells (Richardson, TX: Foundation for Thought and Ethics, 2004), pp. iv–vi.

10. For documentation of parallels, see B. Forrest, Expert Witness Report, Plaintiffs, *Kitzmiller v Dover*, Middle District of Pennsylvania, 04-CV-2688: 1-49, http://www2.ncseweb.org/kvd/all_legal/2005-03_expert_witnesses/2005-04-01_Forrest_Ps_expert_report_readable.pdf; N.J. Matzke and P.R. Gross, "Analyzing Critical Analysis: The Fallback Antievolutionist Strategy," in *Not in Our Classrooms: Why Intelligent Design Is Wrong for Our Schools*, ed. E.C. Scott and G. Branch (Boston: Beacon Press, 2006), pp. 28–56; and B. Forrest, "From 'Creation Science' to 'Intelligent Design': Tracing Id's Creationist Ancestry" (2006), http://www.creationismstrojanhorse.com/Tracing_ID_Ancestry.pdf (accessed April 2007).

11. F. Sonleitner, "What's Wrong with *Pandas*?" (1994), http://www.ncseweb.org/article.asp?category=21 (accessed March 30, 2007).

12. P. W. Davis, D. H. Kenyon, and C. B. Thaxton, *Of Pandas and People: The Central Question of Biological Origins* (Dallas, TX: Haughton Publication Co., 1993), p. 92.

13. Ibid., p. 77.

14. Genesis 2:7, King James Version.

15. Davis et al., *Of Pandas and People*, pp. 99–100.

16. K. Padian, "Gross Misrepresentation," *Bookwatch Reviews* 2, no. 11 (1989), http://www.ncseweb.org/resources/articles/9767_22_padian_1989_gross_misr_10_26_2004.asp.

17. E. C. Scott, "New Creationist Book on the Way," *NCSE Reports* 9, no. 2 (1989): 21, http://www.ncseweb.org/resources/articles/2279_31_scott_1989_new_creatio_11_23_2004.asp.

18. M. Ruse, "They're Here!" *Bookwatch Reviews* 2, no. 11 (1989), http://www.ncseweb.org/resources/articles/2216_23_ruse_1989_they39re__10_26_2004.asp.

19. R. T. Pennock, *Tower of Babel: The Evidence against the New Creationism* (Cambridge, MA: MIT Press, 1999), p. 293.

20. B. Forrest and P. R. Gross, *Creationism's Trojan Horse: The Wedge of Intelligent Design* (Oxford: Oxford University Press, 2004).

21. L. Krishtalka, "Don't Let Creationists Corrupt Science Standards," *Pittsburgh Post Gazette*, 2001, http://www.post-gazette.com/forum/20010107edkristalka9.asp.

22. The experts' witness statements were submitted in the spring of 2005 and are available at http://www2.ncseweb.org/kvd/index.php?path=experts/.

23. N. J. Matzke, "Design on Trial: How NCSE Helped Win the *Kitzmiller* Case," *Reports of the National Center for Science Education* 26, nos. 1–2 (2006): 37–44.

24. The exhibit numbers are taken from Clerk of the Court, E. Rothschild, S. G. Harvey, A. H. Wilcox, C. J. Lowe, T. B. Schmidt III, W. J. Walczak, P. K. Knudsen, A. Khan, R. B. Katskee, and A. J. Luchenitser, Clerk's Exhibit Listing for Plaintiffs, *Kitzmiller v. Dover*, Middle District of Pennsylvania, 04-CV-2688: 1–57, http://www2.ncseweb.org/kvd/all_legal/2005-12-08_Ps_exhibit_list.pdf.

25. Plaintiffs' Exhibit-562, chap. 2, p. 14; chap. 2, p. 15. Emphasis added. Similar versions of this sentence can be found back to the 1986 *Biology and Creation.*

26. The charts are available online at http://www2.ncseweb.org/kvd/index.php?path=exhibits/.

27. Plaintiffs' Exhibit-567, chap. 3, p. 41.

28. Larson, *Trial and Error: The American Controversy over Creation and Evolution*; R. L. Numbers, *The Creationists: From Scientific Creationism to Intelligent Design* (Cambridge, MA: Harvard University Press, 2006).

29. There was both conflict and cooperation between Clark and the ICR, and the details are hard to resolve, since Bird and the ICR blamed Clark for the loss after the trial. On the other hand, Gish attended the trial and passed notes to the defense, and some of the *McLean* experts were later used by Bird in the Louisiana case. See N. L. Geisler, *Creation and the Courts: Eighty Years of Conflict in the Classroom and the Courtroom (With Never before Published Eyewitness Testimony from the Scopes Trial)* (Wheaton, IL: Crossway Books, 2007) for a summary of the Clark/Bird dispute.

30. Ibid., p. 179.

31. Ibid., p. 159; italics added.

32. Ibid., p. 168.

33. Ibid.

34. Ibid.

35. Ibid., p. 169.

36. Ibid.; bracketed "an" original.

37. R. Lewin, "Creationism on the Defensive in Arkansas," *Science* 215, no. 4258 (1982): 33–34.

38. Court transcripts of the defense case were never completed, because the decision was issued before the transcripts were finished, and the decision was not

appealed. Therefore, only the plaintiffs' testimony (the first half of the trial) is currently available (see the *McLean* Documentation Project at http://www.antievolution.org/projects/mclean/new_site/index.htm). For the defense testimony, N. L. Geisler, A. F. Brooke II, and M. J. Keough, *The Creator in the Courtroom "Scopes II": The 1981 Arkansas Creation-Evolution Trial* (Milford, MI: Mott Media, 1982) is the most detailed account available, with the addition of Geisler, *Creation and the Courts*, which published for the first time the text of Geisler's testimony.

39. Geisler et al., *The Creator in the Courtroom "Scopes II,"* pp. 139–41; Geisler's paraphrase.

40. Morrow's puzzling position seems to have been that of a general contrarian. When cross-examined and asked about the flat-earth theory, Morrow said it "would be a very interesting model to teach" (R. M. Baum, "Science Confronts Creationist Assault," *Chemical & Engineering News* 60, no. 3 [1982]: 20).

41. Geisler et al., *The Creator in the Courtroom "Scopes II,"* pp. 125–26; Geisler's paraphrase.

42. In one contemporary account of the trial in *Creation/Evolution* (F. Edwords, "Victory in Arkansas: The Trial, Decision, and Aftermath," *Creation/Evolution* 3, no. 1 [1982]: 36, http://www.ncseweb.org/resources/articles/8661_issue_07_volume_3_number_1__3_4_2003.asp#Victory%20in%20Arkansas), Helder is described as the vice president of the CRS. However, a search of the *CRS Quarterly* for December 1981 and surrounding dates indicates that others held this position. In the 1990s, Answers in Genesis and other creationist organizations describe Helder, who is still writing creationist articles, as the vice president of the CSSAA.

43. Geisler et al., *The Creator in the Courtroom "Scopes II,"* p. 142; Geisler's paraphrase.

44. Ibid., p. 146.

45. Ibid.; Geisler's paraphrase. The ellipsis before "How did" is original to Geisler.

46. Ibid., p. 147; Geisler's paraphrase.

47. Matzke and Gross, "Analyzing Critical Analysis: The Fallback Antievolutionist Strategy," pp. 28–56.

48. F. Hoyle and N. C. Wickramasinghe, *Evolution from Space: A Theory of Cosmic Creationism* (New York: Simon & Schuster, 1982). The original edition, with only the short title, was published in England in 1981.

49. The speck of science behind the bacteria-in-space claim was spectroscopic evidence of organic molecules in interstellar dust clouds. The existence of these organic molecules, which can form through simple reactions, is well accepted. How-

ever, there are numerous problems for bacteria-in-space, including the fact that cosmic radiation will kill anything not well shielded—a particularly severe problem over the millions of years required for interstellar drifting.

50. *McLean v. Arkansas* (1982).

51. Barry Gross was one of the philosophers who consulted on the *McLean* case. His article (B. R. Gross, "Commentary: Philosophers at the Bar—Some Reasons for Restraint," *Science, Technology & Human Values* 8, no. 4 [1983]: 30–38) gives some useful background on the behind-the-scenes discussions that resulted in the philosophy testimony that was presented to Overton. Strangely, this article is almost never cited in the numerous subsequent discussions of Laudan's critique of Ruse and Overton.

52. R. B. Bliss, *Origins: Two Models: Evolution, Creation* (San Diego, CA: Creation-Life Publishers, 1976); J. N. Moore and H. S. Slusher, *Biology: A Search for Order in Complexity* (Grand Rapids, MI: Zondervan Publishing House, 1970); H. M. Morris, *Scientific Creationism* (San Diego, CA: Creation-Life Publishers, 1974); see also *McLean v. Arkansas* (1982) for Overton's assessment.

53. *McLean v. Arkansas* (1982); emphasis in original.

54. Larson, *Trial and Error: The American Controversy over Creation and Evolution*, p. 166.

55. Ibid. The lawsuit with creationists as plaintiffs was *Keith v. Louisiana Department of Education.* The ACLU lawsuit was initially *Aguillard v. Treen.* Governor Edwards eventually succeeded Governor Treen, and the names are reversed on appeal, so the Supreme Court case is known as *Edwards v. Aguillard.*

56. "Unbiased Biology Textbook Planned," *Origins Research* 4, no. 2 (1981): 1.

57. "Life in the Bias Sphere," *Newsletter of the American Scientific Affiliation* 23, no. 4 (1981), http://www.asa3.org/ASA/topics/NewsLetter80s/AUGSEP81.html.

58. As a Campus Crusade for Christ trainee, Buell shared a room with the visiting guest lecturer Henry Morris at one point in the 1960s. H. M. Morris, *A History of Modern Creationism* (San Diego, CA: Master Book Publishers, 1984), p. 159.

59. E. Hatfield, "Probing the Campus," *Christianity Today* 20, no. 16 (1976): 42–44.

60. L. P. Lester, R. G. Bohlin, and V. E. Anderson, *The Natural Limits to Biological Change* (Grand Rapids, MI: Probe Ministries International, 1984). This book contains several instances of the term "intelligent design" and makes some specific arguments later taken up in *Pandas*; for example, a rebuttal to Stephen Jay Gould's argument about the panda's thumb appears in the book. The panda's thumb argument became the source of the title of *Pandas.*

61. "Scholarly Opportunities Abound," *Newsletter of the American Scientific Affiliation* 22, no. 2 (1980), http://www.asa3.org/ASA/topics/NewsLetter80s/APRMAY80.html.

62. "Personals," *Newsletter of the American Scientific Affiliation* 23, no. 3 (1980), http://www.asa3.org/ASA/topics/NewsLetter80s/JUNJUL81.html.

63. "First-Order Reflections of an X-Ray Diffractionist," *Newsletter of the American Scientific Affiliation* 12, no. 5 (1980), http://www.asa3.org/ASA/topics/NewsLetter70s/OCT70.html.

64. J. Budziszewski and D. L. Weeks, *Evangelicals in the Public Square: Four Formative Voices on Political Thought and Action* (Grand Rapids, MI: Baker Academic, 2006), p. 73.

65. Ibid.

66. Ibid., p. 74.

67. C. B. Thaxton, "Presuppositions and Catastrophism," *Pensée* 2, no. 2 (1972), http://www.catastrophism.com/online/pubs/journals/pensee/ivr01/index.htm.

68. Thaxton's scientific output in the 1970s consisted of several coauthored articles on physical chemistry; see C. B. Thaxton, "Curriculum Vitae," http://www3.ksde.org/outcomes/sceptcvthaxton.pdf (accessed July 2005).

69. L. Witham, *By Design: Science and the Search for God* (San Francisco, CA: Encounter Books, 2003), p. 116.

70. Implied in L. Witham, *Where Darwin Meets the Bible: Creationists and Evolutionists in America* (New York: Oxford University Press, 2002), p. 220.

71. C. B. Thaxton, W. L. Bradley, and R. L. Olsen, *The Mystery of Life's Origin: Reassessing Current Theories* (New York: Philosophical Library, 1984).

72. C. Thaxton, A. C. Wilcox, D. E. Boyle, and E. L. White III, July 19, 2005 Deposition of Charles Thaxton, *Kitzmiller v Dover*, Middle District of Pennsylvania, 04-CV-2688: 62, http://www2.ncseweb.org/kvd/depo/2005-07-19_deposition_Thaxton_escript.pdf.

73. FTE, "Articles of Incorporation of the Foundation for Thought and Ethics," document submission to the Texas Secretary of State, *NCSE archives*: 1–4.

74. J. D. Spearing, E. J. Rothschild, T. B. Schmidt III, W. J. Walczak, P. T. Gillen, N. S. Benn, T. J. Barna, D. E. Boyle, L. G. Brown III, and J. A. Buell, Transcript of Oral Argument on July 14, 2005, on Reporters' Motion to Quash and FTE's Motion to Intervene, *Kitzmiller v. Dover*, Middle District of Pennsylvania, 04_CV-2688: 86, http://www2.ncseweb.org/kvd/all_legal/2005-05_FTE-related/2005-06_FTE_intervention/2005-07-14_transcript_pretrial_hearing_on_reporters_and_FTE-Buell.pdf.

75. FTE, "What Is the Foundation for Thought and Ethics?" document submission to Internal Revenue Service, *NCSE archives*: 1.

76. Thaxton et al., July 19, 2005 Deposition of Charles Thaxton, *Kitzmiller v Dover*, pp. 52–53.

77. Spearing et al., Transcript of Oral Argument on July 14, 2005, on Reporters' Motion to Quash and FTE's Motion to Intervene, *Kitzmiller v. Dover*, p. 99.

78. Ibid., p. 101.

79. Geisler et al., *The Creator in the Courtroom "Scopes II,"* p. 194.

80. An e-mail to Wendell Bird asking for recollections on this point, sent in March 2007, was unanswered.

81. Geisler, *Creation and the Courts*, pp. 23–24.

82. *Newsletter of the American Scientific Affiliation* 24, no. 3 (1982), http://www.asa3.org/ASA/topics/NewsLetter80s/JUNJUL82.html.

83. C. B. Thaxton, "Creationism/Evolution: Contrasting Views (no. 3)," *Crossroads: Science Meets Society* 2 (1982): 7–8.

84. Ibid. Crossroads was a series of monographs edited by Robert C. Barkman, Springfield College, rbarkman@spfldcol.edu. A copy was obtained courtesy of Dr. Barkman, and it can now be found in the NCSE archives.

85. W. Frair, "Report of 1974 Board of Directors Meeting," *Creation Research Society Quarterly* 11, no. 2 (1974): 126–27.

86. P. W. Davis, "Land-Dwelling Vertebrates and the Origin of the Tetrapod Limb," *Creation Research Society Quarterly* 2, no. 1 (1965): 27–31; G. F. Howe and P. W. Davis, "Natural Selection Reexamined," *Creation Research Society Quarterly* 8, no. 1 (1971): 30–43, http://creationresearch.org/crsq/abstracts/sum8_1.html.

87. P. W. Davis, "Biographical Sketch," http://www.clearwater.edu/faculty/BillDavis.pdf (accessed February 2007).

88. W. Frair and P. W. Davis, *The Case for Creation* (Chicago: Moody Press, 1967; 1983).

89. E.g., S. C. Meyer, "Danger: Indoctrination. A Scopes Trial for the '90s," *Wall Street Journal*, 1993, http://www.discovery.org/scripts/viewDB/index.php?command=view&id=93.

90. R. Salner, "Professor Teaches a Supernatural Creation of World," *San Francisco Examiner*, 1980, p. A9.

91. R. Lewin, "Creationism on the Defensive in Arkansas," *Science* 215, no. 4258 (1982): 33–34.

92. J. Weatherly, "Creationists Lose in Arkansas: Missing Witnesses and a Divided Defense Muddled the Issue," *Christianity Today* (1982): 29.

93. W. Bird, Plaintiffs' Summaries of Expert Testimony, Plaintiffs, *Keith v. Louisiana Department of Education*: 1–76. This document is not dated, but since the creationists were plaintiffs and the description of expert testimony is used to inform the parties of planned testimony at trial, the date must be 1982. The document was provided by the ACLU archives at Mudd Library, Princeton.

94. W. Bradley, foreword to *Origin Science: A Proposal for the Creation-Evolution Controversy* (Grand Rapids, MI: Baker Book House, 1987), p. 7.

95. Larson, *Trial and Error: The American Controversy over Creation and Evolution*, p. 168.

96. D. H. Kenyon, Affidavit of Expert Witness Dean Kenyon, *Edwards v. Aguillard*, Eastern District of Louisiana, Civil Action No. 81-4787.

97. Ibid.

98. For further analysis of the drafts, see B. Forrest, Supplement to Expert Witness Report, Plaintiffs, *Kitzmiller v. Dover*, Middle District of Pennsylvania, 04-CV-2688: 1–14, http://www2.ncseweb.org/kvd/all_legal/2005-03_expert_witnesses/2005-07-29_Forrest_Ps_expert_supplemental_readable.pdf.

99. J. Buell, "Intelligent Design and the Dover Case," http://www.fteonline.com/buell-dallas-blog.html (accessed January 29, 2006).

100. C. Luskin, "Response to Barbara Forrest's *Kitzmiller* Account Part V: Phillip Johnson and *Of Pandas and People*," http://www.evolutionnews.org/2006/09/response_to_barbara_forrests_k_4.html (accessed April 5, 2006).

101. Plaintiffs' Exhibit-653, chap. 6, p. 28.

102. Ibid., chap. 2, p. 2 (insert).

103. Ibid.

104. Ibid., pp. 6–20.

105. Ibid., chap. 6, pp. 1–2; emphases added.

106. P. W. Davis et al., *Of Pandas and People*, p. 78.

107. Bradley is the key figure here; in various 1980s interviews he basically states that he performed this conversion in rebuttal to the evolutionist criticisms of the creation scientists' "Second Law" argument. He might have been doing this directly in response to the *McLean* plaintiffs' witness Harold Morowitz (whose main job was to rebut the Second Law argument), especially because Bradley was signed up as a creationist expert for the Louisiana case. However, I have not yet fully investigated this issue.

108. N. L. Geisler and J. K. Anderson, *Origin Science: A Proposal for the Creation-Evolution Controversy* (Grand Rapids, MI: Baker Book House, 1987), p. 11.

109. M. D. Hartwig and S. C. Meyer, "A Note to Teachers," in *Of Pandas and People: The Central Question of Biological Origins* (Dallas, TX: Haughton Publishing), p. 159. Meyer also made this argument in a paper given at the 1992 ID conference at Southern Methodist University (Robert T. Pennock, personal communication), published as S. C. Meyer, "Laws, Causes and Facts: Response to Michael Ruse. Darwinism, Science or Philosophy?" Proceedings of a symposium titled "Darwinism, Scientific

Inference or Philosophical Preference?" Southern Methodist University, Dallas, TX, March 26–28, 1992, *Foundation for Thought and Ethics* (1994): 29–40.

110. For the origins and significance of this favorite snippet of rhetoric, used unblinkingly by all creationists including ID advocates, see E. C. Scott, Expert Witness Statement by Eugene C. Scott, Plaintiffs, *Selman v. Cobb*, Northern District of Georgia, 1:02-CV-2325-CC: 1–26, http://www2.ncseweb.org/selman/2006-11-16_Scott_expert_report.pdf.

111. N. L. Geisler, "A Scientific Basis for Creation: The Principle of Uniformity," *Creation/Evolution* 4, no. 3 (1983): 1–6.

112. N. L. Geisler and R. C. Sproul, *Miracles and Modern Thought* (Grand Rapids, MI: Zondervan, 1982), p. 58. As noted previously, Thaxton was an editor on this book and so could still have suggested the Mount Rushmore idea to Geisler.

113. B. Forrest, Supplement to Expert Witness Report, Plaintiffs, *Kitzmiller v. Dover*; see also the revised edition of B. Forrest and P. Gross, *Creationism's Trojan Horse* (New York: Oxford University Press, 2007).

114. T. Woodward, *Doubts about Darwin: A History of Intelligent Design* (Grand Rapids, MI: Baker Books, 2003), p. 85.

115. E.g., by Stephen Meyer in an interview with Tony Snow of Fox News on August 6, 2005. Transcript available at http://www.pandasthumb.org/archives/2006/04/my_encounter_wi.html.

116. "Some Summer Summary," *Newsletter of the American Scientific Affiliation* 30, no. 4 (1988), http://www.asa3.org/ASA/topics/NewsLetter80s/AUGSEP88.html.

117. M. J. Woodruff, K. W. Colby, S. E. Ericsson, and F. D. Montgomery, Brief of the Christian Legal Society and National Association of Evangelicals as Amici Curiae Supporting Appellants, *Edwards v. Aguillard*, US Supreme Court, No. 85-1513. This cannot be argued here, but the CLS/NAE amicus brief is another proto-ID document, citing Thaxton and even using the term "intelligent design" twice.

118. J. Cole. "More Patter of Little Pandas," *NCSE Reports* 15, no. 1 (1995): 21, http://www.ncseweb.org/resources/articles/5844_41_cole_1995_more_patter__11_24_2004.asp.

119. Numbers, *The Creationists*.

120. This is documented at http://www.pandasthumb.org/archives/2005/09/why_didnt_they.html.

CHAPTER 21

WHAT IS DARWINISM?

PHILLIP E. JOHNSON

There is a popular television game show called *Jeopardy*, in which the usual order of things is reversed. Instead of being asked a question to which they must supply the answer, the contestants are given the answer and asked to provide the appropriate question. This format suggests an insight that is applicable to law, to science, and indeed to just about everything. The important thing is not necessarily to know all the answers, but rather to know what question is being asked.

That insight is the starting point for my inquiry into Darwinian evolution and its relationship to creation, because Darwinism is the answer to two very different kinds of questions. First, Darwinian theory tells us how a certain amount of diversity in life-forms can develop once we have various types of complex living organisms already in existence. If a small population of birds happens to migrate to an isolated island, for example, a combination of inbreeding, mutation, and natural selection may cause this isolated population to develop different characteristics from those possessed by the ancestral pop-

This paper was originally delivered as a lecture at a symposium at Hillsdale College in November 1992. Papers from the symposiuim were published in the collection *Man and Creation: Perspectives on Science and Theology*, ed. M. Bauman (Hillsdale, MI: Hillsdale College Press, 1993). Reprinted with permission.

ulation on the mainland. When the theory is understood in this limited sense, Darwinian evolution is uncontroversial, and has no important philosophical or theological implications.

Evolutionary biologists are not content merely to explain how variation occurs within limits, however. They aspire to answer a much broader question—which is how complex organisms like birds, flowers, and human beings came into existence in the first place. The Darwinian answer to this second question is that the creative force that produced complex plants and animals from single-celled predecessors over long stretches of geological time is essentially the same as the mechanism that produces variations in flowers, insects, and domestic animals before our very eyes. In the words of Ernst Mayr, the dean of living Darwinists, "transspecific evolution [i.e., macroevolution] is nothing but an extrapolation and magnification of the events that take place within populations and species." Neo-Darwinian evolution in this broad sense is a philosophical doctrine so lacking in empirical support that Mayr's successor at Harvard, Stephen Jay Gould, once pronounced it in a reckless moment to be "effectively dead." Yet neo-Darwinism is far from dead; on the contrary, it is continually proclaimed in the textbooks and the media as unchallengeable fact. How does it happen that so many scientists and intellectuals, who pride themselves on their empiricism and open-mindedness, continue to accept an unempirical theory as scientific fact?

The answer to that question lies in the definition of five key terms. The terms are *creationism*, *evolution*, *science*, *religion*, and *truth*. Once we understand how these words are used in evolutionary discourse, the continued ascendancy of neo-Darwinism will be no mystery and we need no longer be deceived by claims that the theory is supported by "overwhelming evidence." I should warn at the outset, however, that using words clearly is not the innocent and peaceful activity most of us may have thought it to be. There are powerful vested interests in this area which can thrive only in the midst of ambiguity and confusion. Those who insist on defining terms precisely and using them consistently may find themselves regarded with suspicion and hostility, and even accused of being enemies of science. But let us accept that risk and proceed to the definitions.

The first word is *creationism*, which means simply a belief in creation. In Darwinist usage, which dominates not only the popular and professional scientific literature but also the media, a creationist is a person who takes the creation account in the book of Genesis to be true in a very literal sense. The earth was created in a single week of six twenty-four-hour days no more than ten thousand years ago; the major features of the geological were produced by Noah's flood; and there have been no major innovations in the forms of life since the beginning. It is a major theme of Darwinist propaganda that the only persons who have any doubts about Darwinism are young-earth creationists of this sort, who are always portrayed as rejecting the clear and convincing evidence of science to preserve a religious prejudice. The implication is that citizens of modern society are faced with a choice that is really no choice at all. Either they reject science altogether and retreat to a premodern worldview, or they believe everything the Darwinists tell them.

In a broader sense, however, a creationist is simply a person who believes in the existence of a *creator*, who brought about the existence of the world and its living inhabitants in furtherance of a *purpose*. Whether the process of creation took a single week or billions of years is relatively unimportant from a philosophical or theological standpoint. Creation by gradual processes over geological ages may create problems for biblical interpretation, but it creates none for the basic principle of theistic religion. And creation in this broad sense, according to a 1991 Gallup poll, is the creed of 87 percent of Americans. If God brought about our existence for a purpose, then the most important kind of knowledge to have is knowledge of God and of what He intends for us. Is creation in that broad sense consistent with evolution?

The answer is absolutely not, when "evolution" is understood in the Darwinian sense. To Darwinists evolution means *naturalistic* evolution, because they insist that science must assume that the cosmos is a closed system of material causes and effects, which can never by influenced by anything outside of material nature—by God, for example. In the beginning, an explosion of matter created the cosmos, and undirected, naturalistic evolution produced everything that followed. From this philosophical standpoint it follows deductively that from the beginning no intelligent purpose guided evolution. If intelligence exists today, that is only because it has itself evolved through purposeless material processes.

A materialistic theory of evolution must inherently invoke two kinds of processes. At bottom the theory must be based on chance, because that is what is left when we have ruled out everything involving intelligence or purpose.

Theories which invoke *only* chance are not credible, however. One thing that everyone acknowledges is that living organisms are enormously complex—far more so than, say, a computer or an airplane. That such complex entities came into existence simply by chance is clearly less credible than that they were designed and constructed by a creator. To back up their claim that this appearance of intelligent design is an illusion, Darwinists need to provide some complexity-building force that is mindless and purposeless. Natural selection is by far the most plausible candidate.

If we assume that random genetic mutations provided the new genetic information needed, say, to give a small mammal a start toward wings, and if we assume that each tiny step in the process of wing-building gave the animal an increased chance of survival, then natural selection ensured that the favored creatures would thrive and reproduce. It follows as a matter of logic that wings can and will appear as if by the plan of a designer. Of course, if wings or other improvements do not appear, the theory explains their absence just as well. The needed mutations didn't arrive, or "developmental constraints" closed off certain possibilities, or natural selection favored something else. There is no requirement that any of this speculation be confirmed by either experimental or fossil evidence. To Darwinists just being able to imagine the process is sufficient to confirm that something like that must have happened.

Richard Dawkins calls the process of creation by mutation and selection "the blind watchmaker," by which label he means that a purposeless, materialistic designing force substitutes for the "watchmaker" deity of natural theology. The creative power of the blind watchmaker is supported only by very slight evidence, such as the famous example of a moth population in which the percentage of dark moths increased during a period when the birds were better able to see light moths against the smoke-darkened background trees. This may be taken to show that natural selection can do something, but not that it can create anything that was not already in existence. Even such slight evidence is more than sufficient, however, because evidence is not really necessary to prove something that is practically self-evident. The existence of a potent blind watchmaker follows deductively from the philosophical premise that nature had to do its own creating. There can be argument about the details, but if God was not in the picture something very much like Darwinism simply has to be true, regardless of the evidence.

That brings me to my third term, *science.* We have already seen that Darwinists assume as a matter of first principle that the history of the cosmos and its life-forms is fully explicable on naturalistic principles. This reflects a philosophical doctrine called scientific naturalism, which is said to be a necessary consequence of the inherent limitations of science. What scientific naturalism does, however, is to transform the limitations of science into limitations upon reality, in the interest of maximizing the explanatory power of science and its practitioners. It is, of course, entirely possible to study organisms scientifically on the premise that they were all created by God, just as scientists study airplanes and even works of art without denying that these objects are intelligently designed. The problem with allowing God a role in the history of life is not that science would cease, but rather that scientists would have to acknowledge the existence of something important which is outside the boundaries of natural science. For scientists who want to be able to explain everything—and "theories of everything" are now openly anticipated in the scientific literature—this is an intolerable possibility.

The second feature of scientific naturalism that is important for our purpose is its set of rules governing the criticism and replacement of a paradigm. A paradigm is a general theory, like the Darwinian theory of evolution, which has achieved general acceptance in the scientific community. The paradigm unifies the various specialties that make up the research community, and guides research in all of them. Thus, zoologists, botanists, geneticists, molecular geologists, and paleontologists all see their research as aimed at filling out the details of the Darwinian paradigm. If molecular biologists see a pattern of apparently neutral mutations, which have no apparent effect on an organism's fitness, they must find a way to reconcile their findings with the paradigm's requirement that natural selection guides evolution. This they can do by postulating a sufficient quantity of invisible adaptive mutations, which are deemed to be accumulated by natural selection. Similarly, if paleontologists see new fossil species appearing suddenly in the fossil record, and remaining basically unchanged thereafter, they must perform whatever contortions are necessary to force this recalcitrant evidence into a model of incremental change through the accumulation of micromutations.

Supporting the paradigm may even require what in other contexts would be called deception. As Niles Eldredge candidly admitted, "We paleontologists have said that the history of life supports [the story of gradual adaptive change], all the while knowing it does not."[1] Eldredge explained that this pattern of misrepresentation occurred because of "the certainty so characteristic of evolutionary ranks since the late 1940s, the utter assurance not only that natural selection operates in nature, but that we know precisely how it works." This certainty produced a degree of dogmatism that Eldredge says resulted in the relegation to the "lunatic fringe" of paleontologists who reported that "they saw something out of kilter between contemporary evolutionary theory, on the one hand, and patterns of change in the fossil record on the other."[2] Under the circumstances, prudent paleontologists understandably swallowed their doubts and supported the ruling ideology. To abandon the paradigm would be to abandon the scientific community; to ignore the paradigm and just gather the facts would be to earn the demeaning label of "stamp collector."

As many philosophers of science have observed, the research community does not abandon a paradigm in the absence of a suitable replacement. This means that negative criticism of Darwinism, however devastating it may appear to be, is essentially irrelevant to the professional researchers. The critic may point out, for example, that the evidence that natural selection has any creative power is somewhere between weak and nonexistent. That is perfectly true, but to Darwinists the more important point is this: If natural selection did not do the creating, what did? "God" is obviously unacceptable, because such a being is unknown to science. "We don't know" is equally unacceptable, because to admit ignorance would be to leave science adrift without a guiding principle. To put the problem in the most practical terms: it is impossible to write or evaluate a grant proposal without a generally accepted theoretical framework.

The paradigm rule explains why Gould's acknowledgment that neo-Darwinism is "effectively dead" had no significant effect on the Darwinist faithful, or even on Gould himself. Gould made that statement in a paper predicting the emergence of a new general theory of evolution, one based on the macromutational speculations of the Berkeley geneticist Richard Goldschmidt.[3] When the new theory did not arrive as anticipated, the alternatives were either to stick with Ernst Mayr's version of neo-Darwinism, or to concede that biologists do not after all know of a naturalistic mechanism that can produce biological complexity. That was no choice at all. Gould had to beat a hasty retreat back to classical Darwinism to avoid giving aid and comfort to the enemies of scientific naturalism, including those disgusting creationists.

Having to defend a dead theory tooth and nail can hardly be a satisfying activity, and it is no wonder that Gould lashes out with fury at people such as myself, who call attention to his predicament.[4] I do not mean to ridicule Gould, however, because I have a genuinely high regard for the man as one of the few Darwinists who has recognized the major problems with the theory and reported them honestly. His tragedy is that he cannot admit the clear implications of his own thought without effectively resigning from science.

The continuing survival of Darwinist orthodoxy illustrates Thomas Kuhn's famous point that the accumulation of anomalies never in itself falsifies a paradigm, because "To reject one paradigm without substituting another is to reject science itself."[5] This practice may be appropriate as a way of carrying on the professional enterprise called science, but it can be grossly misleading when it is imposed upon persons who are asking questions other than the ones scientific naturalists want to ask. Suppose, for example, that I want to know whether God really had something to do with creating living organisms. A typical Darwinian response is that there is no reason to invoke supernatural action because Darwinian selection was capable of performing the job. To evaluate that response, I need to know whether natural selection really has the fantastic creative power attributed to it. It is not a sufficient answer to say that scientists have nothing better to offer. The fact that scientists don't like to say "we don't know" tells me nothing about what they really do know.

I am not suggesting that scientists have to change their rules about retaining and discarding paradigms. All I want them to do is to be candid about the disconfirming evidence and admit, if it is the case, that they are hanging on to Darwinism only because they prefer a shaky theory to having no theory at all. What they insist upon doing, however, is to present Darwinian evolution to the public as a fact that every rational person is expected to accept. If there are reasonable grounds to doubt the theory, such dogmatism is ridiculous, whether or not the doubters have a better theory to propose.

To believers in creation, the Darwinists seem thoroughly intolerant and dogmatic when they insist that their own philosophy must have a monopoly in the schools and the media. The Darwinists do not see themselves that way, of course. On the contrary, they often feel aggrieved when creationists (in either the broad or narrow sense) ask to have their own arguments heard in public and fairly considered. To insist that schoolchildren be taught that Darwinian evolution is a fact is in their minds merely to protect the integrity of science education; to present the other side of the case would be to allow fanatics to force their opinions on others. Even college professors have been forbidden to express their doubts about Darwinian evolution in the classroom, and it seems to be widely believed that the Constitution not only permits but actually requires such restriction on academic freedom. To explain this bizarre situation, we must define our fourth term: *religion.*

Suppose that a skeptic argues that evidence for geological creation by natural selection is obviously lacking, and that in the circumstances we ought to give serious consideration to the possibility that the development of life required some input from a preexisting, purposeful creator. To scientific naturalists this suggestion is "creationist" and therefore unacceptable in principle, because it invokes an entity unknown to science. What is worse, it suggests the possibility that this creator may have communicated in some way with humans. In that case there could be real prophets—persons with a genuine knowledge of God who are neither frauds nor dreamers. Such persons could conceivably be dangerous rivals for the scientists as cultural authorities.

Naturalistic philosophy has worked out a strategy to prevent this problem from arising: It labels naturalism as science and theism as religion. The former is then classified as *knowledge,* and the latter as mere *belief.* The distinction is of critical importance, because only knowledge can be objectively valid for everyone; belief is valid only for the believer, and should never be passed off as knowledge. The student who thinks that two and two make five, or that water is not made up of hydrogen and oxygen, or that the theory of evolution is not true, is not expressing a minority viewpoint. He or she is ignorant, and the job of education is to cure that ignorance and to replace it with knowledge. Students in the public schools are thus to be taught at an early age that "evolution is a fact," and as time goes by they will gradually learn that evolution means naturalism.

In short, the proposition that God was in any way involved in our creation is effectively outlawed and implicitly negated. This is because naturalistic evolution is by definition in the category of scientific knowledge. What contradicts knowledge is implicitly false, or imaginary. That is why it is possible for scientific naturalists in good faith to claim on the one hand that their science says nothing about God, and on the other to claim that they have said everything that can be said about God. In naturalistic philosophy both propositions are at bottom the same. All that needs to be said about God is that there is nothing to be said of God, because on that subject we can have no knowledge.

Our fifth and final term is *truth*. Truth as such is not a particularly important concept in naturalistic philosophy. The reason for this is that "truth" suggests an unchanging absolute, whereas scientific knowledge is a dynamic concept. Like life, knowledge evolves and grows into superior forms. What was knowledge in the past is not knowledge today, and the knowledge of the future will surely be far superior to what we have now. Only naturalism itself and the unique validity of science as the path to knowledge are absolutes. There can be no criterion for truth outside of scientific knowledge, no mind of God to which we have access.

This way of understanding things persists even when scientific naturalists employ religious-sounding language. For example, the physicist Stephen Hawking ended his famous book *A Brief History of Time* with the prediction that man might one day "know the mind of God." This phrasing caused some friends of mine to form the mistaken impression that he had some attraction to theistic religion. In context Hawking was not referring to a supernatural eternal being, however, but to the possibility that scientific knowledge will eventually become complete and all-encompassing because it will have explained the movements of material particles in all circumstances.

The monopoly of science in the realm of knowledge explains why evolutionary biologists do not find it meaningful to address the question whether the Darwinian theory is true. They will gladly concede that the theory is incomplete and that further research into the mechanisms of evolution is needed. At any given point in time, however, the reigning theory of naturalistic evolution represents the state of scientific knowledge about how we came into existence. Scientific knowledge is by definition the closest approximation of absolute truth available to us. To ask whether this knowledge is true is therefore to miss the point, and to betray a misunderstanding of "how science works."

So far I have described the metaphysical categories by which scientific naturalists have excluded the topic of God from rational discussion, and thus ensured that Darwinism's fully naturalistic creation story is effectively true by definition. There is no need to explain why atheists find this system of thought control congenial. What is a little more difficult to understand, at least at first, is the strong support Darwinism continues to receive in the Christian academic world. Attempts to investigate the credibility of the Darwinist evolution story are regarded with little enthusiasm by many leading Christian professors of science and philosophy, even at institutions that are generally regarded as conservative in theology. Given that Darwinism is inherently naturalistic and therefore antagonistic to the idea that God had anything to do with the history of life, and that it plays the central role in ensuring agnostic domination of the intellectual culture, one might have supposed that Christian intellectuals (along with religious Jews) would be eager to find its weak spots.

Instead, the prevailing view among Christian professors has been that Darwinism—or "evolution," as they tend to call it—is unbeatable, and that it can be interpreted to be consistent with Christian belief. And in fact Darwinism is unbeatable as long as one accepts the thought categories of scientific naturalism that I have been describing. The problem is that those same thought categories make Christian theism, or any other theism, absolutely untenable. If science has exclusive authority to tell us how life was created, and if science is committed to naturalism, and if science never discards a paradigm until it is presented with an acceptable naturalistic alternative, then Darwinism's position is impregnable within science. The same reasoning that makes Darwinism inevitable, however, also bans God from taking any action within the history of the Cosmos, which means that it makes theism illusory. Theistic naturalism is self-contradictory.

Some hope to avoid the contradiction by asserting that naturalism rules only within the realm of science, and that there is a separate realm called "religion" in which theism can flourish. The problem with this arrangement, as we have already seen, is that in a naturalistic culture scientific conclusions are considered to be knowledge, or even fact. What is outside of fact is fantasy, or at best subjective belief. Theists who accommodate with scientific naturalism therefore may never affirm that their God is *real* in the same sense that evolution is real. This rule is essential to the entire mind-set that produced Darwinism in the first place. If God exists, He could certainly work through mutation and selection if that is what He wanted to do, but He could also create by some means totally outside the ken of our science. Once we put God into the picture, however, there is no good reason to attribute the creation of biological complexity to random mutation and natural selection. Direct evidence that these mechanisms have substantial creative power is not to be found in nature, the laboratory, or the fossil record. An essential step in the reasoning that establishes that Darwinian selection created the wonders of biology, therefore, is that nothing else was available. Theism is by definition the doctrine that something else was available.

Perhaps the contradiction is hard to see when it is stated at an abstract level, so I will give a more concrete example. Persons who advocate the compromise position called "theistic evolution" are in my experience always vague about what they mean by "evolution." They have good reason to be vague. As we have seen, Darwinian evolution is by definition unguided and purposeless, and such evolution cannot in any meaningful sense be theistic. For evolution to be genuinely theistic it must be guided by God, whether this means that God programmed the process in advance or stepped in from time to time to give it a push in the right direction. To Darwinists evolution guided by God is a soft form of creationism, which is to say it is not evolution at all. To repeat, this understanding goes to the very heart of Darwinist thinking. Allow a preexisting supernatural intelligence to guide evolution, and this omnipotent being can do a whole lot more than that.

Of course, theists can think of evolution as God-guided whether naturalistic Darwinists like it or not. The trouble with having a private definition for theists, however, is that the scientific naturalists have the power to decide what that term "evolution" means in public discourse, including the science classes in the public schools. If theistic evolutionists broadcast the message that evolution as *they* understand it is harmless to theistic religion, they are misleading their constituents unless they add a clear warning that the version of evolution advocated by the entire body of mainstream science is something else altogether. That warning is never clearly delivered, however, because the main point of theistic evolution is to preserve peace with the mainstream scientific community. The theistic evolutionists therefore unwittingly serve the purposes of the scientific naturalists, by helping persuade the religious community to lower its guard against the incursion of naturalism.

We are now in a position to answer the question with which this [chapter] began. What is Darwinism? Darwinism is a theory of empirical science only at the level of microevolution, where it provides a framework for explaining such things as the diversity that arises when small populations become reproductively isolated from the main body of the species. As a general theory of biological creation Darwinism is not empirical at all. Rather, it is a necessary implication of a philosophical doctrine called scientific naturalism, which is based on the a priori assumption that God was always absent from the realm of nature. As such evolution in the Darwinian sense in inherently antithetical to theism, although evolution in some entirely different and nonnaturalistic sense could conceivably have been God's chosen method of creation.

In 1874, the great Presbyterian theologian Charles Hodge asked the question I have asked: What is Darwinism? After a careful and thoroughly fair-minded evaluation of the doctrine, his answer was unequivocal: "It is Atheism." Another way to state the proposition is to say that Darwinism is the answer to a specific question that grows out of philosophical naturalism. To return to the game of *Jeopardy* with which we started, let us say that Darwinism is the answer. What, then, is the question? The question is: "How must creation have occurred if we assume that God had nothing to do with it?" Theistic evolutionists accomplish very little by trying to Christianize the answer to a question that comes straight out of the agenda of scientific naturalism. What we need to do instead is to challenge the assumption that the only questions worth asking are the ones that assume that naturalism is true.

NOTES

1. Niles Eldredge, *Time Frames* (Portsmouth, NM: Heinemann, 1986), p. 144.

2. Ibid., p. 93.

3. Stephen Jay Gould, "Is a New and General Theory of Evolution Emerging?" *Paleobiology* 6 (1980): 119–30, reprinted in *Evolution Now: A Century After Darwin*, ed. Maynard Smith (San Francisco: W. H. Freeman, 1982).

4. See Stephen Jay Gould, "Impeaching a Self-Appointed Judge," *Scientific American* (July 1992): 118–22. *Scientific American* refused to publish my response to this attack, but the response did appear in the March 1993 issue of *Perspectives on Science and Christian Faith*, the journal of the American Scientific Affiliation.

5. Thomas S. Kuhn, *The Structure of Scientific Revolutions*, 2nd ed. (Chicago: University of Chicago Press, 1970), p. 79.

CHAPTER 22

IS IT SCIENCE YET? INTELLIGENT DESIGN, CREATIONISM, AND THE CONSTITUTION

MATTHEW J. BRAUER, BARBARA FORREST, AND STEVEN G. GEY

IS IT SCIENCE YET?

Much of what proponents write about the legal issues surrounding the theory of intelligent design is based on the assumption that if the theory can be demoninated "science," then no legal barrier can be erected to including the theory in public school classrooms.[1] In part this is a response to a section of the Arkansas creationism decision *McLean v. Arkansas*,[2] which intelligent design proponents routinely denounce.[3] *McLean* contains the most thorough substantive consideration of creationist theory yet to appear in any judicial opinion and therefore continues to influence the debate about various antievolution efforts, including intelligent design.... *McLean* held unconstitutional an Arkansas statue that required "balanced treatment to creation science and evolution-science."[4] One portion of the statute described the details of each theory to be given balanced treatment, and included in the definition of "creation science" a number of specifically

---

biblical references to phenomena such as the occurrence of a worldwide flood and the "relatively recent inception of the earth."[5] In the course of holding that the statute violated the Establishment Clause by injecting religion into public school classrooms, the court noted that creation science as defined by the statute "is simply not science."[6] The court then set forth five criteria that, it said, typically define a scientific theory:

(1) It is guided by natural law;
(2) It has to be explanatory by reference to nature law;
(3) It is testable against the empirical world;
(4) Its conclusions are tentative, i.e., are not necessarily the final word; and
(5) It is falsifiable.[7]

Intelligent design proponents have several inconsistent responses to this list of criteria. On one hand, they argue that the entire effort to define "science" is impossible because "many philosophers of science have generally abandoned attempts to define science by reference to abstract demarcation criteria."[8] On the other hand, they cite some of the same criteria to argue that a definitive category called "science" does exist and creationism fits the definition as well as evolutionary theory. "[N]aturalistic and non-naturalistic origins theories (including both Darwinism and design theory) are 'methodically equivalent,' both in their ability to meet various demarcation criteria and as historical theories of origin."[9] When these conflicting claims are reduced to their essence, the basic intelligent design argument is that no one has the ability to assess the scientific validity of their theory—especially experts in particular scientific fields who are not aligned with the intelligent design movement. To put it bluntly, they assert that the simple pretense of scientific validity should be enough to satisfy the legal standard for including the theory in the public school curriculum.

> Since . . . no ruling body in science can determine when a minority scientific interpretation has attracted sufficient support to warrant discussion in the science classroom, the pedagogical debate will necessarily, and properly, devolve to individual teachers and local school boards. In any case, defining

> permissible science as co-extensive with majority scientific opinion erects a more restrictive standard than the law itself now recognizes in deciding the admissibility of expert scientific opinion.[10]

Political assessments of scientific validity, in other words, should trump assessments by the scientific community itself.

Intelligent design advocates have no choice but to take this odd route to academic acceptance. As the quote above indicates, intelligent design advocates implicitly acknowledge that their theory has virtually no standing among mainstream scholars. Their only option, therefore, is to argue that the conclusions of a fringe movement is just as valid as the conclusions of the overwhelming majority of scholars in the field. They argue, in effect, that so long as a handful of advocates with plausible academic credentials announce their support for the theory, then that theory is sufficiently "scientific" to be granted access to public school classrooms. All science is created equal, they argue, so let students hear both sides.

There are serious flaws in this argument, and these flaws illustrate how intelligent design proponents mischaracterize *McLean* and other cases dealing with scientific matters. The central problem is that intelligent design proponents greatly overstate the extent to which the law is willing to recognize any theory—no matter how implausible—as "science." It may be true that there is no infallible and universally applicable test for when a particular theory constitutes "science." But it is also true that some theories are so inconsistent with current scientific understandings of the world that they cannot be reasonably construed as scientifically valid. It is irrelevant whether one calls such theories unscientific or merely "bad science."[11] There is simply no legitimate reason to include them in scientific discussions.

The very cases intelligent design proponents use to bolster their position in fact demonstrate why their theory should not be given the credibility they demand. Intelligent design proponents are fond of quoting in support of their position *Daubert v. Merrell Pharmaceuticals, Inc.*[12]—the Supreme Court's recent decision regarding the standard for admitting scientific and other expert testimony.[13] In *Daubert* the Court modified its previous rules regarding the admissibility of scientific evidence under the Federal Rules of Civil Procedure. The Court interpreted the standard of Rule 401—"All relevant evidence is admissible"[14]—as liberalizing the prevailing standard prior to the adoption of the rule. The pre-*Daubert* rule required scientific evidence introduced at trial to be "sufficiently established to have gained general acceptance in the particular field in which it belongs."[15] Intelligent design proponents argue that *Daubert* casts doubt on *McLean* because in *Daubert* the Court adopted a kind of anything-goes approach to accepting any theory that can claim any empirical basis:

> This trend makes reliance upon the demarcation criteria in *McLean v. Arkansas* even more questionable. Since *Daubert* has made the question of scientific legitimacy turn on "evidentiary reliability," the courtroom should be hospitable to competing theories provided each theory has an empirical basis. To exclude an interpretation simply because it has not yet achieved majority support usurps the function that juries ought to serve. By analogy, the debate over origins theory should not exclude a viewpoint at the outset because of the inability to command a majority of scientists; it should be the function of scientific inquiry itself to permit competing theories to argue, on the basis of empirical data, for wider acceptance.[16]

However, this badly misreads the Court's holding in *Daubert.* The Court emphasized in *Daubert* that its opinion "does not mean . . . that the Rules themselves place no limits on the admissibility of purportedly scientific evidence. Nor is the trial judge disabled from screening such evidence. To the contrary, under the Rules, the trial judge must ensure that any and all scientific testimony or evidence admitted is not only relevant, but reliable."[17] The Court emphasized that expert testimony must relate to "scientific knowledge," which the Court emphasized "connotes more than subjective belief or unsupported speculation."[18] The Court then listed a series of pertinent considerations for judges to keep in mind when considering the proffer of scientific evidence. None of these should make intelligent design proponents comfortable. One "key question," the Court noted, "will be whether [a theory] can be (and has been) tested."[19] Like the judge in *McLean,* the *Daubert* Court emphasized the critical element of falsification.[20] Intelligent design proponents should keep this "key question" in mind when devising the methodology for testing their central thesis that a Supreme Being created the world in more or less its present form. A second consideration noted by the Court is "whether the theory or technique has been subjected to peer review and publication."[21] The Court found the reason for relying on peer review obvious:

> [S]ubmission to the scrutiny of the scientific community is a component of "good science," in part because it increases the likelihood that substantive flaws in methodology will be detected.... The fact of publication (or lack thereof) in a peer reviewed journal thus will be a relevant, though not dispositive, consideration in assessing the scientific validity of a particular technique or methodology on which an opinion is premised.[22]

Finally, the Court emphasized that "general acceptance" is still very important in assessing the reliability of a scientific theory: "Widespread acceptance can be an important factor in ruling particular evidence admissible, and 'a known technique which has been able to attract only minimal support within the community,' may properly be viewed with skepticism."[23]

Daubert presents proponents of intelligent design creationism with a major dilemma: They advocate a theory whose central precept cannot be tested or falsified; they seldom if ever have their theoretical papers accepted for publication in peer-reviewed science journals; and their theory is rejected by virtually the entire scientific community. As science, therefore, their theory—in the words of a case they themselves frequently cite—"may properly be viewed with skepticism."[24] For what it is worth, this is precisely the point made by the district court in *McLean.* During its discussion of the definition of science and descriptions of "what scientists do," the *McLean* court noted: "The obvious implication of this description is that, in a free society, knowledge does not require the imprimatur of legislation in order to become science."[25] Or to put the matter another way, intelligent design creationism cannot use the political process to overcome its failures as science.

CONCLUSION

Much of the new battle between intelligent design and evolutionary theory is reminiscent of the old battle between creationism and evolution. This is not surprising, since intelligent design is merely a stripped-down version of its more explicitly biblical predecessors. God is at the center of all versions of the theory, whether He is denominated as such, or is identified merely as the Supreme Being or intelligent designer. Given the similarities between all versions of the theory, the demarcation lines of the battle are already well drawn, and the conclusion to the legal aspect of the conflict is not in serious doubt. In sum, the proposal to incorporate intelligent design theory into the public school science curriculum cannot be reconciled with a consistent application of relevant Supreme Court precedents on the subject of creationism, and none of the alternative First Amendment theories intelligent design proponents offer in response can withstand even cursory analysis.

There is little question that intelligent design proponents have a serious dispute with the scientific community's virtually unanimous support for the proposition that evolution happens—in both micro and macro forms. But this dispute is at bottom a religious, not a scientific dispute. Both scientists and the government must respect the rights of private individuals to reject scientific conclusions on religious grounds in favor of intelligent design and other theocentric approaches to humanity's origins. But at the same time scientists must be allowed to do science and science teachers must be allowed to teach it—unconstrained by the objections of those who find science inconsistent with their religious beliefs. As Bertolt Brecht's Galileo noted, "the sum total of the angles in a triangle can't be changed to suit the requirements of the curia."[26] The Court's Establishment Clause jurisprudence makes it clear that modern governments can't alter basic scientific conclusions to suit the requirements of politically powerful religious groups, either.

NOTES

1. "[I]f, arguably, design theory has both a theoretical basis and evidential support, and if it meets abstract definitional criteria of scientific status equally as well as its main theoretical rivals, then it seems natural to ask: on what grounds can design theory now be excluded from public school science curriculum?" David K. DeWolf, Stephen C. Meyer, and Mark E. DeForrest, *Intelligent Design in Public School Science Curricular: A Legal Guidebook* (1999), p. 74, http://www.arn.org/docs/dewolf/guidebook.htm.

2. 529 F. Supp. 1255 (D. Ark. 1982).

3. See Jeffrey F. Addicott, "Storm Clouds on the Horizon of Darwinism: Teaching the Anthropic Principle and Intelligent Design in the Public Schools," *Ohio State Law Journal* 63 (2002): 1568 (concluding that "Judge Overton's simplistic definition of science has been soundly refuted by numerous legal and scientific commentators as woefully inadequate and unrealistic"); Francis J. Beckwith, "Science and Religion Twenty Years after *McLean v. Virginia*: Evolution, Public Education, and the New Challenge of Intelligent Design," *Harvard Journal of Law and Public Policy* 26 (2003): 455, 494 (criticizing *McLean* and arguing that "Judge Overton's criteria, at least as applied to creation science, are seriously flawed"); DeWolf, Meyer, and DeForrest, *Intelligent Design in Public School Science Curricular: A Legal Guidebook*, pp. 66–78 (critiquing *McLean* and concluding that its definition of science is "questionable").

4. *McLean*, 529 F. Supp. 1256.

5. Ibid., p. 1264.

6. Ibid., p. 1267.

7. Ibid.

8. DeWolf, Meyer, and DeForrest, *Intelligent Design in Public School Science Curricular: A Legal Guidebook*, p. 69; see also Addicott, "Storm Clouds on the Horizon of Darwinism: Teaching the Anthropic Principle and Intelligent Design in the Public Schools," p. 1568 (arguing that the *McLean* definition has been "refuted by numerous legal and scientific commentators"); Francis J. Beckwith, "Science and Religion Twenty Years after *McLean v. Virginia*: Evolution, Public Education, and the New Challenge of Intelligent Design" (describing *McLean*'s definition of science as "anachronistic" and "self-refuting").

9. DeWolf, Meyer, and DeForrest, *Intelligent Design in Public School Science Curricular: A Legal Guidebook*, p. 72.

10. Ibid., p. 75.

11. See ibid., pp. 73–74.

> [E]ven many of those who previously wielded demarcation arguments as a way of protecting the Darwinist hegemony in public education, including the most prominent advocates of these arguments, have either abandoned or repudiated them. For example, Eugenie Scott of The National Center for Science Education (an advocacy group for an exclusively Darwinist curriculum) no longer seeks to dismiss creation science as pseudoscience or as unscientific; instead, she argues that it constitutes "bad science." Ibid.

12. 509 U. S. 579 (1993).

13. See Addicott, "Storm Clouds on the Horizon of Darwinism: Teaching the Anthropic Principle and Intelligent Design in the Public Schools," pp. 1569–70; Beckwith, "Science and Religion Twenty Years after *McLean v. Virginia*: Evolution, Public Education, and the New Challenge of Intelligent Design," p. 491; DeWolf, Meyer, and DeForrest, *Intelligent Design in Public School Science Curricular: A Legal Guidebook*, p. 75.

14. Fed. R. Evid. 402.

15. *Frye v. United States*, 293 F. 1013, 1014 (D.C. Ct. App. 1923).

16. DeWolf, Meyer, and DeForrest, *Intelligent Design in Public School Science Curricular: A Legal Guidebook*, pp. 77–78; Addicott, "Storm Clouds on the Horizon of Darwinism: Teaching the Anthropic Principle and Intelligent Design in the Public Schools," pp. 1569–70 ("Under *Daubert*, the test for scientific legitimacy will be evaluated not on a bandwagon approach or by the fulfillment of a *McLean*-style set of arbitrary criteria. Instead, the Court will now evaluate the legitimacy of a new theory—even if a minority view—on the basis of a variety of factors, with emphasis on the actual empirical research."); Beckwith, "Science and Religion Twenty Years after *McLean v. Virginia*: Evolution, Public Education, and the New Challenge of Intelligent Design," p. 491 ("[The test of scientific legitimacy] is, very simply, now a matter of arguments and their soundness, not a matter of popularity.").

17. *Daubert*, 509 U.S. 589 (internal footnote omitted).

18. Ibid., pp. 589–90.

19. Ibid., p. 593.

20. See ibid. (quoting Karl Popper, *Conjectures and Refutations: The Growth of Scientific Knowledge*, 5th ed. [1989], p. 37). ("[T]he criterion of the scientific status of a theory is its falsifiability, or refutability, or testability.")

21. Ibid., p. 593.

22. Ibid., pp. 593–94 (internal citations omitted).

23. Ibid., p. 594 (quoting *United States v. Downing*, 753 F.2d 1224, 1238 [3d Cir. 1985]).

24. Ibid.

25. *McLean v. Arkansas Board of Education* 529 F. Supp. 1255, 1267 (E.D. Ark. 1982).

26. Bertolt Brecht, *Life of Galileo, Collected Plays* 5 (1972): 58.

CHAPTER 23

KITZMILLER V. DOVER AREA SCHOOL DISTRICT EXPERT WITNESS TESTIMONY

MICHAEL BEHE

DIRECT EXAMINATION (EXCERPTS)*

Q. Sir, what is intelligent design?

A. Intelligent design is a scientific theory that proposes that some aspects of life are best explained as the result of design, and that the strong appearance of design in life is real and not just apparent.

Q. Now Dr. Miller defined intelligent design as follows: Quote, Intelligent design is the proposition that some aspects of living things are too complex to have been evolved and, therefore, must have been produced by an outside creative force acting outside the laws of nature, end quote. Is that an accurate definition?

A. No, it's a mischaracterization.

Q. Why is that?

A. For two reasons. One is understandable—that Professor Miller is viewing intelligent design from the perspective of his own views and

---

*Direct examination questions were posed by defense attorney Robert Muise. From *Tammy Kitzmiller, et al.* v. *Dover Area School District, et al.*, 400 F. Supp. 2d 707 (M.D. Pa. 2005).

sees it simply as an attack on Darwinian theory. And it is not that. It is a positive explanation.

And the second mischaracterization is that intelligent design is a scientific theory. Creationism is a religious, theological idea. And that intelligent design is—relies rather on empirical and physical and observable evidence plus logical inferences for its entire argument.

Q. Is intelligent design based on any religious beliefs or convictions?

A. No, it isn't.

Q. What is it based on?

A. It is based entirely on observable, empirical, physical evidence from nature plus logical inferences.

Q. Dr. Padian testified that paleontologists make reasoned inferences based on comparative evidence. For example, paleontologists know what the functions of the feathers of different shapes are in birds today. They look at those same structures in fossil animals and infer that they were used for a similar purpose in the fossil animal. Does intelligent design employ similar scientific reasoning?

A. Yes, that's a form of inductive reasoning, and intelligent design uses similar inductive reasoning.

Q. Now I want to review with you the intelligent design argument. Have you prepared a slide for this?

A. Yes, I have. On the next slide is a short summary of the intelligent design argument. The first point is that we infer design when we see that parts appear to be arranged for a purpose. The second point is that the strength of the inference, how confident we are in it, is quantitative. The more parts that are arranged, and the more intricately they interact, the stronger is our confidence in design. The third point is that the appearance of design in aspects of biology is overwhelming.

The fourth point then is that, since nothing other than an intelligent cause has been demonstrated to be able to yield such a strong appearance of design, Darwinian claims notwithstanding, the conclusion that the design seen in life is real design is rationally justified.

Q. Now when you use the term design, what do you mean?

A. Well, I discussed this in my book *Darwin's Black Box*, and a short description of design is shown in this quotation from chapter 9. Quote, What is design? Design is simply the purposeful arrangement of parts. When we perceive that parts have been arranged to fulfill a purpose, that's when we infer design.

Q. Can you give us a biochemical example of design?

A. Yes, that's on the next slide. I think the best, most visually striking example of design is something called the bacterial flagellum. This is a figure of the bacterial flagellum taken from a textbook by authors named Voet and Voet, which is widely used in colleges and universities around the country. The bacterial flagellum is quite literally an outboard motor that bacteria use to swim. And in order to accomplish that function, it has a number of parts ordered to that effect.

This part here, which is labeled the filament, is actually the propeller of the bacterial flagellum. The motor is actually a rotary motor. It spins around and around and around. And as it spins, it spins the propeller, which pushes against the liquid in which the bacterium finds itself and, therefore, pushes the bacterium forward through the liquid.

The propeller is attached to something called the drive shaft by another part which is called the hook region which acts as a universal joint. The purpose of a universal joint is to transmit the rotary motion of the drive shaft up from the drive shaft itself through the propeller. And the hook adapts the one to the other.

The drive shaft is attached to the motor itself which uses a flow of acid from the outside of the cell to the inside of the cell to power the turning of the motor, much like, say, water flowing over a dam can turn a turbine. The whole apparatus, the flagellum, has to be kept stationary in the plane of the bacterial membrane, which is represented by these dark curved regions.

As the propeller is turning, much as an outboard motor has to be clamped onto a boat to stabilize it while the propeller is turning. And there are regions, parts, protein parts which act as what is called a stator to hold the apparatus steady in the cell.

The drive shaft has to traverse the membrane of the cell. And there are parts, protein parts, which are, which act as what are called bushing materials to allow the drive shaft to proceed through. And I should add that, although this looks complicated, the actual—this is really only a little illustration, a kind of cartoon drawing of the flagellum. And it's really much more complex than this.

But I think this illustration gets across the point of the purposeful arrangement of parts. Most people who see this and have the function explained to them quickly realize that these parts are ordered for a purpose and, therefore, bespeak design.

***

Q. Do sciences recognize evidence of design in nature?

A. Yes, they do.

Q. And do you have some examples to demonstrate that point?

A. Yes, I do. On the next slide is the cover of a book written by a man named Richard Dawkins, who is a professor of biology at Oxford University and a very strong proponent of Darwinian evolution. In 1986, he wrote a book entitled *The Blind Watchmaker: Why the Evidence of Evolution Reveals a Universe without Design.* Nonetheless, even though he is, in fact, a strong Darwinist, on the first page of the first chapter of his book, he writes the following.

Quote, Biology is the study of complicated things that give the appearance of having been designed for a purpose, close quote. So let me just emphasize that here's Richard Dawkins saying, this is the very definition of biology, the study of complicated things that give the appearance of having been designed for a purpose.

Q. Does he explain why they appear designed; how it is that we can detect design?

A. Yes, he does. And that is shown on the next slide. It is not because of some emotional reaction. It is not due to some fuzzy thinking. It's due to the application of an engineering point of view. He writes on page 21 of the first chapter, quote, We may say that a living body or organ is well designed if it has attributes that an intelligent and knowledgeable engineer might have built into it in order to achieve some sensible purpose, such as flying, swimming, seeing. Any engineer can recognize an object that has been designed, even poorly designed, for a purpose, and he can usually work out what that purpose is just by looking at the structure of the object, close quote.

So let me just emphasize that he, in other words, is stating that we recognize design by the purposeful arrangement of parts. When we see parts arranged to achieve some sensible purpose, such as flying, swimming, and seeing, we perceive design.

Q. Now is it fair to say that he's looking at, and intelligent design proponents look at physical structures similar to like the paleontologist does and then drawing reasonable inferences from those physical structures?

A. That's exactly right. What intelligent design does is look at the physical, observable features and use logic to infer deductions from that.

Q. Now you, as well as Dawkins in the slides that we've just been looking at, refer to purpose. Now when you use—when you were using purpose, are you making a philosophical claim by using that term?

A. No. The word purpose, like many other words, can have different meanings. And the purpose here used by Professor Dawkins and in intelligent design does not refer to some fuzzy purpose of life or some such thing as that. It's purpose in the sense of function.

And I think on the next slide, I emphasize that Dawkins is using some sensible purpose, such as flying, swimming, seeing. An engineer can work out the purpose of an object by looking at its structure. He's talking about purpose in the sense of function.

Q. Now this appearance of design, is this a faint appearance?

A. No, indeed. This is not just some marginal vague impression. Richard Dawkins, a strong proponent of Darwinian evolution, insists, he says, quote, Yet the living results of natural selection overwhelmingly impress us with the appearance of design, as if by a master watchmaker, impress us with the illusion of design and planning, close quote.

Let me make two points with this. He thinks that this is an illusion because he thinks he has an alternative explanation for what he sees. Nonetheless, what he sees directly gives him the overwhelming impression of design.

Q. Have other scientists made similar claims regarding the evidence of design in nature?

A. Yes. On the next slide is a quotation from a book written by a man named Francis Crick. Francis Crick, of course, is the Nobel laureate with James Watson who won the Nobel Prize for their discovery of the double helical structure of DNA.

In a book published in 1998, he wrote, quote, Biologists must constantly keep in mind that what they see was not designed, but rather evolved. So apparently, in the view of Francis Crick, biologists have to make a constant effort to think that things that they studied evolved and were not designed.

Q. I want to return to Richard Dawkins here for a moment and *The Blind Watchmaker.* Did he borrow his title from somewhere?

A. Yes, the watchmaker of his title has an allusion which he explained on page 4 of his book. He says, quote, The watchmaker of my title is borrowed from a famous treatise by the eighteenth-century theologian William Paley. And he starts to quote William Paley. So he is using his book as an answer to, or an argument to, William Paley's discussions of these issues. And he treats William Paley with the utmost respect.

Q. I believe we have a slide to highlight that.

A. Yes, here's a quotation from William Paley. Paley is best known for what is called his watchmaker argument. And that is briefly this. He says that, when we walk—if we were walking across a field, and we hit our foot against a stone, well, we wouldn't think much of it. We would think that the stone might have been there forever.

But if we stumble across a watch and we pick it up, then Paley goes on to say, when we come to inspect the watch, we perceive that its several parts are framed and put together for a purpose; for example, that they so formed and adjusted as to produce motion, and that motion so regulated as to point out the hour of the day. Let me close quote here, and say that, he is talking about the purposeful arrangement of parts.

Let me continue with a quotation from William Paley. Quote, he says, The inference we think is inevitable, that the watch must have had a maker, close quote. So he is inferring from the physical structure of the watch to an intelligent designer.

Q. Is that a theological argument?

A. No, this is a scientific argument based on physical facts and logic. He's saying nothing here about any religious precept, any theological notion. This is a scientific argument.

Q. Does Richard Dawkins himself recognize it as an argument based on logic?

A. Yes, he does, and he goes to great lengths to address it in his book *The Blind Watchmaker.*

Q. What sort of reasoning or argument is this that we're talking about, this scientific argument that you're referring to?

A. This is an instance of what is called inductive reasoning when we—

Q. I'm sorry. We have a slide here to demonstrate this point?

A. Yes, thank you. Just to help illustrate this point, I just grabbed an article from the *Encyclopedia Britannica* online entitled "Inductive Reasoning." And the *Encyclopedia Britannica* says, quote, When a person uses a number of established facts to draw a general conclusion, he uses inductive reasoning. This is the kind of logic normally used in the sciences.

Let me skip the middle of the quotation and say, "It is by this process of induction and falsification that progress is made in the sciences." So this William Paley's argument, the kind of argument that, say, Professor Padian made about bird feathers and so on are all examples of inductive reasoning, and they are all examples of scientific reasoning.

Q. This is the sort of reasoning that is employed in science quite readily?

A. Yes. As the article makes clear, this is the normal mode of thinking in science.

Q. Is that the sort of reasoning you employ to conclude design, for example, in your book *Darwin's Black Box?*

A. Yes, this is exactly the kind of reasoning that I used in *Darwin's Black Box.* On this slide here, which includes an excerpt from chapter 9 entitled Intelligent Design, I say the following.

Quote, Our ability to be confident of the design of the cilium or intracellular transport rests on the same principles as our ability to be confident of the design of anything, the ordering of separate components to achieve an identifiable function that depends sharply on the components, close quote. In other words, the purposeful arrangement of parts.

***

Q. Dr. Behe, is intelligent design science?

A. Yes, it certainly is.

Q. And why is that?

A. Because it relies completely on the physical, observable, empirical facts about nature plus logical inferences.

Q. And that again is a scientific method?

A. That is the way science proceeds.

Q. I want to ask you if you agree with this testimony provided by Dr. Miller. He testified that it is a standard scientific practice for scientists to point to the scientific literature, to point to observations and experiments that have been done by other people in other laboratories, have been peer reviewed, have been published, and to cite to that evidence, cite to those data, and to cite to those experiments in their arguments. Do you agree with that?

A. Yes, I agree completely.

Q. Is that what you have done, and intelligent design has done in presenting its arguments?

A. That's what I have done. That's what the scientists that wrote those books I showed earlier have done. That's a very common practice in science.

Q. Did Crick and Watson employ the same procedure?

A. Yes, that's correct. Francis Crick and James Watson, whose names I have mentioned earlier, who won the Nobel Prize for determining the double helical structure of DNA, actually did not do the experimental work upon which their conclusions were based.

The experimental work, which consisted of doing x-ray fiber defraction studies on DNA, was actually done by a woman named Rosalyn Franklin, and they used her data to reach their conclusions.

Q. I want to ask you if you also agree with Dr. Miller that the question is not whether you or any other scientist has done experiments in your own laboratories that have produced evidence for a particular claim, the question is whether or not the inferences that you and the scientists draw on your analysis from that data are supported?

A. Yes, I agree completely. Again, those books that I showed in the beginning, that is exactly what those scientists did. They looked very widely for all relevant scientific information that would bear on the argument that they were making.

Q. Again, is that what Crick and Watson employed?

A. Yes, that's what Crick and Watson did, too. Scientists do it all the time.

Q. Is that what you're doing in support of your claim for intelligent design?

A. Yes, that's exactly right.

Q. And have you argued that intelligent design is science in your writings?

A. Yes, I have.

Q. Is intelligent design falsifiable?

A. Yes, it is.

Q. And I want to get to that in a little bit more detail later. Now just to summarize. When you say you are relying on logical inferences, you're referring to inductive reasoning, correct?

A. Yes, inductive reasoning.

Q. And other than intelligent design, as you discussed, and you discussed a little bit about paleontology, do you have an example of this sort of reasoning, inductive reasoning that's used in sciences?

A. Well, I think an excellent example of inductive reasoning is the big bang theory. Most people forget that in the early part of the twentieth century that physicists thought the universe was timeless, eternal, and unchanging.

Then in the late 1920s, observations were made which led astronomers to think that galaxies that they could observe were rushing away from each other and rushing away from the Earth as if in the aftermath of some giant explosion.

So they were using inductive reasoning of their experience of explosions to, and applying that to their astronomical observations. And let me emphasize that they were—the inductive method, as philosophers will tell you, always extrapolates from what a we know to instances of what we don't know.

So those scientists studying the big bang were extrapolating from their knowledge of explosions as seen in, say, fire crackers, cannon balls, and so on, and extrapolating that to the explosion of the entire universe, which is quite a distance from the basis set from which they drew their induction.

But nonetheless, they were confident that this pattern suggested an explosion based on their experience with more familiar objects.

Q. And basically, we don't have any experience with universes exploding, correct?

A. I do not, no.

Q. And scientists do not?

A. No, scientists don't either.

Q. Again, is this similar to the reasoning used in paleontology? For example, we haven't seen any live prehistoric birds, for example, but they have features that resemble feathers, as we know them from our common experience today, and we infer that they were used for flying or similar functions, again based on our common experience?

A. Yes, that's right. That's another example of induction from what we know to things we don't know.

Q. Again, that's scientific reasoning?

A. Yes, it is.

Q. Can science presently tell us what caused the bang?

A. No. I'm not a physicist, but I understand the cause of the big bang is still unknown.

Q. Is that similar to intelligent design's claim that science presently cannot tell us the source of design in nature?

A. Yes, that's very similar. All theories, when they're proposed, have outstanding questions, and intelligent design is no exception. And I'd like to make a further point that I just thought of and was going to make earlier, but that, that induction from explosions of our experience to explosions of the universe is analogous to, similar to the induction that intelligent design makes from our knowledge of objects, the purposeful arrangements of parts in our familiar world and extrapolating that to the cell as well. So that, too, is an example of an induction from what we know to what we have newly discovered.

Q. Now was the big bang theory controversial when it was first proposed?

A. Yes, it turns out that the big bang theory was, in fact, controversial because—not because of the scientific data so much, but because many people, including many scientists, thought that it had philosophical and even theological implications that they did not like.

And on the next slide, I have a quotation of a man named Arthur Eddington, which is quoted in a book by a philosopher of science, Susan Stebbing. Arthur Eddington wrote, quote, Philosophically, the notion of an abrupt beginning to the present order of nature is repugnant to me, as I think it must be to most. And even those who would welcome a proof of the intervention of a creator will probably consider that a single winding up at some remote epoch is not really the kind of relation between God and his world that brings satisfaction to the mind, close quote.

Let me say a couple things. I don't think I mentioned that Arthur Eddington was a very prominent astronomer of that age. The second point is that, notice that the reason that he does not like this theory, this scientific proposal, is not because of scientific reasons, but because of philosophical and theological reasons.

But nonetheless, that does not affect the status of the big bang proposal, which was based completely on physical, observable evidence plus logical inferences. And because of that, it was strictly a scientific theory, even though Arthur Eddington saw other ramifications that he did not like.

Q. I believe you have another quote to demonstrate that point?

A. Yes. Here's a passage from a book by a man named Karl von Weizsacker. Karl von Weizsacker was again an astronomer in the middle part of the twentieth century, and he wrote a book in 1964 entitled *The Relevance of Science* where he recalled his interactions with other scientists when the big bang theory was being proposed.

Let me quote from that passage. Quote, He (and he's referring to Walter Nernst, who was a very prominent chemist of that time) said, the view that there might be an age of the universe was not science. At first, I did not understand him. He explained that the infinite duration of time was a basic element of all scientific thought, and to deny this would mean to betray the very foundations of science.

I was quite surprised by this, and I ventured the objection that it was scientific to form hypotheses according to the hints given by experience, and that the idea of an age of the universe was such a hypothesis. He retorted that we could not form a scientific hypothesis which contradicted the very foundations of science.

He was just angry, and thus the discussion, which was continued in his private library, could not lead to any result. What impressed me about Nernst was not his arguments. What impressed me was his anger. Why was he angry? Close quote.

Let me make a couple comments on this passage. This is an example of when people are arguing about what science is. To Walter Nernst, the very idea that there could be a beginning to the universe was unscientific, and we could not entertain that.

On the other hand, von Weizsacker said that science has to take its hints from what evidence is available. We have to form hypotheses according to the hints given by experience. And to me, this is very similar to what I see going on in the debate over intelligent design today.

Many people object that this can't be science, this violates the very definition of science, whereas other people, myself including, say that we have to form hypotheses according to the hints given by experience.

Q. Does the big bang continue to be controversial in more modern times?

A. Yes. Surprisingly, it's still controversial and still mostly because of its extrascientific implications. For example, here is an image of an editorial which appeared in the journal *Nature* in the year 1989 with the surprising title "Down with the Big Bang." And if you advance to the next slide, we can see it more easily.

The subtitle of the article, where it is written, quote, Apart from being philosophically unacceptable, the big bang is an over-simple view of how the universe began. So let me point out that this was written by a man named John Maddox. John Maddox was the editor of *Nature*, the most prestegious science journal in the world.

For twenty years, he was the editor, and he wrote an editorial entitled "Down with the Big Bang," at least partly because he viewed the idea of the big bang as philosophically unacceptable.

Q. Do you have another quote from this?

A. Yes, I do. Actually in the test of the Maddox article, he goes on to explain in further detail some of his objections to the big bang. And he says the following. Quote, creationists and those of similar persuasion seeking support for their opinions have ample justification in the doctrine of the big bang. That, they might say, is when and how the universe was created, close quote.

Let me make a couple of points here. Again, he does not like this theory apparently because of its extrascientific implications, because he sees theological implications in the theory. He says that creationists have ample justification, and he objects to that justification.

Let me make another point. He's using the word creationist here in a very broad sense to mean anybody who thinks that the very beginning of the universe might have been a—an extra—a supernatural act, that the laws of the universe might have been made, have been set from somewhere beyond nature.

And he uses the word creationist in a very pejorative sense to incite the disapprobation of the readers against people who would hold this view.

Q. Do the implications that Maddox refers to here, does this make the big bang theory creationism?

A. No, it certainty does not. One has to be very careful in looking at scientific ideas, because many scientific ideas do have interesting philosophical or other ramifications, and the big bang is one of those. Nonetheless, the big bang is an entirely scientific proposal, because again, it is based simply on the observable, empirical, physical evidence that we find in nature plus logical inferences.

Q. Do you see similarity between the big bang theory and intelligent design?

A. Yes, I do. I see a number of similarities. First, some people have seen controversial philosophical and perhaps even theological implications of those two proposals. But in both cases, they are based entirely on the physical, empirical evidence of nature plus logical inferences.

Q. Is it true that the big bang bracket [*sic*] can be a question of cause?

A. Yes, that's a good point to consider. The big bang hypothesis struck many people, such as John Maddox and Arthur Eddington and so on, as perhaps having pretty strong, even theological implications. Maybe this was a creation event.

But nonetheless, physicists were able to work within the big bang model that the question of what caused the big bang was just left as an open question and work proceeded on other issues within the big bang.

Q. Do you see any similarity in that regard with intelligent design?

A. Yes, I do. The design in life can be readily apprehended by the purposeful—by the purposeful arrangement of parts. However, identifying a designer or identifying how the design was accomplished, they are different questions which might be much more difficult and much harder to address. Questions such as that can be left aside and other sorts of questions could be asked.

Q. Does this make intelligent design a, quote, unquote, science stopper, as we heard in this case?

A. No more than it makes the big bang a science stopper. The big bang posits a beginning to nature which some people thought was the very antithesis of science. It presented a question, the cause of the big bang, which could not be answered, and which has not been answered to this very day, and nonetheless, I think most people would agree that a large amount of science has been done within the big bang model.

Q. So after the big bang theory was proposed, we didn't shut down all our science departments and close up all the laboratories and just stop scientific exploration?

A. Not to my knowledge.

Q. I believe you have a quote from one of your articles making the point regarding the scientific nature of intelligent design, is that correct?

A. Yes, that's right. I think it's on the next slide in the article "Reply to my Critics," which I published in the journal *Biology and Philosophy.* I pointed this out explicitly. Let me just go to the underlined part, the bold part. I wrote, quote, The conclusion of intelligent design in biochemistry rests exclusively on empirical evidence, the structures and functions of the biochemical systems, plus principles of logic. Therefore, I consider design to be a scientific explanation, close quote.

Q. Now another complaint that we've heard in the course of this trial is that intelligent design is not falsifiable. Do you agree with that claim?

A. No, I disagree. And I think I further in slides from my article in *Biology and Philosophy* in which I wrote on that. If you get to the next slide—oh, I'm sorry. Thank you. You got that. In this, I address it. I'm actually going to read this long quotation, so let me begin.

Quote, In fact, intelligent design is open to direct experimental rebuttal. Here is a thought experiment that makes the point clear. In Darwin's *Black Box*, I claimed that the bacterial flagellum was irreducibly complex and so required deliberate intelligent design. The flip side of this claim is that the flagellum can't be produced by natural selection acting on random mutation, or any other unintelligent process.

To falsify such a claim, a scientist could go into the laboratory, place a bacterial species lacking a flagellum under some selective pressure, for mobility, say, grow it for ten thousand generations, and see if a flagellum, or any equally complex system, was produced. If that happened, my claims would be neatly disproven. Close quote.

So let me summarize that slide. It says that if, in fact, by experiment, by growing something or seeing that in some organism such as a bacterium grown under laboratory conditions, grown for and examined before and afterwards, if it were seen that random mutation and natural selection could indeed produce the purposeful arrangement of parts of sufficient complexity to mimic things that we find in the cell, then, in fact, my claim that intelligent design was necessary to explain such things would be neatly falsified.

Q. I got a couple questions about the proposal that you make. First of all, when you say you place something under selective pressure, what does that mean?

A. Well, that means you grow it under conditions where, if a mutation—a mutant bacterium came along which could more easily grow under those conditions, then it would likely propagate faster than other cells that did not have that mutation.

So, for example, if you grew a flask of bacteria and let them sit in a beaker that was motionless, and the bacteria did not have a flagellum to help it swim around and find food, they could only eat then the materials that were in their immediate vicinity.

But if some bacterium, some mutant bacterium were produced that could move somewhat, then it could gather more food, reproduce more, and be favored by selection.

Q. Is that a standard technique that's used in laboratories across the country?

A. Yes, such experiments are done frequently.

Q. And I just want to ask you a question about this grow it for ten thousand generations. Does that mean we have to wait ten thousand years of some sort to prove this or disprove this?

A. No, not in the case of bacteria. It turns out that the generation time for bacteria is very short. A bacterium can reproduce in twenty minutes. So ten thousand generations is actually, I think, just a couple years. So it's quite doable.

Q. Have scientists, in fact, grown bacteria out to ten thousand generations?

A. Yes, there are experiments going on where bacteria have been grown for forty thousand generations. So again, this is something that can be done.

Q. So this is a readily doable experiment?

A. That's correct.

Q. Sir, do you believe that natural selection is similarly falsifiable?

A. No. Actually, I think that, in fact, natural selection and Darwinian claims are actually very, very difficult to falsify. And let me go back to my article, "Reply to my Critics," from the journal *Biology and Philosophy*.

And I don't think I'm actually going to read this whole thing, because it refers to things that would take a while to explain. But let me just try to give you the gist of it. Let me read the first sentence. Quote, Let's turn the tables and ask, how could one falsify a claim that a particular biochemical system was produced by Darwinian processes? Close quote.

Now let me just kind of try to explain that in my own—well, verbally here. Suppose that we did that same experiment as I talked about earlier. Suppose a scientist went into a laboratory, grew a bacterium that was missing a flagellum under selective pressure for motion, waited ten thousand; twenty thousand; thirty thousand; forty thousand generations, and at the end of that time, examined it and saw that, well, nothing much had been changed, nothing much had changed.

Would that result cause Darwinian biologists to think that their theory could not explain the flagellum? I don't think so. I think they would say, number 1, that we didn't wait long enough; number 2, perhaps we started with the wrong bacterial species; number 3, maybe we applied the wrong selective pressure, or some other problem.

Now leaving aside the question of whether those are reasonable responses or not, and some of them might be reasonable, nonetheless, the point is that, it's very difficult to falsify Darwinian claims. What experiment could be done which would show that Darwinian processes could not produce the flagellum?

And I can think of no such experiment. And as a matter of fact, on the next slide, I have a quotation, kind of putting a point on that argument. In that same article, "Reply to my Critics," I wrote that I think Professor Coyne and the National Academy of Sciences have it exactly backwards. And Professor Jerry Coyne is an evolutionary biologist who said that intelligent design is unfalsifiable, and in a publication of the National Academy, they asserted the same thing.

I wrote that a strong point of intelligent design is its vulnerability to falsification. A weak point of Darwinian theory is its resistance to falsification. What experimental evidence could possibly be found that would falsify the contention that complex molecular machines evolved by a Darwinian mechanism? I can think of none, close quote.

So again, the point is that, I think the situation is exactly opposite of what much—of what many arguments assume, that ironically intelligent design is open to falsification, but Darwinian claims are much more resistant to falsification.

CROSS-EXAMINATION (EXCERPT)*

Q. Now, you claim that intelligent design is a scientific theory.
A. Yes.

Q. But when you call it a scientific theory, you're not defining that term the same way that the National Academy of Sciences does.
A. Yes, that's correct.

Q. You don't always see eye to eye with the National Academy?
A. Sometimes not.

Q. And the definition by the National Academy, as I think you testified is, a well-substantiated explanation of some aspect of the natural world that can incorporate facts, laws, inferences, and tested hypotheses, correct?
A. Yes.

Q. Using that definition, you agree intelligent design is not a scientific theory, correct?
A. Well, as I think I made clear in my deposition, I'm a little bit of two minds of that. I, in fact, do think that intelligent design is well substantiated for some of the reasons that I made clear during my testimony. But again, when you say well substantiated, sometimes a person would think that there must be a large number of people then who would agree with that. And so, frankly, I, like I said, I am of two minds of that.

Q. And actually you said at your deposition, "I don't think intelligent design falls under this definition." Correct?
A. Yeah, and that's after I said—if I may see where in my deposition that is? I'm sorry.

Q. It's on pages 134 and 135.
A. And where are you—where are you reading from?

---

*Cross-examination questions were posed by plaintiff attorney Eric Rothschild.

Q. I'll be happy to read the question and answer to you. I asked you whether intelligent design—I asked actually on the top of 133, I asked you whether intelligent design qualifies as a scientific theory using the National Academy of Sciences definition.

A. What line is that, I'm sorry?

Q. That's 133, line 18.

A. Is that going—question beginning, "Going back to the National Academy of Science?"

Q. Yes. And you first said, "I'm going to say that I would argue that in fact it is." And that's 134, line 10.

A. Yes.

Q. Okay. And I said, "Intelligent design does meet that?" And you said, "It's well substantiated, yes." And I said, "Let's be clear here, I'm asking—looking at the definition of a scientific theory in its entirety, is it your position that intelligent design is a scientific theory?" And you said, going down to line 23, "I think one can argue these a variety of ways. For purposes of an answer to the—relatively brief answer to the question, I will say that I don't think it falls under this." And I asked you, "What about this definition; what is it in this definition that ID can't satisfy to be called a scientific theory under these terms?" And you answer, "Well, implicit in this definition it seems to me that there would be an agreed upon way to decide something was well substantiated. And although I do think that intelligent design is well substantiated, I think there's not—I can't point to external—an external community that would agree that it was well substantiated."

A. Yes.

Q. So for those reasons you said it's not—doesn't meet the National Academy of Sciences definition.

A. I think this text makes clear what I just said a minute or two ago, that I'm of several minds on this question. I started off saying one thing and changing my mind and then I explicitly said, "I think one can argue these things a variety of ways. For purposes of a relatively brief answer to the question, I'll say this." But I think if I were going to give a more complete answer, I would go into a lot more issues about this.

So I disagree that that's what I said—or that's what I intended to say.

Q. In any event, in your expert report, and in your testimony over the last two days, you used a looser definition of "theory," correct?

A. I think I used a broader definition, which is more reflective of how the word is actually used in the scientific community.

Q. But the way you define scientific theory, you said it's just based on your own experience; it's not a dictionary definition, it's not one issued by a scientific organization.

A. It is based on my experience of how the word is used in the scientific community.

Q. And as you said, your definition is a lot broader than the NAS definition?

A. That's right, intentionally broader to encompass the way that the word is used in the scientific community.

Q. Sweeps in a lot more propositions.

A. It recognizes that the word is used a lot more broadly than the National Academy of Sciences defined it.

Q. In fact, your definition of scientific theory is synonymous with hypothesis, correct?

A. Partly—it can be synonymous with hypothesis, it can also include the National Academy's definition. But in fact, the scientific community uses the word "theory" in many times as synonymous with the word "hypothesis," other times it uses the word as a synonym for the definition reached by the National Academy, and at other times it uses it in other ways.

Q. But the way you are using it is synonymous with the definition of hypothesis?

A. No, I would disagree. It can be used to cover hypotheses, but it can also include ideas that are in fact well substantiated and so on. So while it does include ideas that are synonymous or in fact are hypotheses, it also includes stronger senses of that term.

Q. And using your definition, intelligent design is a scientific theory, correct?

A. Yes.

Q. Under that same definition astrology is a scientific theory under your definition, correct?

A. Under my definition, a scientific theory is a proposed explanation which focuses or points to physical, observable data and logical inferences. There are many things throughout the history of science which we now think to be incorrect which nonetheless would fit that—which would fit that definition. Yes, astrology is in fact one, and so is the ether theory of the propagation of light, and many other—many other theories as well.

CHAPTER 24

KITZMILLER V. DOVER AREA SCHOOL DISTRICT EXPERT REPORT

ROBERT T. PENNOCK

I. BASIS OF MY EXPERTISE

I am associate professor of science and technology studies at Michigan State University's Lyman Briggs School of Science and associate professor of philosophy in the Department of Philosophy. I'm also a faculty member in MSU's Ecology & Evolutionary Biology and Behavior Program and in the Department of Computer Science.

My PhD was in history and philosophy of science from the University of Pittsburgh. My dissertation advisor was Wesley Salmon, whose causal-mechanical account is the most important contemporary analysis of the nature of scientific explanation. My dissertation was on the nature of evidence in science, especially on causal reasoning. I have also done research on naturalism as a National Endowment for the Humanities Summer Institute Fellow.

I have studied the creationist movement for over twenty years, focusing especially on the intelligent design creationists since the early 1990s. I have published over a dozen articles on philosophical issues in the creationism debate and a book, *Tower of Babel: The Evidence against the New Creationism.* I also

From *Tammy Kitzmiller, et al. v. Dover Area School District, et al.*, 400 F. Supp. 2d 707 (M.D. Pa. 2006).

edited *Intelligent Design Creationism and Its Critics: Philosophical, Theological and Scientific Perspectives*, which is the most complete source book on the topic, and am currently editing a collection of new articles that rebut the claims of Jonathan Wells's book *Icons of Evolution.*

I also do scientific research on experimental evolution and evolutionary design using evolving computer organisms, including work showing how evolutionary mechanisms can produce the kinds of complex features creationists say is impossible. I am a member of the education committee of the Society for the Study of Evolution, the international professional organization for evolutionary biologists.

I have published numerous papers and given well over a hundred invited talks on these subjects at universities and professional conferences nationally and internationally. With regard to creationism in particular, I have testified on the subject before state boards of education, assisted legislators dealing with proposed intelligent design legislation, and helped in school districts in cases where individual teachers have taught intelligent design. I am the founder and current board president of Michigan Citizens for Science, which works to defend and promote sound science education in Michigan.

I have won two awards given by the Templeton Foundation for my writing and teaching on issues in science and religion. I am on the National Advisory Board of Americans United for Separation of Church and State.

I grew up and attended public schools in central Pennsylvania.

II. GENERAL OPINION

In my considered opinion, allowing so-called intelligent design (ID) to be included as part of a science class would have the effect of introducing material that is not only unscientific, but is essentially religious in nature. Like other kinds of creationism, the ID movement rejects the scientific findings of evolution and posits instead creation by a supernatural entity. This is a truly radical proposition. To teach such a view, under whatever name, is not only to dismiss well-established scientific findings that are a fundamental part of biology in favor of an unsupported religious belief, but also to reject the very nature of science. In what follows, I will explain the reasons and evidence for this general opinion in detail, but here is a brief abstract of my opinion.

Science as it is understood by practicing scientists is not so much a list of conclusions as it is a set of methods for investigating the physical world and thereby adding or revising conclusions. ID departs from the acceptable methodological practice of science from the very first step, in appealing to a realm beyond nature. The concept of "design" as it is used by ID theorists is inherently supernatural. Science does not reject the supernatural dogmatically, but rather because such claims cannot be tested by empirical evidence. One will look in vain in the peer-reviewed scientific literature for any method that could be used to confirm supernatural hypotheses. By its own admission, ID wants to "change the ground rules of science" by allowing supernatural "explanations." However, without any acceptable method to test such hypotheses, ID has no positive evidence for its core claim. Like earlier forms of creationism, it can do no more than try to win by default by claiming that there are "weaknesses" in evolutionary theory and pointing to "problems" that science purportedly cannot explain, such as what ID proponents call "irreducible complexity." ID theory also makes specific commitments to theological propositions that identify it as a theistic view, and even a narrowly sectarian view. However, even if one were to overlook these aspects of the view, ID remains at base inherently supernaturalistic. By virtue of that fact alone it is not science, but religion.

III. THE NATURE AND SUBSTANCE OF THE INTELLIGENT DESIGN MOVEMENT

I base my opinion on the nature of the intelligent design movement and its substantive claims upon reading and analyzing hundreds of articles, books, films, interviews, and Internet postings by its leaders and members, and from listening to them give talks to both supporters and opponents. With some fifteen years of material to draw upon, much of it highly repetitive, one can easily document their views from a wide range of sources, but I will support my opinions here with just some representative quotations drawn mostly from the core leaders of the movement. In this and in subsequent sections, I will point out ways that ID concepts are exemplified in their text *Of Pandas and People* (the ID "reference book" cited in the Dover curriculum), but it will be equally important to pay attention to other sources of ID materials since it is likely that students will search the Internet as well as the library if they are asked to research the topic. I will not base my opinion upon any assessment of ID creationists' religious motivations or their connections with other religious organizations, although these are easily documented and relevant to the case at hand, but will confine my analysis to their substantive claims.

3.1. The defining concepts of the ID movement

Phillip Johnson is the main pioneer, strategist, and intellectual leader of the ID movement. Others in the movement recognize him as the "leading edge" of what they call "the wedge" of intelligent design, so to understand what ID is one should always look first to him. As Johnson articulates it, **the defining concept of the ID movement is theistic realism.**

> My colleagues and I speak of "theistic realism"—or sometimes, "mere creation"—as the defining concept of our movement. This means that we affirm that God is objectively real as Creator, and that the reality of God is tangibly recorded in evidence accessible to science, particularly in biology. (Johnson 1996)

More specifically, this is the God of the New Testament. According to Johnson, "**Either the gospel of Christ is the centerpiece of a new order or it's nothing**" (Johnson 2002).[1] He regularly points out that the starting point for intelligent design is John 1, which says: "In the beginning was the word." William Dembski, another ID leader, cites the same scripture in giving a definition of the basic content of intelligent design: "**Intelligent design is the Logos of John's Gospel restated in the idiom of information theory**" (Dembski 1999b, 84).

This sets forth in a nutshell the defining presuppositions of intelligent design. Their other fundamental claims are built upon this foundation. Here are four key elements of the view:

(1) ID asserts that a transcendent, immaterial, supernatural designer purposefully created biological and physical complexities.

(2) ID asserts that the naturalistic ground rules of science should be rejected and replaced with a revolutionary "theistic science."
(3) ID asserts that the scientific, naturalistic theory of evolution and all other parts of science that deal with functional complexity are false, as is any possible naturalistic theory of functional complexity.
(4) ID asserts that theological views under which evolution and religion are compatible (such as theistic evolution) are unacceptable.

Although the terminology is slightly different in a few points, these central claims are identical to those of creation science, the earlier version of creationism that attempted to bypass the wall of separation by not explicitly mentioning the Bible and claiming to be science. The ID movement aims to provide a big tent for a wide range of creationist views, and for this reason it refuses to take an explicit stand on some claims that are found in creation science, which divide "young earth" creationists from "old earth" creationists.[2] I will focus mostly on the minimal set of common commitments, since that is a sufficient basis for the conclusion that ID is not science but religion. However, it is important to note at the outset, and keep in mind throughout, that all the characteristic young earth creation science claims would be brought into the classroom under the general heading of intelligent design, in the following manner.

The ID movement attempts to unite various creationist factions against their common enemy under a banner of "mere creation" or "design" by temporarily setting aside internal differences. As Johnson told *Christianity Today*, "People of differing theological views should learn who's close to them, form alliances and put aside divisive issues 'til later." Aiming to quell the battle between young- and old-earthers to redirect their energies in tandem against evolutionists, he continued: "I say after we've settled the issue of a Creator, we'll have a wonderful time arguing about the age of the Earth" (quoted in Walker 1998).

The critical point with regard to "teaching the controversy" in the schools is that, when intelligent design creationists ("IDCs") differ, they believe that their differences may not be excluded from the curriculum. Since IDC includes young-earthers, this means that all the standard young-earth creation science arguments are to be included under the ID heading. I will illustrate this briefly with the example of how they handle *common descent*, which refers to the basic fact of evolution, namely, that all organisms have descended with modification from common ancestors, forming a great "tree of life."

In public settings, IDCs will often make it sound as if they object only to the Darwinian evolutionary mechanism, and otherwise accept that evolution occurred. Michael Behe wrote recently in the *New York Times* "Intelligent design proponents do question whether random mutation and natural selection completely explain the deep structure of life. But they do not doubt that evolution occurred" (Behe 2005). However, Dean Kenyon and Percival Davies, the authors of *Pandas*, explicitly reject common descent, as do other core ID leaders such as Paul Nelson. William Dembski allows that organisms have undergone some change through natural history, but like proponents of creation science says this occurred only within strict limits, and holds that human beings were specially created (Dembski 1995). Indeed, Behe is the only ID leader who gets mentioned as not necessarily rejecting common descent. Dembski acknowledges this disagreement within the big tent and then says that their disputes should be included in the schools.

> **Design theorists themselves are divided on [the question of common descent]**. Dean Kenyon and Percival Davis, for instance, argue against common descent.... Michael Behe provisionally accepts common descent. Nonetheless design theorists agree that **discussion of this question must not be shut down simply because a majority of biologists happen to embrace common descent**. The limits of evolutionary change form a legitimate topic of scientific inquiry. **It is therefore illegitimate to exclude this topic from public school science curricula**. (Dembski 1999a, 250)

In *Pandas* the authors actually go further *and explicitly contrast intelligent design with common descent* (what they call "natural descent") claiming, for instance, that homologous structures found in different organisms do not indicate common ancestry. Rather they assert that living things are a "mosaic" of fixed patterns that have been "assembled in various patterns, not unlike subroutines in a computer program" (Davis and Kenyon 1993, 33).

The IDC big tent principle of inclusion means that any of the standard creation science claims may be taught in the public schools under the ID banner as a "legitimate topic of scientific inquiry" no matter how at odds it is with the settled findings of science. Besides being divided on the issue of common descent, IDCs are similarly divided on whether the earth is billions of years old or only six to ten thousand years old and whether or not there was a global, catastrophic flood. *Pandas* author Percival Davis rejects the accepted geological age of the earth in favor of the young-earth view. He rejects plate tectonics. He argues that a global, catastrophic flood caused the major geographical features of the planet (Frair and Davis 1983). These and other such claims are basic creation science views and many are shared by Paul Nelson and other young-earthers in the ID movement. Allowing ID into the schools thus allows all these views perforce.

Anecdotally, this appears to be what happens in practice. In recent cases I have dealt with in two school districts in Michigan, three individual teachers taught intelligent design in their science classes on their own, and included such standard creation science views and materials under that heading.

Returning now to their central assertions, the basic commitment of ID theory is its rejection of naturalism. IDCs assert that natural processes are in principle incapable of explaining the complexity of the biological world. As Dembski puts it:

> The fundamental claim of intelligent design is straightforward and easily intelligible, namely, **there are natural systems that are in principle incapable of being explained in terms of natural causes** and that exhibit features that in any other circumstance we would attribute to intelligence. (Dembski 2004)

According to ID theory, no natural causes—neither natural law nor chance nor any combination thereof—can produce complex functional systems, which they claim are cases of "irreducible complexity" or "complex specified information." Dembski insists that only an immaterial intelligent cause can create such complexity "ex nihilo" (Dembski 2002, 162). Since ID holds that the world does exhibit such biological complexity, **they are fundamentally committed to supernatural creation *ex nihilo*.** In line with their defining commitment to theistic realism, their speeches and writings always try to focus attention on this issue of naturalism. Asked what are the major issues in the debate over naturalism, Phillip Johnson explained:

> The most important question is whether God is real or imaginary. Did God create man or did man create God? The latter is the teaching of evolutionary naturalism, and even many Christian thinkers tacitly assume that position. (Christianbook.com 2000)

IDCs reject what they call science's "naturalistic creation story." Johnson elsewhere points out the importance of this:

> If you have a biblical creation story, then getting the right relationship with God and getting to heaven are the most important things. If you throw that overboard and you have a naturalistic creation story, those things become unimportant and what becomes important is how we apply scientific knowledge to make a heaven here on earth. (Quoted in Goode 1999)

IV. ID IS NOT SCIENCE

Although IDCs, like creation scientists, use the term science to describe their view, their "alternative theory" is fundamentally at odds with what is currently understood as the nature of science. As noted previously, their defining commitment is to explanation in terms of immaterial, supernatural agents. In proposing this, IDCs are departing from what they themselves acknowledge are "the ground rules of science." Their self-proclaimed revolutionary theistic science aims to overturn the way that science has been practiced in modern times and to "fundamentally change our conception" of science.

4.1. ID rejects basic methodological constraints of science

If science is understood as its set of conclusions, then ID theory clearly does not count as science, because it rejects central elements of evolutionary theory that are well established and fundamental. But even more important than specific conclusions (which in science are open to revision on the basis of new evidence) are the methods of science. IDC's so-called theistic science rejects science's methodology and therefore does not belong within the subject. ID theory, like creation science, abrupt appearance theory, and so on are attempts to put a scientific veneer on a narrow religious view to try to garner the prestige of science and the forum of the science classroom (Pennock 2002).

As we have seen, the defining element of IDC is its essential reliance upon supernatural beings and powers—entities that are unconstrained by either lawful necessity or chance processes. The ID movement thus rejects a basic element of scientific empirical evidence, namely, that explanations appeal only to natural causal processes. Scientific explanations need not cite a specific law of nature, but they are always understood to be restricted to the physical realm of law-bound cause and effect relations. In science this is a principle of method, not a metaphysical dogma. This is typically spoken of as *methodological naturalism* in contrast to metaphysical naturalism (also sometimes referred to as ontological or philosophical naturalism).[3]

Historically, people did appeal to the supernatural. Epilepsy was a "sacred disease" and said to result from possession. Lightning bolts were thrown by Zeus. Sick livestock were explained by a hex. Even Newton posited that God nudged the planets to keep them from falling into the sun. However, it has been many centuries since science took such "explanations" seriously. Supernaturalism is not allowed, as will be discussed in more detail below, because it is not testable. Methodological naturalism is not unique to evolutionary biology, but holds throughout all contemporary sciences.[4]

Moreover, there is no sign that science is about to redefine itself or return to a premodern view. I did a database search to see if I could find cases where science might be reintroducing appeal to the supernatural. Several journals did include studies that examined people's beliefs in the supernatural—such as people in KwaZula/Natal, South Africa, who attributed certain forms of diarrhea to supernatural causation, or in Gabon, who blamed certain fevers on the supernatural design of malicious spirits or witches—but these studies considered these beliefs only insofar as they affected quality of medical care, for instance. I did find one case, albeit in a journal devoted to alternative medicine, in which the author proposed a supernatural explanation for the assumed healing effects of prayer. Yet even this was significant in that the author explicitly granted that such a possibility was unprovable scientifically.

> [N]onlocal effects can be conceived of as naturalistic; that is, they are explained by physical laws that may be unbelievable or unfamiliar to most physicians but that are nonetheless becoming recognized as operant laws of the natural universe. The concept of the supernatural, however, is something

> altogether different, and is, by definition, outside of or beyond nature. Herein may reside an either wholly or partly transcendent Creator-God who is believed by many to heal through means that transcend the laws of the created universe, both its local and nonlocal elements, and that are thus inherently inaccessible to and unknowable by science. (Levin 1996)

The authors of *Pandas* recognize this as well:

> Archaeology has pioneered the development of methods for distinguishing the effects of natural and intelligent causes. We should recognize, however, that if we go further, and conclude that the intelligence responsible for biological origins is outside the universe (supernatural) or within it, we do so without the help of science. (Davis and Kenyon 1993)

4.2. IDCs themselves recognize that their view is not science

The ID movement itself recognizes that their view stands in opposition to science. One finds this not only in the leaked Discovery Institute "Wedge" document, which discusses overturning what they see as the antitheistic assumptions of modernism, but throughout ID writings. I'll just give a few examples. William Dembski writes:

> **The scientific picture of the world championed since the Enlightenment is not just wrong but massively wrong**. Indeed entire fields of inquiry, especially in the human sciences... need to be rethought from the ground up in terms of intelligent design. (Dembski 1999a, 224)

Another ID theorist, J. P. Moreland, expressed the conviction that ID is not science by coining a new term:

> If (naturalists) want to define science in naturalistic terms, then **we can define a new term, creascience, that allows for the recognition of discontinuities in nature that indicate the intentional, immediate intervention of a first cause that resembles a person**. Note, if God does not exist, or if he has never intervened in the world through primary causality, then science and creascience are empirically equivalent and equally adequate

> approaches to the study of nature. The main difference between science and creascience is that the latter allows for the possibility that primary causality has occurred and can be recognized. (Moreland 1989)

Here we see another conceptual link to creation science even in Moreland's choice for the roots of his coined word. Whatever one calls it, IDCs themselves recognize that it is not science.

Phillip Johnson made the same point in a criticism of Michael Denton, who was an inspiration for early IDCs but who had backed away from his rejection of common descent. Denton still rejects the Darwinian mechanism but thinks that purposeful complexity can be seen as the result of natural laws. Johnson said that "the restrictions of naturalism" would force Denton out of his view:

> The problem is, there is no non-Darwinian natural mechanism available to do the work of building biological complexity. There's no alternative science to be done using Denton's approach. So, if one asks, what are scientists actually going to do with Denton's ideas?—well, **I don't think there is any prospect for their success *as science*.**... There is no intellectually viable midpoint between naturalism and intelligent design. (Nelson, Behe, et al. 1999)

In his latest book, Dembski says that intelligent design is a revolutionary doctrine "that **will fundamentally change our conception of science** and the world." He emphasizes this key point again a moment later, writing that

> there is good reason to think intelligent design fits the bill as a genuine scientific revolution. Indeed, it is challenging not merely the grand idol of evolutionary biology (Darwinism) but it **is also changing the ground rules by which the natural sciences are conducted**. Ever since Darwin, the natural sciences have rejected the idea that intelligent causes could play a substantive, empirically significant role in the natural world. Intelligent causes might emerge out of a blind evolutionary process but were in no way fundamental to the operation of the world. Intelligent design challenges this exclusion of design from the natural sciences. In so doing, **it promises to remake science and the world**. (Dembski 2004, 19)

4.3. IDCs define key terms in unscientific ways

According to Dembski, "Naturalistic explanations by definition exclude appeals to intelligent agency" (Dembski 2002, 162). Again, this would not be so if design is used in ordinary scientific sense, for example, as when archeologists identify something as an artifact—pottery, for instance—from an ancient people. Science understands people as being a part of the natural causal order. IDC rules out any such natural notion. In Dembski's discussion of what he calls his "explanatory filter" he provides the technical definition upon which his "design inference" rests: Design is just "the set-theoretic complement of necessity and/or chance" (Dembski 1998a). That is to say, design is defined by negation in ID theory, as whatever is not constrained by any natural law ("necessity") or chance process. As they sometimes put it, design just means "transcending natural causes." To be accurate, IDCs should just say nonnatural or supernatural and leave it at that.[5]

According to ID theory, not even extraterrestrial or human intelligences are actually natural, but rather are supernatural, immaterial intelligences that are somehow "embodied." Remember, the basic ID claim is that material processes cannot in principle produce novel functional complexity. No bodies or brains, or natural forces of any kind can do this, they argue, so intelligence cannot be even a supervenient or an emergent property of matter.[6] Thus, if human beings were natural beings, we could not truly be intelligent designers in their sense of the term. Under the assumption that we can produce complex information, as they hold, it can only be because our true selves are immaterial spirits that somehow inhabit a body. Dembski, for instance, claims that human intellect can never be explained in natural, scientific terms.

> [T]he facts point resoundingly to a very imperfect understanding of man in purely scientific categories; that sound philosophy is consistent with this finding, indicating that scientific categories may well be inadequate for a complete understanding of man; and the historic Judeo-Christian theology, by looking to transcendence in both man and God, affirms that this state of affairs will continue. (Dembski 1990)

He explains what he calls "the historical Judeo-Christian position on mind and body" as holding that "the human being unites physical body and immaterial spirit into a living soul" and emphasizes that "this position demands an expanded ontology: unlike semi-materialism with its commitment to supervenience, the historic position does not see spirit as a derivative of the complex physical system that makes up the human body" (Dembski 1990). Dembski cites Genesis as well as the Gospels as the basis of this view.[7] The term "design" in the ID vocabulary refers to the supposed third ontological (metaphysical) category.

IDCs also view evolution in a nonscientific manner. In particular, we have already seen how they define evolution as making metaphysical claims, in contrast to the metaphysically neutral way that scientists understand it. Moreoever, like creation scientists, IDCs holds that evolutionary processes can never increase complexity, but can only stay the same or, more likely, lose information. Dembski, for example, writes:

> If we see evolution as progressive in the sense that the capacities of organisms get honed and false starts get weeded out by natural selection over time, then it seems implausible that a wise and benevolent designer might want to guide such a process. But if we think of evolution as regressive, as reflecting a distorted moral structure that takes human rebellion against the designer as a starting point, then it's possible a flawless designer might use a very imperfect evolutionary process as a means of bringing a prodigal universe back to its senses. (Dembski 2004, 62)

This notion might better be called devolution than evolution. Needless to say, such notions are not scientific, but are inherently religious in character.

4.4. ID offers no positive evidence, just alleged "problems" with evolution

For all their protestations that ID is a scientific theory, IDCs seem to be supremely uninterested either in stating specific testable hypotheses or in providing evidence. It is revealing that William Dembski, in a book that purportedly is about "answering the toughest questions about intelligent design," begins with a motto taken from Blaise Pascal: "People almost invariably arrive at their beliefs not on the basis of proof but on the basis of what they find attractive" (Dembski 2004). But to accept a belief because one finds it attractive is wishful thinking, not science.

In *Pandas*, the ID textbook recommended in Dover, the authors say outright that the characteristics of the intelligent designer cannot be discovered scientifically and that this must be left to "religion and philosophy" (Davis and Kenyon 1993, 7). This is another sign of how unscientific is the ID notion of design. Under ordinary conditions, scientists can sometimes draw conclusions that a human being created something because we have considerable background information about the causal abilities and motivations of human beings. We have observed them designing and creating artifacts and know a lot about their purposes so that we are in a position to judge the results. However, even within the constraints of methodological naturalism, we would have no such ground for any judgments whatsoever about an unnamed, undescribed, and completely nebulous natural designer, let alone the supernatural ones that ID posits.

So, what do creationist "explanations" come down to? *Pandas* is typical in that it declines to offer anything beyond the bare, vague claim that the property in question was designed. For instance:

> Is there any alternative explanation for the marsupial bones and pouches other than that they are homologous and therefore evidence for common ancestry? Yes, another theory is that marsupials were all designed with these reproductive structures. (Davis and Kenyon 1993, 125)

Pandas goes on to admit that they can give no reason for why the intelligent designer would give such structures to one group of animals and not another: "Even if it is assumed that an intelligent designer did indeed have a good reason for every decision that was made, and for including every trait in each organism, it does not follow that such reasons will be obvious to us" (Davis and Kenyon 1993, 125). Again, there is no scientific content to this notion of "design"—on their definition, the term just means that the cause was supernatural.

Only occasionally do we get a more specific idea of what ID theorists have in mind. Phillip Johnson gave the example of the peacock, saying that it is something an "uncaring evolutionary process would never allow to develop" but which is "just the kind of creature that a whimsical Creator might favor" (Johnson 1991, 31). It is hard to see how any such "explanation" in terms of **divine whimsy** could be taken seriously by any science as we understand the notion today.

Besides the previous specific examples and similar ones from *Pandas* of purported problems that evolution cannot explain, IDCs make a more general claim about an entire class of systems that they claim evolution cannot explain, specifically ones that involve what they call "specified complexity" (Dembski) and "irreducible complexity" (Behe). These concepts stand at the base of what IDCs claim is how they can detect design, in the same way that they did for creation science.[8] This design inference supposedly works by means of an "explanatory filter," whereby one first tries to explain a phenomenon using law and then by chance. If neither of these works, then one should conclude that the phenomenon was the result of design. In the same way that creation scientists proposed to prove creation simply by showing things evolution purportedly cannot explain (what is known as an argument from ignorance), this tries to get a conclusion by default without ever having to present any positive evidence. Dembski seems to think this is not a problem, saying that "an argument from ignorance is still better than a pipe dream in which you're deluding yourself. I'm at least admitting to ignorance as opposed to pretending that you've solved the problem when you haven't" (quoted in Monastersky 2001).

Among the many flaws with Dembski's argument, his tripartite classification of necessity, chance, and design is neither mutually exclusive nor jointly exhaustive in the ordinary senses of those terms. As noted above, he gives a technical definition of "design" as the "set-theoretic complement" of the other two, but this negative definition certainly does not capture the ordinary content of the concept, which is orthogonal to chance and necessity. That is to say, if design is understood in the ordinary sense, then it is entirely included within chance and necessity, and his design inference thus fails to get off the ground. It is an example of a false dilemma (or here, a false trilemma). Moreover, Dembski's concept of specified complexity or complex specified information (CSI) is not clearly defined or applicable to biological information in the manner he claims. We cannot say whether any real biological pattern exhibits actual CSI. There is no way to assess the probability in any real biological case, and Dembski's notion of after-the-fact specification is similarly problematic. Even the value of the universal probability bound, which is essential in Dembski's inference, is an open question; Dembski dismisses out of hand important hypotheses in physics that suggest the possibility of multiple universes that would completely undermine his set figure. Leaving these problems aside, it is simplest to just directly refute the illustrative cases he has specified.[9] For instance, Dembski says that Behe's notion of irreducible complexity is an example of specified complexity. My colleagues and I have demonstrated experimentally that a Darwinian mechanism can discover irreducible complex system (Lenski, Ofria, et al. 2003). The basic IDC claim that it is impossible in principle for the natural evolutionary mechanisms to produce irreducible complexity is not persuasive when one can observe evolution do just that.

However, if design is understood in the supernatural sense, then it is not amenable to test. There is no way to have a controlled experiment to test the efficacy of the claimed causal factor. Nor can one infer from what is known of natural designers.

4.5. Testability

Supernaturalism is not included within science because it is untestable. Indeed, introducing the supernatural undermines the very basis for empirical testing. The first and most basic characteristic of supernatural agents and powers is that they are above and beyond the natural world and its agents and powers. Indeed, this is the very definition of the term. They are not constrained by natural laws or chance processes. A second characteristic of the supernatural is that it is inherently mysterious to us. As natural beings, our knowledge all comes via natural laws and processes. If we could apply natural knowledge to understand supernatural powers, then, again by definition, they would not be supernatural. The lawful regularities of our experience do not apply to the supernatural world. If there are other sorts of supernatural "laws" that govern that world, they can be nothing like those that we understand. Occult entities and powers are profoundly mysterious to us. The same point holds about divine beings—we cannot know what they would or would not do in any given case. Scientific models must be judged on natural grounds of evidence, for we have no supernatural ground upon which to stand. A final relevant element of the notion of the supernatural is that supernatural beings and powers are not controllable by humans. If we can control the natural world, it is only because the world is governed by physical laws that must be "obeyed" even when we are pulling the strings, whereas the very idea of the supernatural is that it stands above natural laws and thus outside the possibility of our control.

These characteristics of the supernatural show why supernatural explanations are excluded from scientific theorizing. Science operates by empirical principles of observational testing; hypotheses must be confirmed or disconfirmed by reference to intersubjectively accessible empirical data. One supports a hypothesis by showing that certain consequences obtain, which would follow if what is hypothesized were to be so in fact. Darwin spent most of the *Origin of Species* applying this procedure, demonstrating how a wide variety of biological phenomena could have been produced by (and thus explained by) the simple causal processes he discovered. But supernatural theories can give us no guidance about what follows or does not follow from their supernatural components.

The appeal to supernatural forces is always available for we can cite no necessary constraints upon the powers of supernatural agents. This is just the picture of God that Johnson presents. He says that God could create out of nothing or use evolution if He wanted (Johnson 1991, 14, 113); God is "omnipotent" (Johnson 1991, 113). He says God creates in the "furtherance of a purpose" (Johnson 1991, 4), but that God's purposes are "inscrutable" (Johnson 1991, 71) and "mysterious" (Johnson 1991, 67). A god that is all-powerful and whose will is inscrutable can be called upon to "explain" any event in any situation, and this is one reason for science's methodological prohibition against such appeals. Leaving the designer unnamed and undescribed has the same effect. Given this feature, supernatural hypotheses remain immune from disconfirmation or meaningful testing.

Experimentation requires observation and control of the variables. We confirm causal laws by performing controlled experiments in which the hypothesized independent variable is made to vary while all other factors are held constant so that we can observe the effect on the dependent variable. But we have no control over supernatural entities or forces; hence these cannot be scientifically studied.

Finally, if we were to allow science to appeal to supernatural powers even though they could not be tested, then the scientist's task would become just too easy. One would always be able to call upon the gods for quick theoretical assistance in any circumstance. Once such supernatural explanations are permitted, they could be used in chemistry and physics as easily as creationists have used them in biology, geology, and linguistics. Indeed, all empirical investigation could cease, for scientists would have a ready-made answer for everything. For example, consider *Pandas* author Davis's alternative creationist explanation of the many general similarities among animals (such as common reactions of humans, rats, and monkeys to drugs). These, he and his coauthor, Wayne Frair, say, "can be explained as originating in basic design given by the Creator. Evolution is not needed to account for the similarities" (Frair and Davis 1983, 14). In short the "explanation" does not go beyond claiming that this pattern is so because the Creator designed it so. There is no way to test this kind of one-size-fits-all explanation.[10]

Regarding the charge that ID theory is not testable, *Pandas* first tries to shrug off the problem, saying that ID theory is "not unique in its flexibility" but then it tries to claim that ID is falsifiable in one way:

> [T]he concept of ID predicts that complex information ... never arises from purely chemical or physical antecedents.... Experience will show that only intelligent agency gives rise to functional information. All that is necessary to falsify the hypothesis of ID is to show confirmed instances of purely physical or chemical antecedents producing such information. (Davis and Kenyon 1993, 160)

IDCs repeat this single example whenever challenged, but it shows just how hollow and far-removed from science the IDC view is. Once again, it makes clear that "design" is not being used in the ordinary sense, for it does not follow as a prediction from the hypothesis (or even the fact) that something was created by design that purely chemical and physical antecedents could not also do so. Moreover, even if we grant the conditional using ID theorists' supernatural notion of design, an example of the sort mentioned hardly functions as a test. If it did, then we could as easily say we have tested hypotheses such as "Complex information never arises without the help of elves," or "Lightning bolts never form unless Zeus throws them," or "You can never be in good health unless your chakras are aligned" by showing natural processes that account for lightning and good health. For the same reasons noted previously, all such supernatural hypotheses are untouched by any possible observation. The electrical experiments that demonstrate electromagnetic theory (under the normal constraints of methodological naturalism) tell us nothing about whether Zeus was or was not pitching from beyond the natural realm. A person could be in good heath or bad health with charkas aligned or misaligned for all anyone can tell. If we do point to cases of functional information arising through natural processes (as I and others have already done), can we really be sure that the elves were not secretly helping?

Again, the point here is that the scientific methodological principle of restricting appeals to natural causal processes is perfectly reasonable. Allowing appeals to the supernatural undermines the very notion of an empirical test.

Do IDCs have any other method to offer? It is hard to tell whether Phillip Johnson is serious about a couple of suggestions he has made:

> Science is committed by definition to... find[ing] truth by observation, experiment, and calculation rather than by studying sacred books or achieving mystical states of mind. It may well be, however, that there are certain questions... that cannot be answered by the methods available to our science. These may include not only broad philosophical issues such as whether the universe has a purpose, but also questions we have become accustomed to think of as empirical, such as how life first began or how complex biological systems were put together.

It is unclear why we should think that sacred books and mystical states of mind may be able to answer such questions. If IDCs are serious about such "methods," they do not say how this may be done, so there seems to be no good reason to join their revolution to overturn natural science.

4.6. Calling ID a science does not make it one

A famous philosopher posed the following question: If you call a tail a leg, how many legs does a dog have? The answer, he said, is four; calling a tail a leg doesn't make it one. Calling intelligent design creationism an "alternative scientific theory" and using scientific-sounding terminology does not make it a science now any more than it did for creation science. ID theory rejects both fundamental conclusions and basic methodological constraints of science. It posits an unnamed and undescribed supernatural designer as its sole explanatory principle. It provides no positive evidence for its extraordinary claims. And because it cannot stand on the evidential ground that science requires, it tries to change the ground rules of science. Even by its own lights, ID theory is not science.

V. ID IS RELIGION

The ID movement is not just a religious view in a general, nondenominational sense, but also in a sectarian sense. Just as creation science took Old Testament claims as fundamental, the ID movement takes New Testament claims as the basis for its "theistic science." Its advocates claim that design theory puts Christianity into the realm of objective fact. Nor is this even a mainstream version of Christianity, but rather a specific theological view that explicitly rules out various other religions as well as other standard Christian views. To admit ID into the public schools would be to advance one particular religious view over others. However, even if ID was not based on a particular sectarian view and even if it did not explicitly mention God, it would still be religious simply because of its base reliance upon supernatural beings.

When speaking to a small audience of apparent supporters, IDCs can be straightforward in describing their theory as religious. In a talk I heard him give to a campus evangelical Christian group, ID leader Walter Bradley got a good laugh when he said that intelligent design is "the politically correct way to say God." In a talk I heard William Dembski give, he too seemed to recommend this political advantage of ID, pointing out at the end how he had been able to explain their ideas—for instance, how the designer doesn't create the world like a watchmaker and let it run on its own like a watch, but plays it like a violinist—"without using the G-word." In the public square, however, IDCs will deny that their view is religious. Michael Behe, for instance, recently wrote: "The theory of intelligent design is not a religiously based idea, even though devout people opposed to the teaching of evolution cite it in their arguments.... [I]ntelligent design itself says nothing about the religious concept of a creator" (Behe 2005). However, as we have already seen in some of the ID literature quoted above, such denials are not credible. In fact, ID does make essential commitments to God, both directly and indirectly.

5.1. ID is inherently theistic

Behe himself, almost immediately after making the above denial, goes on to explain irreducible complexity as an example of William Paley's design argument, which is a classical argument for the existence of God. Stephen Meyer, one of the authors of the appendix in *Pandas*, explains ID as being "the return of the god hypothesis" in one of his articles (Meyer 1999). William Dembski writes that "the crucial breakthrough of the intelligent design movement has been to show that this great theological truth—that God acts in the world by dispersing information—also has scientific content" (Dembski 1999a, 233). Phillip Johnson has revealed the theistic nature of intelligent design in many places, including a recent interview on the American Family Radio:

> Our strategy has been to change the subject a bit so that we can get the issue of intelligent design, which really means the reality of God, before the academic world and into the schools. (Phillip Johnson, American Family Radio, January 10, 2003, broadcast)

Elsewhere he explains how the ID movement's defining concept of theistic realism "assumes that the universe and all its creatures were brought into existence for a purpose by God. Theistic realists expect this 'fact' of creation to have empirical, observable consequences that are different from the consequences one would observe if the universe were the product of nonrational causes" (Johnson 1995, 208–209).

Indeed, IDCs regularly claim that accepting evolution and naturalism takes more faith than religion and that such thinking is superstitious and irrational. For instance, Johnson writes: "I've had an opportunity to see how influential naturalistic thinking is, I want to show people how to identify it and contrast it with genuinely theistic reasoning" (Johnson 1995). According to Johnson, "[R]ational thinking is God-based thinking" (Johnson 1995). This, purportedly, is what intelligent design provides.

5.2. ID is not just general theism, but sectarian religion

We have already noted how Dembski explains that "intelligent design is the Logos of John's Gospel restated in the idiom of information theory" (Dembski 1999b, 84). He elsewhere advances a "Law of Priority in Creation" and explains it this way:

> The creator is always strictly greater than the creature. It is not possible for the creature to equal the creator, much less surpass the creator. The Law of Priority in Creation is a conservation law. It states in the clearest possible terms that you can't get something for nothing. There are no free lunches. (Dembski 1990, 222)

Dembski writes that he would like to see this law elevated to a status comparable with the laws of thermodynamics, but says that he cannot take credit for the law:

> The law is not new with me. It is found in Scripture: "Jesus has been found worthy of greater honor than Moses, just as the builder of a house has greater honor than the house itself." (Dembski 1990, 222)

In his later writings for secular audiences he again puts this in "the idiom of information theory" and renames it the "Law of Conservation of Information," claiming that it is a "4th law" of thermodynamics.

Johnson also emphasizes the New Testament basis of intelligent design. Here is how one sympathetic writer reported about an intelligent design event:

> A conference on "Mere Creation" at Biola University in suburban Los Angeles brought together an unprecedented cross-disciplinary gathering of 200 men and women—mostly academics and mostly Christians—interested in building a credible origins model based on "theistic design."
>
> "This isn't really, and never has been, a debate about science.... It's about religion and philosophy." Mr. Johnson also insists the real issue in the century-old debate isn't even about the early chapters of Genesis. "I turn instead to John 1," says the astute Presbyterian layman, "where we're told that 'In the beginning was the word.'" (Belz 1996)

One can easily extend such a list documenting the religious nature of ID. Stephen Meyer, citing Paley, argues that the astonishing functional complexity of the world "could not originate strictly through the blind forces of nature," and claims that ID theory supports "a Judeo-Christian understanding of Creation." ID advocate Nancy Pearcey writes, "By uncovering evidence that natural phenomena are best accounted for by Intelligence, Mind, and Purpose, the theory of intelligent design reconnects religion to the realm of public knowledge. It takes Christianity out of the sphere of noncognitive value and restores it to the realm of objective fact, so that it can once more take a place at the table of public discourse" (Pearcey 2004, 72–73). And in his blurb for Pearcey's book on intelligent design *Total Truth*, Michael Behe is forthright about the religious nature of ID writing: "With marvelous clarity of thought and prose, Pearcey explains how modern science reinforces Christianity—and why more Christians should be aware of it."

5.3. ID explicitly rejects other theological views

To allow intelligent design in the public schools would be to privilege a narrow theological view over other religious beliefs. Most significantly, ID theory explicitly rejects theistic evolution, a theologically mainstream view that takes evolution and belief in God to be compatible. In the introduction to the ID anthology *Mere Creation*, Dembski writes:

> Intelligent design is logically compatible with everything from utterly discontinuous creations (e.g., God intervening at every conceivable point to create new species) to the most far-ranging evolution (e.g., God seamlessly melding all organisms together into one great tree of life). (Dembski 1998b)

However, in his next breath, he says,

> That said, intelligent design is incompatible with what typically is meant by theistic evolution . . . theistic evolution is no different from atheistic evolution. (Dembski 1998b)

Elsewhere Dembski has written:

> *Design theorists are no friends of theistic evolution.* As far as design theorists are concerned, theistic evolution is American evangelicalism's ill-conceived accommodation to Darwinism. (Dembski 1995, 3 [emphasis in original])

Phillip Johnson is equally blunt, saying that theistic evolution is a "disastrous accommodation" to Darwinism that provides "a veneer of biblical and Christian interpretation ... to camouflage a fundamentally naturalistic creation story" (Johnson 2002b, 137). He elsewhere said that theistic evolutionary views are "bogus intellectual systems" that read the Bible "figuratively rather than literally" (Johnson 1997, 111).

ID also explicitly rejects deistic and various other religious views. For instance, Dembski says that deism is only a "logical possibility" but that ID rejects it because "there is no evidence for it" and interactive design is a better fit with the fact that "information tends to appear discretely at particular times and places.... [Deism] restricts design to structuring the laws of nature and thereby precludes design from violating those laws and thus violating nature's causal structure" (Dembski 2002, 344–47). He dismisses Hinduism as "religious naturalism," writing that "there can be no transcendent God within such a framework. These gods of the Vedas are not prior to nature but intrinsic to it." Dembski says that both the Hindu and the Greek gods "are pathetic because nature's fundamental laws can always overrule them" (Dembski 1999a, 101–102).

VI. CONCLUSION

ID's claims that it isn't religious are not credible. Standard definitions of religion focus on belief in the supernatural and in some transcendent realm. God need not be mentioned explicitly. Thus, even if ID were taught "without mentioning the G-word" it would still be religious. ID's defining element of theistic realism, its basic inclusion of transcendent entities and powers, and its explicit rejection of even the possibility of theistic evolution all make explicitly religious commitments, as do many other of its claims.

Science, on the other hand, is neutral and nondogmatic with regard to metaphysical possibilities. As I have noted, naturalism in science is a methodological view, not a metaphysical dogma. Science takes no stand with regard to the metaphysical possibility of supernatural entities or powers. Neither evolution nor any other science affirms or denies the possible existence of God or other transcendent beings. If someone were to find a way to empirically confirm the existence of an immaterial designer or any other supernatural being, science should change its methodology. However, there is no evidence in the scientific literature that supernatural hypotheses are being considered and it is hard to imagine how that could possibly change without undermining the very notion of empirical evidence. ID and other creationist views want to appeal to transcendent beings, but have not provided an acceptable alternative method to test such appeals. ID remains at base inherently supernaturalistic. By virtue of that fact alone it is not science, but religion.

NOTES

1. Here and elsewhere, I use bold typeface to highlight salient points, whereas terms in quotations that are italicized for emphasis are in the original.

2. The general term "creationism" refers to any view that rejects evolution in favor of the action of some personal, supernatural creator. Creationism is not limited to Bible-based views. Other religions have their own creation accounts that may be in conflict with evolution. For instance, some fundamentalist Hindu sects, like the Hare Krishnas, reject evolution in favor of their own specific theistic account. Many Native American tribal groups do as well, as do various Pagan religions.

Even Bible-based creationism comes in many varieties. For instance, some creationists hold that creation occurred in six literal days just six to ten thousand years ago ("young-earth" creationism). Other creationists read Genesis in a different literal manner to allow room for billions of years ("old-earth" creationism). Some creationists insist on a global, catastrophic flood, while others hold to a global but tranquil flood, while others think the flood was local. Intelligent design is an alliance of young-earth and old-earth creationists.

Although ID leaders sometimes try to suggest that creationism refers only to young-earth creationism in order to claim that ID is not creationism, they have elsewhere granted this definition of creationism. For instance, Johnson defines creationism:

> "Creationism" means belief in creation in a . . . general sense. Persons . . . are "creationists" if they believe that a supernatural Creator not only initiated

> this process but in some meaningful sense controls it in furtherance of a purpose. (Johnson 1991)

This is just what ID holds. William Dembski, in a section he devotes to getting the terminology straight, writes:

> The only thing one can say for certain is that to reject fully naturalistic evolution is to accept some form of creationism broadly construed, i.e., the belief that God or some intelligent agent has produced life with a purpose in mind. Young earth creationism certainly falls under such a broad construal of creationism, but is hardly coextensive with creationism in this broad sense. (Dembski 1995)

ID theorists have at times explicitly referred to each other as creationists. I will use intelligent design (ID) and intelligent design creationism (IDC) interchangeably.

It is also important to be clear that not all religions are creationist. Many religions and theological traditions accept the scientific understanding of evolution and so are not forms of creationism. The Catholic Church and most mainline Protestant denominations, for instance, do not take evolution to be in conflict with Christian faith, holding that God could have ordained the evolutionary mechanism as the process for creating the biological world. These views are not creationist, but rather are forms of theistic evolution. As noted, ID explicitly rejects theistic evolution.

3. ID theorists regularly conflate methodological and metaphysical naturalism and charge science with itself being a religion. This is a mistake. Because this is a principle of method only, science is neutral with regard to the metaphysical existence of God.

4. Indeed, this principle in science is no different from modern understanding of legal reasoning. A supernatural theistic science is no more reasonable than a supernatural theistic legal system (Pennock 1999). The fact that courts prosecuted witchcraft in the past does not mean that such claims would or should be countenanced today. Courts do not take seriously any claims or defenses based on supernatural intervention, nor should they. If, as IDCs propose, science should be redefined to allow in the supernatural, then the law would have to be redefined as well.

5. Again, it is important to recognize that the supernatural does not mean what is "outside" the universe, which might still be a part of nature and covered by physical law, but rather what is unconstrained by the lawful causal structure of nature. Supernatural spirits are believed to operate "inside" the universe as well.

6. Put simply, these are properties that arise at higher levels of organization.

7. He writes, for example: "The position as I have stated it is but a straightforward restatement of the Genesis account of man's creation: 'The LORD God formed the man from the dust of the ground [body] and breathed into his nostrils the breath of life [spirit], and the man became a living being [soul]'" (Dembski 1990, 202–26). He also cites Paul and James from the New Testament.

8. For instance, all the main elements of Dembski's argument for design and even some details, like the example of Mount Rushmore and the SETI project, were previously made by young-earth creation scientists such as A. E. Wilder-Smith and, especially, Norman Giesler, who was a creationist expert witness at the Arkansas "balanced-treatment" trial (Pennock 1999). Henry Morris, the young-earth creationist who developed the notion of creation science, has recently chided Dembski for giving arguments that he had previously made but just under a different name.

> [I]t is not really a new approach, using basically the same evidence and arguments used for years by scientific creationists but made to appear more sophisticated with complex nomenclature and argumentation.... Dembski uses the term "specified complexity" as the main criterion for recognizing design. This has essentially the same meaning as "organized complexity," which is more meaningful and which I have often used myself. He refers to the Borel number (1 in 10^{50}) as what he calls a "universal probability bound," below which chance is precluded. He himself calculates the total conceivable number of specified events throughout cosmic history to be 10^{150} with one chance out of that number as being the limit of chance. In a book written a quarter of a century ago, I had estimated this number to be 10^{110}, and had also referred to the Borel number for comparison. His treatment did add the term "universal probability bound" to the rhetoric. (Morris 2005, 194a)

Morris even points out that Behe's example of the bacterial flagellum was already given by creation scientists.

9. I have shown this for several specific cases Dembski and others have given, such as phone numbers (Pennock 1999), Dürer woodprints (Pennock 2001, 645–67), and so on.

10. The above summary is excerpted from a more detailed discussion of testability in Pennock 1995, chap. 6.

REFERENCES

Behe, M. "Design for Living." *New York Times*, 2005.

Belz, J. "Witnesses for the Prosecution: Darwin on Trial Author Brings Together Anti-Darwin Coalition to Bring Down Evolution." *World Magazine* 11 (1996): 18.

Christianbook.com. Interview with Phillip Johnson, 2000.

Davis, P., and D. H. Kenyon. *Of Pandas and People.* Dallas, TX: Haughton Publishing, 1993.

Dembski, W. A. "Converting Matter into Mind: Alchemy and the Philosopher's Stone in Cognitive Science." *Perspectives on Science and Christian Faith* 42 (1990): 202–26.

———. "What Every Theologian Should Know about Creation, Evolution, and Design." *Center for Interdisciplinary Studies Transactions* 3 (1995): 1–8.

———. *The Design Inference: Eliminating Chance through Small Probabilities.* New York: Cambridge University Press, 1998a.

———. *Mere Creation: Science, Faith & Intelligent Design.* Downers Grove, IL: InterVarsity Press, 1998b.

———. *Intelligent Design: The Bridge between Science & Theology.* Downers Grove, IL: InterVarsity Press, 1999a.

———. "Signs of Intelligence: A Primer on the Discernment of Intelligent Design." *Touchstone* 12 (1999b): 76–84.

———. *No Free Lunch: Why Specified Complexity Cannot Be Purchased without Intelligence.* Lanham, MD: Rowman & Littlefield, 2002.

———. *The Design Revolution: Answering the Toughest Questions about Intelligent Design.* Downers Grove, IL: InterVarsity Press, 2004.

Frair, W., and P. Davis. *A Case for Creation.* Chicago: Moody Press, 1983.

Goode, S. "Johnson Challenges Advocates of Evolution." *Insight*, 1999.

Johnson, P. "Reason in the Balance: An Interview with Phil Johnson about His Forthcoming Book." *Real Issue* (1995).

Johnson, P. E. *Darwin on Trial.* Washington, DC: Regnery Gateway, 1991.

———. *Reason in the Balance: The Case against Naturalism in Science, Law & Education.* Downers Grove, IL: InterVarsity Press, 1995.

———. *Starting a Conversation about Evolution: A review of* The Battle of the Beginnings: Why Neither Side Is Winning the Creation-Evolution Debate *by Del Ratzsch.* Access Research Network, 1996.

———. *Defeating Darwinism.* Downers Grove, IL: InterVarsity Press, 1997.

———. "Recognizing the Power of Religion." *Touchstone* 15 (2002a): 12.

———. *The Right Questions: Truth, Meaning and Public Debate.* Downers Grove, IL: Intervarsity Press, 2002b.

Lenski, R. E., C. Ofria, R. Pennock, and C. Adami. "The Evolutionary Origin of Complex Features." *Nature* 423 (2003): 139–44.

Levin, J. S. "How Prayer Heals: A Theoretical Model." *Alternative Therapies in Health Medicine* 2 (1996): 66–73.

Meyer, S. C. "The Return of the God Hypothesis." *Journal of Interdisciplinary Studies* 11 (1999): 1–38.

Monastersky, R. "Seeking Deity in the Details." *Chronicle of Higher Education* 48 (2001).

Moreland, J. P. *Christianity and the Nature of Science: A Philosophical Investigation.* Grand Rapids, MI: Baker Book House, 1989.

Morris, H. "The Design Revelation." *Back to Genesis*, 194a., 2005.

Nelson, P., and M. Behe, et al. "A Roundtable on Nature's Destiny." *Origins & Design* 19 (1999): 1–11.

Pearcey, N. R. "Darwin Meets the Barenstain Bears: Evolution as a Total Worldview." In W. A. Dembski, *Uncommon Dissent: Intellectuals Who Find Darwinism Unconvincing*, 53–73. Wilmington, DE: ISI Books, 2004.

Pennock, R. T. *Tower of Babel: The Evidence against the New Creationism.* Cambridge, MA: MIT Press, 1999.

———. "The Wizards of ID: Reply to Dembski." In R. T. Pennock, *Intelligent Design Creationism and Its Critics: Philosophical, Theological and Scientific Perspectives*, 645–67. Cambridge, MA: MIT Press, 2001.

———. "Should Creationism Be Taught in the Public Schools?" *Science & Education* 11 (2002).

Walker, K. "Young-Earth Theory Gains Advocates." *Christianity Today* (1998).

CHAPTER 25

A STEP TOWARD THE LEGALIZATION OF SCIENCE STUDIES

STEVE FULLER

In February 2005, the Thomas More Law Center of Ann Arbor, Michigan, sought my help as a "rebuttal witness" for the defense in *Kitzmiller v. Dover Area School District*, a trial scheduled to begin early in the autumn of that year, which would be the first case to test the eligibility of "intelligent design theory" (IDT) for inclusion alongside the neo-Darwinian theory of evolution in high school biology classes. As a rebuttal witness, my charge was to contradict the claims made by the plaintiffs' witnesses, all of whom were seasoned veterans of related trials involving creationism. However, I learned of their prior experience only once I started preparing for the trial.[1] I decided to participate simply after having read the expert witness reports as filed by the plaintiffs' lawyers. These struck me as based on tendentious understandings of the nature of science that would not have survived scrutiny on an informed listserv such as HOPOS-L, let alone the peer-review process of a relevant journal. My critical eye was clearly informed by knowledge gained from the science studies disciplines, since I am not a known advocate of—or expert in—either IDT or Neo-Darwinism.[2]

I may be the first person to declare under oath that knowledge of the his-

---

tory, philosophy, and sociology of science provides a better basis for evaluating the scientific standing of a field of inquiry than someone formally trained in science. However, I am not the first whose expertise conformed to this declaration. The testimony of Michael Ruse, a scientific amateur, was the intellectual centerpiece in the verdict delivered in *McLean v. Arkansas* (1982).[3] It was based on a Popper-inspired criterion, according to which the proposed version of creation science failed to be a science by virtue of its reliance on unfalsifiable—indeed, infallible—biblical pronouncements about nature. At the time, Ruse was criticized by fellow philosophers for having advanced a largely discredited conception of science.[4] Nevertheless, the significance of his testimony is that the judge did not simply defer to the experts in the contested field to authorize his decision. He sought an independent standard, which Ruse provided, insofar as the falsifiability principle does not beg the question in favor of a science's dominant paradigm. Of course, it helped that Ruse's independent standard also coincided with expert judgment, but the legal precedent had been set. Future candidates for inclusion in the science curriculum could satisfy criteria that do not require the approval of those who claimed to speak on behalf of the entire scientific community. I take this point in the spirit of job qualifications that a candidate may meet, despite the personal judgment of a prospective employer—against whom then a charge of discrimination may be legitimately raised.[5]

It would be easy to treat the precedent set by Ruse's testimony cynically as an opportunity for creationists to make the minimal adjustments needed to convey the impression that they have crossed the threshold into science. Moreover, the events that precipitated *Kitzmiller* suggest as much, since the IDT textbook on offer—*Of Pandas and People*—began life as a "young earth" creationist text, and the Dover school board members keen on it were more interested in IDT as a fig leaf for their creationist beliefs than in the details of the theory itself. As a result, the Discovery Institute, the Seattle-based think tank most responsible for promoting IDT, refused to cooperate with the defense lawyers.[6] It also didn't help that the person who turned out to be (after other defections) the "star witness" for IDT, Michael Behe, is a contributor to the book's new edition. The defense lawyers, perhaps anticipating these problems, instructed me not to speak with anyone on the Dover school board, nor read *Pandas*, the content of which I only gleaned during testimony given at the trial. While *Pandas* is not the book I would write to introduce IDT in a scientific light, it does imply that social constructivism might itself be considered a version of IDT.[7] In any case, my own testimony was limited to the general matter of whether IDT counts as science.

All things considered, the precedent set by Ruse's testimony was the right one, and it is unfortunate that the judge in *Kitzmiller* made no reference to it, though he referred to the case in which it figured. In this regard, his ruling took a step back from the historical impetus for proposing a "scientific method," from Francis Bacon's *experimentum crucis*, through the versions of the demarcation criteria advanced by the positivists and the Popperians, up to recent sociological preoccupations with boundary construction and maintenance in knowledge production. All of these projects have shared an interest in finding a neutral ground for the adjudication of contesting knowledge claims. To be sure, the projects define "neutrality" rather differently, but they agree that it means something other than simply deferring to those currently called scientists to decide who else can count as scientists. A major exception to this consensus of opinion about the scientific method is provided by the work of Thomas Kuhn and others, also in science studies, who share his late Wittgensteinian anthropological sensibility, whereby science is simply as scientists do. The judge's ruling in *Kitzmiller* can be easily understood in these terms, given the weight that the judgment of the US National Academy of Sciences (NAS) carried in undermining IDT's scientific credentials.

It was precisely the NAS's judgment that impelled me to serve as a rebuttal witness. The National Academy of Sciences (1999) has declared that the conduct of science is governed by a principle of "methodological naturalism," a point repeatedly made by the plaintiffs' experts throughout the trial. This expression has been promoted for several years by a think tank based in Oakland, California, the misleadingly named National Center for Science Education, as a safeguard against the kind of threat that IDT supposedly poses to the science curriculum. The phrase "methodological naturalism" is a neologism designed to capture two things at once that the history of the scientific method has tended to keep separate: the source of hypotheses and the conditions under which they are testable.[8] This separation explains the studious neutrality that philosophers of the scientific method have tended to adopt toward "metaphysics," in which both naturalism and its opposite, supernaturalism, are normally included: neither metaphysics offers a royal road to scientific validity but both have had significant heuristic value.

In particular, the history of science is full of hypotheses of "supernatural" inspiration, in that the hypothesized entities are not observable in the normal run of experience, but only under specially crafted conditions that are unavailable when the hypotheses are first proposed. In short, the right clever experiment has yet to be designed. Typically, these supernatural hypotheses —expressions of what is less prejudicially called "metaphysical realism"— receive their initial grounding in a mathematically significant pattern that points to a deeper level of explanation. In the case of, say, Newton's appeal to gravitational attraction or Mendel's to hereditary factors, these hypotheses have had theistic origins that survive in contemporary IDT—namely, ideas of a divine plan and special creation. Science, on this view, is part of the project of creating "a heaven on earth," to recall the old Enlightenment ideal, specifically by realizing in matter things that first exist as products of the (here mathematical) imagination, which is itself presumed—as Newton did of his own mind—divinely inspired.[9]

However, the role of the supernatural in the advancement of science comes to be erased once experimental proof is found for these originally mysterious entities. The context of justification is effectively read back into the context of discovery, which then makes methodological naturalism look persuasive. Newton's and Mendel's theistic inspirations are relegated to superfluity, since subsequent experiments and attendant reasoning are sufficient to demonstrate gravity and genes. This makes methodological naturalism a natural concomitant of a Whig historiography of science, whereby the past is seen in retrospect (from where it ended up), not in prospect (from where it emerged). The interesting exception, as one might expect, is neo-Darwinism itself, where "natural selection" has figured in less than 1 percent of the abstracts of all biology papers indexed since 1960. Even "evolution," in all its protean uses, has figured in only 12 percent of them.[10] Does this mean that the reported research is sufficiently self-contained that it can be explained without appeal to such higher-order concepts? Apparently not. Instead we are told that neo-Darwinism has become so fundamental to the conduct of the life sciences that it is always already assumed.[11]

In the end, these wranglings over the history and philosophy of science must return to the requirements of high school science education. Presumably one wishes to promote an interest in science within the confines of the law. Public opinion surveys suggest that there should be a ready market for textbooks that creatively synthesize scientific and religious interests, given that two-thirds of Americans believe in both divine creation and evolution. However, this market is curbed by a broad interpretation of the so-called Establishment Clause of the US Constitution's First Amendment, which prevents any single religion from monopolizing public life. The clause may be seen as part of the nation's birth trauma, since the American colonies were originally settled by people whose dissenting religious views prevented their full participation in public life in Britain. In the twentieth century, the clause's scope expanded in the context of publicly funded schools, so that all religious expression is now effectively banned as anything other than a formal object of study. The emergence of a scientific establishment as distinct from religious authorities, backed by the American Civil Liberties Union, has been in the forefront of this development.[12]

After successive waves of immigration from southern and eastern Europe starting in the mid-nineteenth century, it became common for self-styled progressive thinkers to claim that science unites and religion divides in the great "melting pot" that the United States had become. Indeed, some, like John Dewey, came close to arguing that the scientific method could function as a secular religion.[13] Without that vision ever quite being realized, religion was nevertheless increasingly driven into the private sphere, on the assumption that it was inherently nonscientific, yet, in an Orwellian sense, still "compatible" with science, just as the "separate but equal" doctrine—prior to the Civil Rights Act of 1964—suggested that blacks and whites could live together, as long as they did not interfere with each other. This assertion of compatibility—and the corresponding judgment that the defense was simply manufacturing a conflict between the scientific establishment and religious believers—was perhaps the most widely reported feature of the *Kitzmiller* ruling. My own view is that the defense did indeed have the weaker case, but equally that the judge did an injustice to the relevant philosophy, politics, and ultimately to science.[14] IDT may be inept in its self-understanding and self-presentation, but it did not deserve to be dismissed outright.[15]

If IDT reads anything "literally," it is not the Bible but the history of biology, in which words implying intelligent agency like "design" and "selection" have figured prominently—though with increasingly ironic inflection: for example, "design without a designer" and "blind selection." ID theorists rightly wonder why neo-Darwinists retain these theistic remnants, other than to avoid marking the clean break with anthropocentrism that is the logical conclusion of their theory.[16] In the philosophically confused world of contemporary secular humanism, it is often overlooked that the mastery of "human nature," which supposedly enables us to "take control" of evolution, trades on differences between *Homo sapiens* (a coinage of the special creationist, Carolus Linnaeus) and other animals that are losing their salience with advances in the neo-Darwinian sciences of life. This point is vividly demonstrated—almost daily in Britain—in the trench warfare between animal rights activists and animal-based experimenters, both of whom claim Darwinian lineage. However, the activists, not the experimenters, are the Darwinian purists—and Peter Singer is their Calvin. It is no accident that IDT enthusiasts tend to hail from the more abstract and laboratory-based areas of science. They may turn out to be the animal experimenters' strongest allies in the long term.[17]

NOTES

1. A summary of each expert's brief is presented at the Web site of the American Civil Liberties Union, which supplied the plaintiffs' legal team: www.aclu.org/religion/intelligent.design/21775res20051123.html. The positions of most of them, repeated in the trial, are also presented in William Dembski and Michael Ruse, eds., *Debating Design: From Darwin to DNA* (Cambridge, MA: Cambridge University Press, 2004).

2. Indeed, the plaintiffs' lawyers had to remind me that I had written an article arguing that IDT should be taught (in universities) alongside Neo-Darwinism, albeit in the interest of cross-fertilization (Steve Fuller, "An Intelligent Person's Guide to Intelligent Design Theory," *Rhetoric and Public Affairs* 1 (1998): 603–10).

3. Although the topic of Ruse's PhD thesis (University of Bristol, 1970) was "The Nature of Biology," all of his degrees have been in philosophy. However, his pedigree typifies the first generation of philosophers of science whose intellectual center of gravity is biology (others include David Hull, Alexander Rosenberg, Elliott Sober, Philip Kitcher, and lately Daniel Dennett). The previous two generations of philosophers of science (spanning the transition from the logical positivists to Kuhn and the historicists) had received formal training in physics and then migrated to philosophy to keep alive the broad cognitive interest in natural philosophy that scientific special-

ization had abandoned. The shift in the direction of migration ("into" vs. "out of" philosophy) reflects a repositioning of the philosopher of science from legislator to underlaborer, a point originally developed in Steve Fuller, *Thomas Kuhn: A Philosophical History for Our Time* (Chicago: University of Chicago Press, 2000). From this standpoint, Ruse is a transitional figure who proffered a physics-based definition of science and whose writing generally brings the whole weight of the history of science—not merely current scientific judgment—to bear on debates over Darwinism. Notwithstanding his pro-Darwin credentials, Ruse has never been a devotee of the "newer is better" sense of intellectual progress sometimes found in contemporary philosophy of science touched by biology (for example, Daniel Dennett, *Darwin's Dangerous Idea: Evolution and the Meanings of Life* [New York: Simon & Schuster, 1995]). Ruse's intellectual independence was recently highlighted in a nasty e-mail exchange with Dennett (Andrew Brown, "When Evolutionists Attack," *Guardian*, London, March 6, 2006).

4. As a graduate student at the University of Pittsburgh's History and Philosophy of Science Department, I recall half-serious discussions about whether Ruse's testimony constituted perjury, though this charge never made it to the published philosophical criticisms. Similar charges have been raised against me by bloggers for my participation in *Kitzmiller*. My own ambivalent support for Ruse was originally voiced in Steve Fuller and James Collier, *Philosophy, Rhetoric and the End of Knowledge: A New Beginning for Science and Technology Studies*, 2nd ed. (Mahwah, NJ: Lawrence Erlbaum Associates, 2004), p. 198.

5. One of the more widely reported aspects of my testimony was my claim that, given the hegemony that neo-Darwinism enjoys in contemporary biology, an "affirmative action" policy would be necessary for IDT to develop its research program sufficiently to mount a credible challenge. Such a policy may include giving "equal time" to the lesser theory at the high school level. Tales of scientific underdogs who eventually triumph typically either took place before the era of Big Science or do not propose as radical a reorientation of science as IDT does. In any case, it should be always an open question for high school educators, even in science, whether the next generation is exposed exclusively to the received wisdom of the current generation or the seeds of epistemic change are planted as well.

6. The uneasy alliance between biblical fundamentalists and IDT supporters is explored in Steve Fuller, "Intelligent Design Theory: A Site for Contemporary Sociology of Knowledge," *Canadian Journal of Sociology* 31 (2006): 277–89.

7. For example, in direct examination of two of the plaintiffs' expert witnesses, John Haught and Brian Alters, the following quote from the textbook was cited as evidence that "intelligent design" is synonymous with "special creation": "Intelligent design means that various forms of life began abruptly through an intelligent agency with their

distinctive features already intact: fish with fins and scales, birds with feathers, beaks, and wings, etc." (Percival Davis and Dean Kenyon, *Of Pandas and People: The Central Questions of Biological Origins*, 2nd ed. [Richardson, TX: Foundation for Thought and Ethics, 1993], pp. 99–100). However, at this level of abstraction, it could equally serve as a somewhat reified account of how, in Kuhn's own words, "the world changes" in a paradigm shift after a scientific revolution, since the paradigm shifter acquires a new worldview as a whole, not in parts. More concretely, this account also captures the student's acquisition of the conceptual framework needed to address problems in her chosen science. The "intelligent agency" in this case is the disciplinary instruction the student receives that enables a whole new domain of objects to come into view.

8. Not surprisingly, the NAS's attempt to legitimate its position by appeal to "methodological naturalism" has sat uncomfortably even with philosophers who oppose IDT (for example, Keith Parsons, "Defending the Radical Center," in *Scientific Values and Civic Virtues*, ed. Noretta Koertge [Oxford: Oxford University Press, 2005], pp. 159–71).

9. The *locus classicus* of this interpretation is Carl Becker, *The Heavenly City of the Eighteenth Century Philosophers* (New Haven, CT: Yale University Press, 1932).

10. On October 1, 2005, I conducted a computer-based search of the two main biology databases, which covered 1,273,417 papers. I was inspired by a similar observation made over a decade ago by Nicolas Rasmussen in "Surveying Evolution," *Metascience* 5 (1994): 55–60. Rasmussen contended that neo-Darwinism was largely a philosophical cottage industry with little bearing on day-to-day biological research.

11. Here one of the most capriciously contextualized quotes in the history of science is often invoked: "Nothing in biology makes sense except in light of evolution" (Theodosius Dobzhansky, "Nothing in Biology Makes Sense Except in Light of Evolution," *American Biology Teacher* 3 [March 1973]: 125–29). The paper ends by referring to divine creation as among the things that only make sense in light of evolution. It is also worth noting that "natural selection" and "evolution" appear more frequently in the abstracts as time goes on. This is the exact opposite of what would be expected, if the corresponding concepts were indeed always already assumed in biology.

12. The ACLU, founded in 1920, first came to national prominence five years later when it paid for Clarence Darrow's defense of John T. Scopes in the infamous 1925 "monkey trial" in Tennessee that began Darwinism's history in the US court system. The ACLU also supplied counsel for the plaintiffs in *Kitzmiller*. See note 1.

13. The rise of science as a unifying element in an increasingly diverse US society in the early twentieth century is discussed in Morton White, *Social Thought in America: The Revolt against Formalism* (Boston, MA: Beacon Press, 1957).

14. The one aspect of the *Kitzmiller* case about which I was truly naïve was my

sharing the defense team's hope that Judge Jones might have ruled against the Dover school board, while taking a more adventurously open-minded attitude toward IDT as such. Pace the Discovery Institute, that would have made for "judicial activism."

15. My testimony included criticism of IDT, as I used the expert witness role as an opportunity to provide a model for how the judge should weigh the merits of both sides of the case. For my trouble, I was quoted authoritatively in the closing arguments by the lawyers on both sides. The judge cited me a dozen times in his ruling, unsurprisingly whenever I said something revealing or critical of IDT. One very valid point that Michael Lynch makes in his unsurprisingly unsympathetic conclusion about my participation in *Kitzmiller* is that the case ended up turning on whether IDT is religion, not science. This forced choice of "either religion or science" was the one constraint to which my testimony had to conform, given the current interpretation of the Establishment Clause. However, my own view—which I believe accords well with those who have studied the matter seriously—is that science and religion are not "separate but equal," as the *Kitzmiller* verdict suggests, but rather are substantially overlapping. (From that standpoint, the explicit exclusion of religion from science promoted by, say, the Lemon Test amounts to institutionalized atheism.) The spirit of that position remained present in my testimony, which made it easy for Judge Jones to use it to show that I believe that ID is really religion.

16. For a development of this critique in terms of Dennett's pan-Darwinism, see Angus Menuge, *Agents under Fire: Materialism and the Rationality of Science* (Lanham, MD: Rowman and Littlefield, 2004), chap. 3 and the conclusion to Steve Fuller, *The New Sociological Imagination* (London: Sage Publications, 2006).

17. Too often the emphasis is placed on the nontheism prevalent among biologists, especially those closest to Darwin's own expertise. However, evidence for accommodation, and perhaps even endorsement, of IDT can be found more widely across the scientific community. Recall that the director of the Human Genome Project at the US National Institute of Health, Francis Collins, a born-again Christian, spoke of aiming to decode "the language of God"—a phrase echoed by President Bill Clinton in 2000, upon completion of the project. Collins described the completed genomic map as a "shop manual with a detailed blueprint for building a human cell." (Whenever asked to comment on "evolution," Collins typically restricts the theory to a demonstration of the interrelatedness of all living things.) The engineering imagery is not accidental but points to the main constituency for IDT among scientists, namely, those who think of themselves as doing on smaller scale (or perhaps bringing to completion) work the Creator has done on a grand scale. Thus, the leading scientific proponent of IDT in the United Kingdom, Andrew McIntosh, professor of thermodynamics at Leeds Uni-

versity, currently leads a major Engineering and Physical Sciences Research Council project on "biomimetics," an emerging field of biotechnology that treats organisms as prototypes for humanly useful things and processes.

REFERENCES

Becker, Carl. *The Heavenly City of the Eighteenth Century Philosophers.* New Haven, CT: Yale University Press, 1932.

Brown, Andrew. "When Evolutionists Attack." *Guardian*, London, March 6, 2006.

Davis, Percival, and Dean Kenyon. *Of Pandas and People: The Central Questions of Biological Origins*, 2nd ed. Richardson, TX: Foundation for Thought and Ethics, 1993.

Dembski, William, and Michael Ruse, eds. *Debating Design: From Darwin to DNA.* Cambridge: Cambridge University Press, 2004.

Dennett, Daniel. *Darwin's Dangerous Idea: Evolution and the Meanings of Life.* New York: Simon & Schuster, 1995.

Dobzhansky, Theodosius. "Nothing in Biology Makes Sense Except in Light of Evolution." *American Biology Teacher* 3 (March 1973): 125–29.

Fuller, Steve. "An Intelligent Person's Guide to Intelligent Design Theory." *Rhetoric and Public Affairs* 1 (1998): 603–10.

———. *Thomas Kuhn: A Philosophical History for Our Time.* Chicago: University of Chicago Press, 2000.

———. "Intelligent Design Theory: A Site for Contemporary Sociology of Knowledge." *Canadian Journal of Sociology* 31, no. 3 (2006): 277–89.

———. *The New Sociological Imagination.* London: Sage Publications, 2006.

Fuller, Steve, and James Collier. *Philosophy, Rhetoric and the End of Knowledge: A New Beginning for Science and Technology Studies*, 2nd ed. Mahwah, NJ: Lawrence Erlbaum Associates, 2004.

Menuge, Angus. *Agents under Fire: Materialism and the Rationality of Science.* Lanham, MD: Rowman and Littlefield, 2004.

National Academy of Sciences. *Science and Creationism: A View from the National Academy of Sciences*, 2nd ed. Washington, DC: NAS Press, 1999.

Parsons, Keith. "Defending the Radical Center." In *Scientific Values and Civic Virtues*, ed. Noretta Koertge, 159–71. Oxford: Oxford University Press, 2005.

Rasmussen, Nicholas. "Surveying Evolution." *Metascience* 5 (1994): 55–60.

White, Morton. *Social Thought in America: The Revolt against Formalism.* Boston, MA: Beacon Press, 1957.

CHAPTER 26

WHAT IS WRONG WITH INTELLIGENT DESIGN?

ELLIOTT SOBER

One striking difference between the intelligent design (ID) position and earlier forms of creationism is that ID is often formulated as a comparatively modest claim. For example, young-earth creationism denied that human beings share common ancestors with other species while affirming that God was the designer of organisms and that life on earth is at most ten thousand years old. ID, at least when stated in a minimalistic form, is officially neutral on these three claims (Behe 1996; 2005). The single thesis of what I will call mini-ID is that the complex adaptations that organisms display (e.g., the vertebrate eye) were crafted by an intelligent designer. Scientists have challenged young-earth creationism by pointing to compelling evidence for common ancestry and ancient life-forms. These challenges do not touch mini-ID. Does that mean that mini-ID is well supported by evidence?

This question about the evidential status of mini-ID differs from the psychological question of why it was developed. Although the rest of this [chapter] will address the first query, a few comments are in order with respect to the second. ID proponents often make assertions that go beyond mini-ID's single

---

claim. For example, they often affirm that the intelligent designer they have in mind is supernatural (Johnson 1991; Dembski 2002), and most deny common ancestry (Davis and Kenyon 1993; Dembski 1999). Why, then, do proponents of ID think that mini-ID is so important? After all, it leaves out so much. One reason is that versions of creationism that mention a supernatural being have a constitutional problem—US courts have deemed them religious, and so they are not permitted in public school science curricula. ID proponents hope that mini-ID can avoid this objection. In addition, mini-ID has the advantage of expressing an idea to which all creationists subscribe; it thus presents a united front, allowing the factions to stop squabbling and to face their common enemy.

Although mini-ID is modest in what it asserts, ID proponents have high hopes for what it will achieve. According to the Discovery Institute's "Wedge Strategy" (available at http://www.antievolution.org/features/wedge.html), which was leaked on the Internet in 2001, "[d]esign theory promises to reverse the stifling dominance of the materialist worldview, and to replace it with a science consonant with Christian and theistic convictions." The Discovery Institute is the flagship ID think tank, and the "Wedge Strategy" is its political manifesto. So much for questions about religious motivation and political context (Forrest and Gross 2004). What about the evidence?

THE "NO DESIGNER WORTH HIS SALT" OBJECTION

Many biologists take the fact that adaptations are often imperfect to provide a decisive objection to creationism and to mini-ID. Charles Darwin presents this type of argument (Burkhardt et al., 1993, 224). More recently, Stephen Jay Gould (1980) made the objection famous in his discussion of the panda's thumb. The "thumb" is a crude spur of bone that enables pandas to laboriously strip the bamboo they eat. Gould contends that if a truly intelligent designer had built the panda, the panda would possess a far more efficient device for preparing its meals. Biologists have cited other examples, but the conclusion drawn is the same—since no designer worth his salt (Raddick 2005) would produce the many imperfect adaptations we observe in nature, creationism is false.

This criticism concedes that creationism is testable. In addition, it assumes that the designer, if he existed, would have wanted pandas to have a more efficient device for stripping bamboo. Creationists have a reply to this criticism. How does Gould (or anyone else) know what God (or some unspecified designer) would have wanted to achieve in building the panda (Nelson 1996; Sober 2005)? This is a good reply by creationists, but it is one that invites an entirely different, but equally serious, criticism of ID.

POPPER'S FALSIFIABILITY CRITERION

If imperfect adaptations do not demonstrate that the mini-ID claim is false, perhaps the right criticism is that this statement cannot be tested. But, what does testability mean? Scientists often answer by using Karl Popper's concept of falsifiability (Popper 1959). According to Popper, a hypothesis is falsifiable precisely when it rules out a possible observational outcome. Popper understood "ruling out" in terms of deductive logic; a falsifiable statement is logically inconsistent with at least one observation statement. Popper further suggested that falsifiability provides a demarcation criterion, separating science from nonscience.

Popper's account entails that some versions of creationism are falsifiable, and hence scientific. Consider, for example, the hypothesis that an omnipotent supernatural being wanted everything to be purple, and had this as his top priority. Of course, no creationist has advocated purple-ID. However, it is inconsistent with what we observe, so purple-ID is falsifiable (the fact that it postulates a supernatural being notwithstanding). The same can be said of other, more modest, versions of ID that do not say whether the designer is supernatural. For example, if mini-ID says that an intelligent designer created the vertebrate eye, then it is falsifiable; after all, it entails that vertebrates have eyes. An even more minimalistic formulation of ID is also falsifiable; the statement that organisms were created by an intelligent designer entails that there are organisms, which is something we observe to be true.

PROBABILITY STATEMENTS ARE NOT FALSIFIABLE

In addition to entailing that many formulations of ID are falsifiable, Popper's criterion also has the consequence that probability statements are unfalsifiable. Consider the statement that a coin has a 50 percent probability of landing heads each time it is tossed. This statement is logically consistent with all possible sequences of heads and tails in any finite run of tosses. Popper attempted to solve this problem by expanding the concept of falsification. Rather than saying that *H* is falsified only when an observation occurs that is logically inconsistent with *H*, Popper suggested that we regard *H* as false when an observation occurs that *H* says is very improbable. But how improbable is improbable enough for us to be warranted in rejecting *H*? Popper thought that there was no objectively correct answer to this question; the choice of cutoff is a matter of convention (Popper 1959, 191).

Popper's idea has much in common with Ronald Fisher's test of significance (Fisher 1959). According to Fisher, if *H* says that an observation *O* is very improbable, and *O* occurs, then a disjunction is true—either *H* is false or something very improbable has occurred. The disjunction does follow, but it does not follow that *H* is false, nor does it follow that we should reject *H*. As many statisticians and philosophers of science have recognized (Hacking 1965; Edwards 1972; Royall 1997), perfectly plausible hypotheses often say that the observations have low probability. This is especially common when a probabilistic hypothesis addresses a large body of data. If we make a large number of observations, it may turn out that *H* confers on *each* observation a high probability, although *H* confers on the *conjunction* of observations a tiny probability. If Fisher's test of significance fails to provide a criterion for when hypotheses should be rejected, it also fails to describe when a hypothesis is falsifiable. Perhaps Popper's *f*-word should be dropped.

The fact that Popperian falsifiability fails to capture what testability is does not mean that we should abandon the latter concept. Rather, a better theory of testability is needed.

TESTING IS COMPARATIVE

To develop an account of testability, we must begin by recognizing that testing is typically a comparative enterprise. If ID is to be tested, it must be tested against one or more competing hypotheses. Creationists now single out evolutionary theory as their stalking horse. Before 1859, the competing theory was the vaguer idea of "chance"—that a mindless random process is responsible for the complex adaptations we observe. The details of these alternative hypotheses do not matter to the problem at hand, but they contribute an insight into the kinds of observational consequences that a formulation of ID needs to have if it is to be tested against its competitors. For example, if mini-ID says that an intelligent designer made the vertebrate eye, and this claim is to be tested against the claim that chance produced the vertebrate eye, we must discover how these two hypotheses disagree about what we should observe. Since both entail that vertebrates have eyes, the observation that this is true does not help. We need to find other predictions that mini-ID makes.

DUHEM'S THESIS

An additional point needs to be taken into account. As the philosopher Pierre Duhem (1954) emphasized, physical theories, on their own, do not make testable predictions. One needs to add "auxiliary propositions" to the theories one wishes to test. For example, the laws of optics do not predict when eclipses will occur. However, if propositions about the positions of the earth, moon, and sun are added to these lows, they do make predictions. Duhem's thesis holds for most theories in most sciences, and it has wide applicability when prediction is understood probabilistically, not just deductively.

Duhem's point applies to mini-ID. Taken alone, the statement that an intelligent designer made the vertebrate eye does not have observational consequences beyond the entailment that vertebrates have eyes. However, mini-ID can be supplemented with further assumptions that allow it to have additional observational entailments. For example, suppose we assume that if an intelligent designer made the vertebrate eye, that we would want it to have the set of features *F*. Mini-ID, when supplemented with this auxiliary assumption, has implications about the detailed features that the eye will have. Just like the laws of optics, mini-ID does not predict much until auxiliary assumptions are added. Does this mean that mini-ID is no worse than the laws of optics?

AUXILIARY PROPOSITIONS MUST BE INDEPENDENTLY SUPPORTED

It is crucial to the scientific enterprise that auxiliary propositions not simply be invented. By inventing assumptions, we can equip a theory with favorable auxiliary propositions that allow it to fit the data. Conversely, a theory also can be equipped with unfavorable auxiliaries that lead it to conflict with the data. An important strategy that scientists use to avoid this nihilistic outcome is to insist that there be independent evidence for the auxiliary propositions that are used. When testing the laws of optics by observing eclipses, we do not arbitrarily invent assumptions about the positions of the earth, moon, and sun. Rather, we use propositions about their positions for which we have independent evidence.

When we test the laws of optics by observing eclipses, the auxiliary propositions we use are "independently justified" in the sense that our reasons for accepting them do not depend on (i) assuming that the theory being tested is true or (ii) using the data on eclipses. The reason to avoid (i) is obvious, since a test of optical theory should not be question-begging. But why avoid (ii)? The reason is that violating this requirement would allow us to show that any theory, no matter how irrelevant it is to the occurrence of eclipses, makes accurate predictions about them. For if *O* describes an observation about the occurrence of an eclipse, and *O* is used to justify the auxiliary propositions we use to test theory *N*, then we can simply construct the auxiliary proposition "not-*N* or *O*"; this disjunction must be true if *O* is, and this auxiliary proposition, when conjoined to *N*, allows *N* to entail *O*.

The important scientific strategy of rendering theories testable by finding independently justified auxiliary propositions does not work for mini-ID. We have no independent evidence concerning which auxiliary propositions about the putative designer's goals and abilities are true (Kitcher 1982). Surprisingly, this is a point that several ID proponents concede. For example, the influential ID textbook *Of Pandas and People: The Central Question of Biological Origins* states that "the message encoded in DNA must have originated from an intelligent cause. What kind of intelligent designer was it? On its own, science cannot answer this question; it must leave it to religion and philosophy" (Davis and Kenyon 1993, 7). In the same vein, Phillip Johnson (1991) says that the designer's motives are "mysterious" (67) and "inscrutable" (71).

WHAT ID PROPONENTS SAY ABOUT TESTABILITY

Proponents of ID have had a variety of reactions to the charge that their position is not testable. Sometimes they embrace the criterion of falsifiability and claim that ID fills the bill:

> The concept of intelligent design entails a strong prediction that is readily falsifiable. In particular, the concept of intelligent design predicts that complex information, such as that encoded in a functioning genome, never arises from purely chemical or physical antecedents.... All that is necessary to falsify the hypothesis of intelligent design is to show confirmed instances of purely physical or chemical antecedents producing such information. (Hartwig and Meyer 1993, p. 160)

We have already seen why Popper's notion of falsifiability fails to capture what testability is. The point of relevance here is that these ID proponents have misapplied Popper's criterion. ID asserts that somewhere on the causal chains leading up to "complex information" there is an intelligent designer at work. If a newspaper contains complex information, ID proponents are not obliged to say that the press used to print the newspaper is intelligent; presumably, the press is just as mindless as the paper it produces. Rather, their claim is that if you look back further along the causal chain, you'll find an intelligent being. And they are right—there is a person setting the type.

If scientists observe that "purely physical antecedents" at time t_9 give rise to complex information at t_{10}, this does not refute the ID claim any more than a mindless printing press does. ID proponents will simply maintain that an intelligent designer was present at an earlier stage. If scientists press their inquiry into the more remote past and discover that mindless physical conditions at t_8 produced the conditions at t_9, ID proponents will have the same reply: an intelligent designer was involved at a still earlier time. If scientists somehow manage to push their understanding of the complex information that exists at t_{10} all the way back to the start of the universe without ever having to invoke an intelligent designer, would that refute the ID position? Undoubtedly, ID proponents will then postulate a supernatural intelligence that exists outside of space and time. Defenders of ID always have a way out. This is not the mark of a falsifiable theory.

In addition, the proponents of ID who make this argument have lost sight of the role of observation in Popper's concept of falsifiability. For a proposition to be falsifiable, it is not enough that it be inconsistent with a possible state of affairs; it must also be inconsistent with a possible observation. Granted, the ID position is inconsistent with the existence of complex information that never had an intelligent designer in its causal history. It is equally true that "all lightning bolts issue from the hand of Zeus" is inconsistent with there existing even one Zeus-like lightning bolt (Pennock 1999). These points fail to address how observations could refute either claim.

Defenders of ID often claim to test their position by another route, by criticizing the theory of evolution. Behe (1996) contends that evolutionary processes cannot produce "irreducibly complex" adaptations; since we observe such traits, evolutionary theory is refuted, leaving ID as the only position standing. Behe (1996) says that a system is irreducibly complex when it is "composed of several well-matched, interacting parts that contribute to the basic function, wherein the removal of any one of the parts causes the system to effectively cease functioning" (39). Before considering whether evolutionary theory really does rule out irreducible complexity, I want to note that this argument does nothing to test ID. For ID to be testable, *it* must make predictions. The fact that a different theory makes a prediction says nothing about whether ID is testable. Behe has merely changed the subject.

One flaw in Behe's argument is his assumption that evolutionary processes must always involve a lockstep increase in fitness. This ignores the fact that contemporary evolutionary theory describes evolution as a probabilistic process. Drift can lead to evolutionary changes that involve no increase in fitness and even to changes that lead fitness to decline. Evolution does not require that each later stage be fitter than its predecessors. At least since the 1930s, biologists have understood that evolution can cross valleys in a fitness landscape.

The most that can be claimed about irreducibly complex adaptations (though this would have to be scrutinized carefully) is that evolutionary theory says that they have low probability. However, that does not justify rejecting evolutionary theory or accepting ID. As noted earlier, many probabilistic theories have the property of saying that a body of observations has low probability. If we reject theories because they say that observations have low probability, all probabilistic theories will be banished from science once they are repeatedly tested.

There is a second problem with Behe's position on irreducible complexity. The fact that a system can be segmented into *n* parts in such a way that it counts as irreducibly complex does not guarantee that the evolution of the system involved a stepwise accumulation of parts, moving from 0 to 1 to . . . to $n - 1$ to *n* parts coming on line. What we call "the parts" may or may not correspond to the historical sequence of accumulating details. Consider the horse and its four legs. A horse with zero, one, or two legs cannot walk or run; suppose the same is true for a horse with three. In contrast, a horse with four legs can walk and run, and it thereby gains a fitness advantage. So far so good—the tetrapod arrangement satisfies the definition of irreducible complexity. The mistake comes from thinking that horses (or their ancestors) had to evolve their tetrapod morphology one leg at a time. In fact, the development of legs is not controlled by four sets of genes, one for each leg; rather, there is a single set that controls the development of appendages. A division of a system into parts that entails that the system is irreducibly complex may or may not correspond to the historical sequence of trait configurations through which the lineage passed. This point is obvious with respect to the horse's four legs, but needs to be borne in mind when other less familiar organic features are considered.

CONCLUSION

It is one thing for a version of ID to have observational consequences, something else for it to have observational consequences that differ from those of a theory with which it competes. The mini-ID claim that an intelligent designer made the vertebrate eye entails that vertebrates have eyes, but that does not permit it to be tested against alternative explanations of why vertebrates have eyes. When scientific theories compete with each other, the usual pattern is that independently attested auxiliary propositions allow the theories to make predictions that disagree with each other. No such auxiliary propositions allow mini-ID to do this.

It is easy enough to construct a version of ID that accommodates a set of observations already known, but it also is easy to construct a version of ID that conflicts with what we have already observed. Neither undertaking results in substantive science, nor is there any point in constructing a version of ID that is so minimalistic that it fails to say much of anything about what we observe. In all its forms, ID fails to constitute a serious alternative to evolutionary theory.

ACKNOWLEDGMENTS

My thanks to Richard Amasino, Alan Attie, Jeremy Butterfield, Michael Cox, Mehmet Elgin, Malcolm Forster, Daniel Hausman, Bret Larget, Gregory Mougin, Ronald Numbers, Robert Pennock, David S. Wilson, and the referees of this journal for useful suggestions.

REFERENCES

Behe, M. J. *Darwin's Black Box: The Biochemical Challenge to Evolution.* New York: Free Press, 1996.

———. "Design for Living." *New York Times*, February 7, 2005, p. A27.

Burkhardt, F. H., J. Browne, D. M. Porter, and M. Richmond, eds. *The Correspondence of Charles Darwin*, vol. 8. Cambridge, MA: Cambridge University Press, 1860.

Davis, P., and D. H. Kenyon *Of Pandas and People: The Central Question of Biological Origins*, 2nd ed. Dallas, TX: Haughton Publishing, 1993.

Dembski, W. A. "Signs of Intelligence: A Primer on the Discernment of Intelligent Design." *Touchstone* 12 (1999): 76–84.

———. *No Free Lunch: Why Specified Complexity Cannot Be Purchased without Intelligence.* Lanham, MD: Rowman and Littlefield, 2002.

Duhem, P. M. M. *The Aim and Structure of Physical Theory.* Princeton, NJ: Princeton University Press, 1954.

Edwards, A. W. F. *Likelihood: An Account of the Statistical Concept of Likelihood and Its Application to Scientific Inference.* Cambridge, MA: Cambridge University Press, 1972.

Fisher, R. A. *Statistical Methods and Scientific Inference*, 2nd ed. Edinburgh: Oliver and Boyd, 1959.

Forrest, B., and P. R. Gross. *Creationism's Trojan Horse: The Wedge of Intelligent Design.* Oxford and New York: Oxford University Press, 2001.

Gould, S. J. *The Panda's Thumb: More Reflections on Natural History.* New York: Norton, 1980.

Hacking, I. *Logic of Statistical Inference.* Cambridge, MA: Cambridge University Press, 1965.

Hartwig, M. D., and S. C. Meyer. "A Note to Teachers." In P. Davis and D. H. Kenyon, *Of Pandas and People: The Central Question of Biological Origins,* 2nd ed., 153–63. Dallas, TX: Haughton Publishing, 1993.

Johnson, P. E. *Darwin on Trial.* Washington, DC: Regnery Gateway, 1991.

Kitcher, P. *Abusing Science: The Case against Creationism.* Cambridge, MA: MIT Press, 1982.

Nelson, P. A. "The Role of Theology in Current Evolutionary Reasoning." *Biology and Philosophy* 11 (1996): 493–517.

Pennock, R. T. *Tower of Babel: The Evidence against the New Creationism.* Cambridge, MA: MIT Press, 1999.

Popper, K. R. *The Logic of Scientific Discovery.* New York: Basic Books, 1959.

Raddick, G. "Deviance, Darwinian-style." *Metascience* 14 (2005): 453–57.

Royall, R. M. *Statistical Evidence: A Likelihood Paradigm.* London and New York: Chapman & Hall, 1997.

Sober, E. "The Design Argument." In *The Blackwell Guide to the Philosophy of Religion,* edited by W. E. Mann, 117–47. Malden, MA: Blackwell Publishing, 2005.

CHAPTER 27

UNITED STATES DISTRICT COURT MEMORANDUM OPINION: *TAMMY KITZMILLER, ET AL. V. DOVER AREA SCHOOL DISTRICT, ET AL.*

JUDGE JOHN E. JONES II

INTRODUCTION

On October 18, 2004, the Defendant Dover Area School Board of Directors passed by a 6–3 vote the following resolution:

> Students will be made aware of gaps/problems in Darwin's theory and of other theories of evolution including, but not limited to, intelligent design. Note: *Origins of Life* is not taught.

On November 19, 2004, the Defendant Dover Area School District announced by press release that, commencing in January 2005, teachers would be required to read the following statement to students in the ninth-grade biology class at Dover High School:

> The Pennsylvania Academic Standards require students to learn about Darwin's Theory of Evolution and eventually to take a standardized test of which evolution is a part.

From *Tammy Kitzmiller, et al. v. Dover Area School District, et al.*, 400 F. Supp. 2d 707 (M.D. Pa. 2005).

Because Darwin's Theory is a theory, it continues to be tested as new evidence is discovered. The Theory is not a fact. Gaps in the Theory exist for which there is no evidence. A theory is defined as a well-tested explanation that unifies a broad range of observations.

Intelligent design is an explanation of the origin of life that differs from Darwin's view. The reference book, *Of Pandas and People*, is available for students who might be interested in gaining an understanding of what intelligent design actually involves.

With respect to any theory, students are encouraged to keep an open mind. The school leaves the discussion of the *Origins of Life* to individual students and their families. As a Standards-driven district, class instruction focuses upon preparing students to achieve proficiency on Standards-based assessments.

A. Background and Procedural History

On December 14, 2004, the plaintiffs filed the instant suit challenging the constitutional validity of the October 18, 2004, resolution and November 19, 2004, press release (collectively, "the ID Policy.")...

For the reasons that follow, we hold that the ID Policy is unconstitutional pursuant to the Establishment Clause of the First Amendment of the United States Constitution and Art. I, § 3 of the Pennsylvania Constitution.

B: The Parties to the Action

...

The trial commenced September 26, 2005, and continued through November 4, 2005. This memorandum opinion constitutes the court's findings of fact and conclusions of law which are based upon the court's review of the evidence presented at trial, the testimony of the witnesses at trial, the parties' proposed findings of fact and conclusions of law with supporting briefs, other documents and evidence in the record, and applicable law.[1] Further orders and judgments will be in conformity with this opinion.

...

E. Application of the Endorsement Test to the ID Policy

. . .

1. An Objective Observer Would Know That ID and Teaching about "Gaps" and "Problems" in Evolutionary Theory Are Creationist, Religious Strategies That Evolved from Earlier Forms of Creationism

The history of the intelligent design movement (hereinafter "IDM") and the development of the strategy to weaken education of evolution by focusing students on alleged gaps in the theory of evolution is the historical and cultural background against which the Dover School Board acted in adopting the challenged ID Policy. As a reasonable observer, whether adult or child, would be aware of this social context in which the ID Policy arose, and such context will help to reveal the meaning of the defendants' actions, it is necessary to trace the history of the IDM.

It is essential to our analysis that we now provide a more expansive account of the extensive and complicated federal jurisprudential legal landscape concerning opposition to teaching evolution, and its historical origins. As noted, such opposition grew out of a religious tradition, Christian fundamentalism, that began as part of evangelical Protestantism's response to, among other things, Charles Darwin's exposition of the theory of evolution as a scientific explanation for the diversity of species. *McLean*, 529 F. Supp. at 1258; see also, e.g., *Edwards*, 482 U.S. at 590–92. Subsequently, as the United States Supreme Court explained in *Epperson*, in an "upsurge of fundamentalist religious fervor of the twenties," 393 U.S. at 98 (citations omitted), state legislatures were pushed by religiously motivated groups to adopt laws prohibiting public schools from teaching evolution. *McLean*, 529 F. Supp. at 1259; see *Scopes*, 289 S.W. 363 (1927). Between the 1920s and early 1960s, antievolutionary sentiment based upon a religious social movement resulted in formal legal sanctions to remove evolution from the classroom. *McLean*, 529 F. Supp. at 1259 (discussing a subtle but pervasive influence that resulted from antievolutionary sentiment concerning teaching biology in public schools).

As we previously noted, the legal landscape radically changed in 1968 when the Supreme Court struck down Arkansas's statutory prohibition against teaching evolution in *Epperson.* 393 U.S. 97. Although the Arkansas statute at issue did not include direct references to the book of Genesis or to the fundamentalist view that religion should be protected from science, the Supreme Court concluded that "the motivation of the [Arkansas] law was the same . . . : to suppress the teaching of a theory which, it was thought, 'denied' the divine creation of man." *Edwards,* 482 U.S. at 590 (quoting *Epperson,* 393 U.S. at 109). (Arkansas sought to prevent its teachers from discussing the theory of evolution as it is contrary to the belief of some regarding the book of Genesis.)

Post-*Epperson,* evolution's religious opponents implemented "balanced treatment" statutes requiring public school teachers who taught evolution to devote equal time to teaching the biblical view of creation; however, such statutes did not pass constitutional muster under the Establishment Clause. See, e.g., *Daniel,* 515 F.2d at 487, 489, 491. In *Daniel,* the Sixth Circuit Court of Appeals held that by assigning a "preferential position for the Biblical version of creation" over "any account of the development of man based on scientific research and reasoning," the challenged statute officially promoted religion, in violation of the Establishment Clause. Ibid. at 489.

Next, and as stated, religious opponents of evolution began cloaking religious beliefs in scientific-sounding language and then mandating that schools teach the resulting "creation science" or "scientific creationism" as an alternative to evolution. However, this tactic was likewise unsuccessful under the First Amendment. "Fundamentalist organizations were formed to promote the idea that the Book of Genesis was supported by scientific data. The terms 'creation science' and 'scientific creationism' have been adopted by these Fundamentalists as descriptive of their study of creation and the origins of man." *McLean,* 529 F. Supp. at 1259. In 1982, the district court in *McLean* reviewed Arkansas's balanced-treatment law and evaluated creation science in light of *Scopes, Epperson,* and the long history of fundamentalism's attack on the scientific theory of evolution, as well as the statute's legislative history and historical context. The court found that creation science organizations were fundamentalist religious entities that "consider[ed] the introduction of creation science into the public schools part of their ministry." Ibid. at 1260. The court in *McLean* stated that creation science rested on a "contrived dualism" that recognized only two possible explanations for life, the scientific theory of evolution and biblical creationism, treated the two as mutually exclusive such that "one must either accept the literal interpretation of Genesis or else believe in the godless system of evolution," and accordingly viewed any critiques of evolution as evidence that necessarily supported biblical creationism. Ibid. at 1266. The court concluded that creation science "is simply not science" because it depends upon "supernatural intervention," which cannot be explained by natural causes, or be proven through empirical investigation, and is therefore neither testable nor falsifiable. Ibid. at 1267. Accordingly, the United States District Court for the Eastern District of Arkansas deemed creation science as merely biblical creationism in a new guise and held that Arkansas's balanced-treatment statute could have no valid secular purpose or effect, served only to advance religion, and violated the First Amendment. Ibid. at 1264, 1272–74.

Five years after *McLean* was decided, in 1987, the Supreme Court struck down Louisiana's balanced-treatment law in *Edwards* for similar reasons. After a thorough analysis of the history of fundamentalist attacks against evolution, as well as the applicable legislative history including statements made by the statute's sponsor, and taking the character of organizations advocating for creation science into consideration, the Supreme Court held that the state violated the Establishment Clause by "restructur[ing] the science curriculum to conform with a particular religious viewpoint." *Edwards*, 482 U.S. at 593.

Among other reasons, the Supreme Court in *Edwards* concluded that the challenged statute did not serve the legislature's professed purposes of encouraging academic freedom and making the science curriculum more comprehensive by "teaching all of the evidence" regarding origins of life because: the state law already allowed schools to teach any scientific theory, which responded to the alleged purpose of academic freedom; and if the legislature really had intended to make science education more comprehensive, "it would have encouraged the teaching of all scientific theories about the origins of humankind" rather than permitting schools to forego teaching evolution, but mandating that schools that teach evolution must also teach creation science, an inherently religious view. Ibid. at 586, 588–89. The Supreme Court further held that the belief that a supernatural creator was responsible for the creation of humankind is a religious viewpoint and that the act at issue "advances a religious doctrine by requiring either the banishment of the theory of evolution from public school classrooms or the presentation of a religious viewpoint that rejects evolution in its entirety." Ibid. at 591, 596. Therefore, as noted, the import of *Edwards* is that the Supreme Court made national the prohibition against teaching creation science in the public school system.

The concept of intelligent design (hereinafter "ID"), in its current form, came into existence after the *Edwards* case was decided in 1987. For the reasons that follow, we conclude that the religious nature of ID would be readily apparent to an objective observer, adult or child.

We initially note that John Haught, a theologian who testified as an expert witness for the plaintiffs and who has written extensively on the subject of evolution and religion, succinctly explained to the Court that the argument for ID is not a new scientific argument, but is rather an old religious argument for the existence of God. He traced this argument back to at least Thomas Aquinas in the thirteenth century, who framed the argument as a syllogism: Wherever complex design exists, there must have been a designer; nature is complex; therefore nature must have had an intelligent designer. (Trial tr. vol. 9, Haught Test., 7–8, September 30, 2005). Dr. Haught testified that Aquinas was explicit that this intelligent designer "everyone understands to be God." Ibid. The syllogism described by Dr. Haught is essentially the same argument for ID as presented by defense expert witnesses Professors Behe and Minnich, who employ the phrase "purposeful arrangement of parts."

Dr. Haught testified that this argument for the existence of God was advanced early in the nineteenth century by Reverend Paley and defense expert witnesses Behe and Minnich admitted that their argument for ID based on the "purposeful arrangement of parts" is the same one that Paley made for design. (9:7–8 [Haught]; Trial tr. vol. 23, Behe Test., 55–57, October 19, 2005; Trial tr. vol. 38, Minnich Test., 44, November 4, 2005). The only apparent difference between the argument made by Paley and the argument for ID, as expressed by defense expert witnesses Behe and Minnich, is that ID's "official position" does not acknowledge that the designer is God. However, as Dr. Haught testified, anyone familiar with Western religious thought would immediately make the association that the tactically unnamed designer is God, as the description of the designer in *Of Pandas and People* (hereinafter "*Pandas*") is a "master intellect," strongly suggesting a supernatural deity as opposed to any intelligent actor known to exist in the natural world. (P-11 at

85). Moreover, it is notable that both Professors Behe and Minnich admitted their personal view is that the designer is God and Professor Minnich testified that he understands many leading advocates of ID to believe the designer to be God. (21:90 [Behe]; 38:36–38 [Minnich]).

Although proponents of the IDM occasionally suggest that the designer could be a space alien or a time-traveling cell biologist, no serious alternative to God as the designer has been proposed by members of the IDM, including the defendants' expert witnesses. (20:102-103 [Behe]). In fact, an explicit concession that the intelligent designer works outside the laws of nature and science and a direct reference to religion is *Pandas*'s rhetorical statement, "what kind of intelligent agent was it [the designer]" and answer: "On its own science cannot answer this question. It must leave it to religion and philosophy." (P-11 at 7; 9:13–14 [Haught]).

A significant aspect of the IDM is that despite the defendants' protestations to the contrary, it describes ID as a religious argument. In that vein, the writings of leading ID proponents reveal that the designer postulated by their argument is the God of Christianity. Dr. Barbara Forrest, one of the plaintiffs' expert witnesses, is the author of the book *Creationism's Trojan Horse.* She has thoroughly and exhaustively chronicled the history of ID in her book and other writings for her testimony in this case. Her testimony, and the exhibits which were admitted with it, provide a wealth of statements by ID leaders that reveal ID's religious, philosophical, and cultural content. The following is a representative grouping of such statements made by prominent ID proponents.[2]

Phillip Johnson, considered to be the father of the IDM, developer of ID's "Wedge Strategy," which will be discussed below, and author of the 1991 book titled *Darwin on Trial*, has written that "theistic realism" or "mere creation" are defining concepts of the IDM. This means "that God is objectively real as Creator and recorded in the biological evidence..." (Trial tr. vol. 10, Forrest Test., 8081, October 5, 2005; P-328). In addition, Phillip Johnson states that the "Darwinian theory of evolution contradicts not just the Book of Genesis, but every word in the Bible from beginning to end. It contradicts the idea that we are here because a creator brought about our existence for a purpose." (11:16–17 [Forrest]; P-524 at 1). ID proponents Johnson, William Dembski, and Charles Thaxton, one of the editors of *Pandas*, situate ID in the book of John in the New Testament of the Bible, which begins, "In the Beginning was the Word, and the Word was God." (11:18–20, 54–55 [Forrest]; P-524; P-355; P-357). Dembski has written that ID is a "ground clearing operation" to allow Christianity to receive serious consideration, and "Christ is never an addendum to a scientific theory but always a completion." (11:50–53 [Forrest]; P-386; P-390). Moreover, in turning to the defendants' lead expert, Professor Behe, his testimony at trial indicated that ID is only a scientific, as opposed to a religious, project for him; however, considerable evidence was introduced to refute this claim. Consider, to illustrate, that Professor Behe remarkably and unmistakably claims that the *plausibility of the argument for ID depends upon the extent to which one believes in the existence of God.* (P-718 at 705) (emphasis added). As no evidence in the record indicates that any other scientific proposition's validity rests on belief in God, nor is the Court aware of any such scientific propositions, Professor Behe's assertion constitutes substantial evidence that in his view, as is commensurate with other prominent ID leaders, ID is a religious and not a scientific proposition.

Dramatic evidence of ID's religious nature and aspirations is found in what is referred to as the "Wedge Document." The Wedge Document, developed by the Discovery Institute's Center for Renewal of Science and Culture (hereinafter "CRSC"), represents, from an institutional standpoint, the IDM's goals and objectives, much as writings from the Institute for Creation Research did for the earlier creation science movement, as discussed in *McLean.* (11:26–28 [Forrest]); *McLean,* 529 F. Supp. at 1255. The Wedge Document states in its "Five Year Strategic Plan Summary" that the IDM's goal is to replace science as currently practiced with "theistic and Christian science." (P-140 at 6). As posited in the Wedge Document, the IDM's "Governing Goals" are to "defeat scientific materialism and its destructive moral, cultural, and political legacies" and "to replace materialistic explanations with the theistic understanding that nature and human beings are created by God." Ibid. at 4. The CSRC expressly announces, in the Wedge Document, a program of Christian apologetics to promote ID. A careful review of the Wedge Document's goals and language throughout the document reveals cultural and religious goals, as opposed to scientific ones. (11:26–48 [Forrest]; P-140). ID aspires to change the ground rules of science to make room for religion, specifically, beliefs consonant with a particular version of Christianity.

In addition to the IDM itself describing ID as a religious argument, ID's religious nature is evident because it involves a supernatural designer. The courts in *Edwards* and *McLean* expressly found that this characteristic removed creationism from the realm of science and made it a religious proposition. *Edwards*, 482 U.S. at 591–92; *McLean*, 529 F. Supp. at 1265–66. Prominent ID proponents have made abundantly clear that the designer is supernatural.

The defendants' expert witness ID proponents confirmed that the existence of a supernatural designer is a hallmark of ID. First, Professor Behe has written that by ID he means "not designed by the laws of nature," and that it is "implausible that the designer is a natural entity." (P-647 at 193; P-718 at 696, 700). Second, Professor Minnich testified that for ID to be considered science, the ground rules of science have to be broadened so that supernatural forces can be considered. (38:97 [Minnich]). Third, Professor Steven William Fuller testified that it is ID's project to change the ground rules of science to include the supernatural. (Trial tr. vol. 28, Fuller Test., 20–24, October 24, 2005). Turning from defense expert witnesses to leading ID proponents, Johnson has concluded that science must be redefined to include the supernatural if religious challenges to evolution are to get a hearing. (11:8–15 [Forrest]; P-429). Additionally, Dembski agrees that science is ruled by methodological naturalism and argues that this rule must be overturned if ID is to prosper. (Trial tr. vol. 5, Pennock Test., 32–34, September 28, 2005).

Further support for the proposition that ID requires supernatural creation is found in the book *Pandas*, to which students in Dover's ninth-grade biology class are directed. *Pandas* indicates that there are two kinds of causes, natural and intelligent, which demonstrate that intelligent causes are beyond nature. (P-11 at 6). Professor Haught, who as noted was the only theologian to testify in this case, explained that in Western intellectual tradition, nonnatural causes occupy a space reserved for ultimate religious explanations. (9:13–14 [Haught]). Robert Pennock, the plaintiffs' expert in the philosophy of science, concurred with Professor Haught and concluded that because its basic proposition is that the features of the natural world are produced by a transcendent, immaterial, nonnatural being, ID is a religious proposition regardless of whether that religious proposition is given a recognized religious label. (5:55–56 [Pennock]). It is notable that not one defense expert was able to explain how the supernatural action suggested by ID could be anything other than an inherently religious proposition. Accordingly, we find that ID's religious nature would be further evident to our objective observer because it directly involves a supernatural designer.

A "hypothetical reasonable observer," adult or child, who is "aware of the history and context of the community and forum" is also presumed to know that ID is a form of creationism. *Child Evangelism*, 386 F.3d at 531 (citations omitted); *Allegheny*, 492 U.S. at 624–25. The evidence at trial demonstrates that ID is nothing less than the progeny of creationism. What is likely the strongest evidence supporting the finding of ID's creationist nature is the history and historical pedigree of the book to which students in Dover's ninth-grade biology class are referred, *Pandas*. *Pandas* is published by an organization called FTE, as noted, whose articles of incorporation and filings with the Internal Revenue Service describe it as a religious, Christian organization. (P-461; P-28; P-566; P-633; Buell Dep. 1:13, July 8, 2005). *Pandas* was written by Dean Kenyon and Percival Davis, both acknowledged creationists, and Nancy Pearcey, a young-earth creationist, contributed to the work. (10:102–108 [Forrest]).

As the plaintiffs meticulously and effectively presented to the Court, *Pandas* went through many drafts, several of which were completed prior to and some after the Supreme Court's decision in *Edwards*, which held that the Constitution forbids teaching creationism as science. By comparing the pre- and post-*Edwards* drafts of *Pandas*, three astonishing points emerge: (1) the definition for creation science in early drafts is identical to the definition of ID; (2) cognates of the word creation (creationism and creationist), which appeared approximately 150 times, were deliberately and systematically replaced with the phrase ID; and (3) the changes occurred shortly after the Supreme Court held that creation science is religious and cannot be taught in public school science classes in *Edwards*. This word substitution is telling, significant, and reveals that a purposeful change of words was effected without any corresponding change in content, which directly refutes FTE's argument that by merely disregarding the words "creation" and "creationism," FTE expressly rejected creationism in *Pandas*. In early pre-*Edwards* drafts of *Pandas*, the term "creation" was defined as "various forms of life that began abruptly through an intelligent agency with their distinctive features intact—fish with fins and scales, birds with feathers, beaks, and wings, etc.," the very same way in which ID is defined in the subsequent published versions. (P560 at 210; P-1 at 2–13; P-562 at 2–14, P-652 at 2–15; P-6 at 99–100; P-11 at 99–100; P-856.2). This definition was described by many witnesses for both parties, notably including defense experts Minnich and Fuller, as "special creation" of kinds of animals, an inherently religious and creationist concept. (28:85–86 [Fuller]; Minnich dep. at 34, May 26, 2005; Trial tr. vol. 1, Miller Test., 141–42, September 26, 2005; 9:10 (Haught); Trial tr. vol. 33, Bonsell Test., 54–56, October 31, 2005). Professor Behe's assertion that this passage was merely a description of appearances in the fossil record is illogical and defies the weight of the evidence that the passage is a conclusion about how life began based upon an interpretation of the fossil record, which is reinforced by the content of drafts of *Pandas*.

The weight of the evidence clearly demonstrates, as noted, that the systemic change from "creation" to "intelligent design" occurred sometime in 1987, after the Supreme Court's important *Edwards* decision. This compelling evidence strongly supports the plaintiffs' assertion that ID is creationism relabeled. Importantly, the objective observer, whether adult or child, would conclude from the fact that *Pandas* posits a master intellect that the intelligent designer is God.

Further evidence in support of the conclusion that a reasonable observer, adult or child, who is "aware of the history and context of the community and forum" is presumed to know that ID is a form of creationism concerns the fact that ID uses the same, or exceedingly similar arguments as were posited in support of creationism. One significant difference is that the words "God," "creationism," and "Genesis" have been systematically purged from ID explanations, and replaced by an unnamed "designer." Dr. Forrest testified and sponsored exhibits showing six arguments common to creationists. (10:140–48 [Forrest]; P-856.5–856.10). Demonstrative charts introduced through Dr. Forrest show parallel arguments relating to the rejection of naturalism, evolution's threat to culture and society, "abrupt appearance" implying divine creation, the exploitation of the same alleged gaps in the fossil record, the alleged inability of science to explain complex biological information like DNA, as well as the theme that proponents of each version of creationism merely aim to teach a scientific alternative to evolution to show its "strengths and weaknesses," and to alert students to a supposed "controversy" in the scientific community. (10:140–48 [Forrest]). In addition, creationists made the same argument that the complexity of the bacterial flagellum supported creationism as Professors Behe and Minnich now make for ID. (P-853; P-845;

37:155–56 [Minnich]). The IDM openly welcomes adherents to creationism into its "Big Tent," urging them to postpone biblical disputes like the age of the earth. (11:3–15 [Forrest]; P-429). Moreover and as previously stated, there is hardly better evidence of ID's relationship with creationism than an explicit statement by defense expert Fuller that ID is a form of creationism. (Fuller dep. at 67, June 21, 2005) (indicated that ID is a modern view of creationism).

Although contrary to Fuller, defense experts Professors Behe and Minnich testified that ID is not creationism, their testimony was primarily by way of bare assertion and it failed to directly rebut the creationist history of *Pandas* or other evidence presented by the plaintiffs showing the commonality between creationism and ID. The sole argument the defendants made to distinguish creationism from ID was their assertion that the term "creationism" applies only to arguments based on the book of Genesis, a young earth, and a catastrophic Noaich flood; however, substantial evidence established that this is only one form of creationism, including the chart that was distributed to the Board Curriculum Committee, as will be described below. (P-149 at 2; 10:129–32 [Forrest]; P-555 at 22–24).

Having thus provided the social and historical context in which the ID Policy arose of which a reasonable observer, either adult or child, would be aware, we will now focus on what the objective student alone would know. We will accordingly determine whether an objective student would view the disclaimer read to the ninth-grade biology class as an official endorsement of religion.

. . .

[*Editor's Note: Sections 2 and 3, which deal with how the ID policy and its implementation would be seen as an official endorsement of religion, are omitted except for these last lines which introduce the next section on whether ID is science.*]

Finally, we will offer our conclusion on whether ID is science not just because it is essential to our holding that an Establishment Clause violation has occurred in this case, but also in the hope that it may prevent the obvious waste of judicial and other resources which would be occasioned by a subsequent trial involving the precise question which is before us.

4. Whether ID Is Science

After a searching review of the record and applicable caselaw, we find that while ID arguments may be true, a proposition on which the Court takes no position, ID is not science. We find that ID fails on three different levels, any one of which is sufficient to preclude a determination that ID is science. They are: (1) ID violates the centuries-old ground rules of science by invoking and permitting supernatural causation; (2) the argument of irreducible complexity, central to ID, employs the same flawed and illogical contrived dualism that doomed creation science in the 1980s; and (3) ID's negative attacks on evolution have been refuted by the scientific community. As we will discuss in more detail below, it is additionally important to note that ID has failed to gain acceptance in the scientific community, it has not generated peer-reviewed publications, nor has it been the subject of testing and research.

Expert testimony reveals that since the scientific revolution of the sixteenth and seventeenth centuries, science has been limited to the search for natural causes to explain natural phenomena. (9:19–22 [Haught]; 5:25–29 [Pennock]; 1:62 [Miller]). This revolution entailed the rejection of the appeal to authority, and by extension, revelation, in favor of empirical evidence. (5:28 [Pennock]). Since that time period, science has been a discipline in which testability, rather than any ecclesiastical authority or philosophical coherence, has been the measure of a scientific idea's worth. (9:21–22 [Haught]; 1:63 [Miller]). In deliberately omitting theological or "ultimate" explanations for the existence or characteristics of the natural world, science does not consider issues of "meaning" and "purpose" in the world. (9:21 [Haught]; 1:64, 87 [Miller]). While supernatural explanations may be important and have merit, they are not part of science. (3:103 [Miller]; 9:19–20 [Haught]). This self-imposed convention of science, which limits inquiry to testable, natural explanations about the natural world, is referred to by philosophers as "methodological naturalism" and is sometimes known as the scientific method. (5:23, 29–30 [Pennock]). Methodological naturalism is a "ground rule" of science today which requires scientists to seek explanations in the world around us based upon what we can observe, test, replicate, and verify. (1:59–64, 2:41–43 [Miller]; 5:8, 23–30 [Pennock]).

As the National Academy of Sciences (hereinafter "NAS") was recognized by experts for both parties as the "most prestigious" scientific association in this country, we will accordingly cite to its opinion where appropriate. (1:94, 160–61 [Miller]; 14:72 [Alters]; 37:31 [Minnich]). NAS is in agreement that science is limited to empirical, observable, and ultimately testable data: "Science is a particular way of knowing about the world. In science, explanations are restricted to those that can be inferred from the confirmable data—the results obtained through observations and experiments that can be substantiated by other scientists. Anything that can be observed or measured is amenable to scientific investigation. Explanations that cannot be based upon empirical evidence are not part of science." (P-649 at 27).

This rigorous attachment to "natural" explanations is an essential attribute to science by definition and by convention. (1:63 [Miller]; 5:29–31 [Pennock]). We are in agreement with plaintiffs' lead expert, Dr. Miller, that from a practical perspective, attributing unsolved problems about nature to causes and forces that lie outside the natural world is a "science stopper." (3:14–15 [Miller]). As Dr. Miller explained, once you attribute a cause to an untestable supernatural force, a proposition that cannot be disproven, there is no reason to continue seeking natural explanations as we have our answer. Ibid.

ID is predicated on supernatural causation, as we previously explained and as various expert testimony revealed. (17:96 [Padian]; 2:35–36 [Miller]; 14:62 [Alters]). ID takes a natural phenomenon and, instead of accepting or seeking a natural explanation, argues that the explanation is supernatural. (5:107 [Pennock]). Further support for the conclusion that ID is predicated on supernatural causation is found in the ID reference book to which ninth-grade biology students are directed, *Pandas. Pandas* states, in pertinent part, as follows:

> Darwinists object to the view of intelligent design *because it does not give a natural cause explanation* of how the various forms of life started in the first place. Intelligent design means that various forms of life began abruptly, through an intelligent agency, with their distinctive features already intact—fish with fins and scales, birds with feathers, beaks, and wings, etc.

P-11 at 99–100 (emphasis added). Stated another way, ID posits that animals did not evolve naturally through evolutionary means but were created abruptly by a nonnatural, or supernatural, designer. The defendants' own expert witnesses acknowledged this point. (21:96–100 [Behe]; P-718 at 696, 700 ["implausible that the designer is a natural entity"]; 28:21–22 [Fuller]) ("... ID's rejection of naturalism and commitment to supernaturalism..."); 38:95–96 (Minnich) (ID does not exclude the possibility of a supernatural designer, including deities).

It is notable that defense experts' own mission, which mirrors that of the IDM itself, is to change the ground rules of science to allow supernatural causation of the natural world, which the Supreme Court in *Edwards* and the court in *McLean* correctly recognized as an inherently religious concept. *Edwards*, 482 U.S. at 591–92; *McLean*, 529 F. Supp. at 1267. First, defense expert Professor Fuller agreed that ID aspires to "change the ground rules" of science and lead defense expert Professor Behe admitted that his broadened definition of science, which encompasses ID, would also embrace astrology. (28:26 [Fuller]; 21:37–42 [Behe]). Moreover, defense expert Professor Minnich acknowledged that for ID to be considered science, the ground rules of science have to be broadened to allow consideration of supernatural forces. (38:97 [Minnich]).

Prominent IDM leaders are in agreement with the opinions expressed by defense expert witnesses that the ground rules of science must be changed for ID to take hold and prosper. William Dembski, for instance, an IDM leader, proclaims that science is ruled by methodological naturalism and argues that this rule must be overturned if ID is to prosper. (5:32–37 [Pennock]); P-341 at 224 ("Indeed, entire fields of inquiry, including especially in the human sciences, will need to be rethought from the ground up in terms of intelligent design").

The Discovery Institute, the think tank promoting ID whose CRSC developed the Wedge Document, acknowledges as "Governing Goals" to "defeat scientific materialism and its destructive moral, cultural and political legacies" and "replace materialistic explanations with the theistic understanding that nature and human beings are created by God." (P-140 at 4). In addition, and as previously noted, the Wedge Document states in its "Five Year Strategic Plan Summary" that the IDM's goal is to replace science as currently practiced with "theistic and Christian science." Ibid. at 6. The IDM accordingly seeks nothing less than a complete scientific revolution in which ID will supplant evolutionary theory.[3]

Notably, every major scientific association that has taken a position on the issue of whether ID is science has concluded that ID is not, and cannot be considered as such. (1:98–99 [Miller]; 14:75–78 [Alters]; 37:25 [Minnich]). Initially, we note that NAS, the "most prestigious" scientific association in this country, views ID as follows:

> Creationism, intelligent design, and other claims of supernatural intervention in the origin of life or of species are not science because they are not testable by the methods of science. These claims subordinate observed data to statements based on authority, revelation, or religious belief. Documentation offered in support of these claims is typically limited to the special publications of their advocates. These publications do not offer hypotheses subject to change in light of new data, new interpretations, or demonstration of error. This contrasts with science, where any hypothesis or theory always remains subject to the possibility of rejection or modification in the light of new knowledge. P-192 at 25

Additionally, the American Association for the Advancement of Science (hereinafter "AAAS"), the largest organization of scientists in this country, has taken a similar position on ID, namely, that it "has not proposed a scientific means of testing its claims" and that "the lack of scientific warrant for so-called 'intelligent design theory' makes it improper to include as part of science education" (P-198). Not a single expert witness over the course of the six-week trial identified one major scientific association, society, or organization that endorsed ID as science. What is more, defense experts concede that ID is not a theory as that term is defined by the NAS and admit that ID is at best "fringe science" which has achieved no acceptance in the scientific community. (21:37–38 [Behe]; Fuller dep. at 98–101, June 21, 2005; 28:47 [Fuller]; Minnich dep. at 89, May 26, 2005).

It is therefore readily apparent to the Court that ID fails to meet the essential ground rules that limit science to testable, natural explanations. (3:101–103 [Miller]; 14:62 [Alters]). Science cannot be defined differently for Dover students than it is defined in the scientific community as an affirmative action program, as advocated by Professor Fuller, for a view that has been unable to gain a foothold within the scientific establishment. Although ID's failure to meet the ground rules of science is sufficient for the Court to conclude that it is not science, out of an abundance of caution and in the exercise of completeness, we will analyze additional arguments advanced regarding the concepts of ID and science.

ID is at bottom premised upon a false dichotomy, namely, that to the extent evolutionary theory is discredited, ID is confirmed. (5:41 [Pennock]). This argument is not brought to this Court anew, and in fact, the same argument, termed "contrived dualism" in *McLean*, was employed by creationists in the 1980s to support "creation science." The court in *McLean* noted the "fallacious pedagogy of the two model approach" and that "[i]n efforts to establish 'evidence' in support of creation science, the defendants relied upon the same false premise as the two model approach... all evidence which criticized evolutionary theory was proof in support of creation science." *McLean*, 529 F. Supp. at 1267, 1269. We do not find this false dichotomy any more availing to justify ID today than it was to justify creation science two decades ago.

ID proponents primarily argue for design through negative arguments against evolution, as illustrated by Professor Behe's argument that "irreducibly complex" systems cannot be produced through Darwinian, or any natural, mechanisms. (5:38–41 [Pennock]; 1:39, 2:15, 2:35–37, 3:96 [Miller]; 16:72–73 [Padian]; 10:148 [Forrest]). However, we believe that arguments against evolution are not arguments for design. Expert testimony revealed that just because scientists cannot explain today how biological systems evolved does not mean that they cannot, and will not, be able to explain them tomorrow. (2:36–37 [Miller]). As Dr. Padian aptly noted, "absence of evidence is not evidence of absence." (17:45 [Padian]). To that end, expert testimony from Drs. Miller and Padian provided multiple examples where *Pandas* asserted that no natural explanations exist, and in some cases that none could exist, and yet natural explanations have been identified in the intervening years. It also bears mentioning that, as Dr. Miller stated, just because scientists cannot explain every evolutionary detail does not undermine its validity as a scientific theory as no theory in science is fully understood. (3:102 [Miller]).

As referenced, the concept of irreducible complexity is ID's alleged scientific centerpiece. Irreducible complexity is a negative argument against evolution, not proof of design, a point conceded by defense expert Professor Minnich. (2:15 [Miller]; 38:82 [Minnich]) (irreducible complexity "is not a test of intelligent design; it's a test of evolution"). Irreducible complexity additionally fails to make a positive scientific case for ID, as will be elaborated upon below.

We initially note that irreducible complexity as defined by Professor Behe in his book *Darwin's Black Box* and subsequently modified in his 2001 article entitled "Reply to My Critics," appears as follows:

> By irreducibly complex I mean a single system which is composed of several well-matched, interacting parts that contribute to the basic function, wherein the removal of any one of the parts causes the system to effectively cease functioning. An irreducibly complex system cannot be produced directly by slight, successive modifications of a precursor system, because any precursor to an irreducibly complex system that is missing a part is by definition nonfunctional.... Since natural selection can only choose systems that are already working, then if a biological system cannot be produced gradually it would have to arise as an integrated unit, in one fell swoop, for natural selection to have anything to act on. P-647 at 39; P-718 at 694

Professor Behe admitted in "Reply to My Critics" that there was a defect in his view of irreducible complexity because, while it purports to be a challenge to natural selection, it does not actually address "the task facing natural selection." (P-718 at 695). Professor Behe specifically explained that "[t]he current definition puts the focus on removing a part from an already functioning system," but "[t]he difficult task facing Darwinian evolution, however, would not be to remove parts from sophisticated pre-existing systems; it would be to bring together components to make a new system in the first place." Ibid. In that article, Professor Behe wrote that he hoped to "repair this defect in future work;" however, he has failed to do so even four years after elucidating his defect. Ibid.; 22:61–65 (Behe).

In addition to Professor Behe's admitted failure to properly address the very phenomenon that irreducible complexity purports to place at issue, natural selection, Drs. Miller and Padian testified that Professor Behe's concept of irreducible complexity depends on ignoring ways in which evolution is known to occur. Although Professor Behe is adamant in his definition of irreducible complexity when he says a precursor "missing a part is by definition nonfunctional," what he obviously means is that it will not function in the same way the system functions when all the parts are present. For example, in the case of the bacterial flagellum, removal of a part may prevent it from acting as a rotary motor. However, Professor Behe excludes, by definition, the possibility that a precursor to the bacterial flagellum functioned not as a rotary motor, but in some other way, for example as a secretory system. (19:88–95 [Behe]).

As expert testimony revealed, the qualification on what is meant by "irreducible complexity" renders it meaningless as a criticism of evolution. (3:40 [Miller]). In fact, the theory of evolution proffers exaptation as a well-recognized, well-documented explanation for how systems with multiple parts could have evolved through natural means. Exaptation means that some precursor of the subject system had a different, selectable function before experiencing the change or addition that resulted in the subject system with its present function (16:146–48 [Padian]). For instance, Dr. Padian identified the evolution of the mammalian middle ear bones from what had been jawbones as an example of this process. (17:6–17 [Padian]). By defining irreducible complexity in the way that he has, Professor Behe attempts to exclude the phenomenon of exaptation by definitional fiat, ignoring as he does so abundant evidence which refutes his argument.

Notably, the NAS has rejected Professor Behe's claim for irreducible complexity by using the following cogent reasoning:

> [S]tructures and processes that are claimed to be "irreducibly" complex typically are not on closer inspection. For example, it is incorrect to assume that a complex structure or biochemical process can function only if all its components are present and functioning as we see them today. Complex biochemical systems can be built up from simpler systems through natural selection. Thus, the "history" of a protein can be traced through simpler organisms.... The evolution of complex molecular systems can occur in several ways. Natural selection can bring together parts of a system for one function at one time and then, at a later time, recombine those parts with other systems of components to produce a system that has a different function. Genes can be duplicated, altered, and then amplified through natural selection. The complex biochemical cascade resulting in blood clotting has been explained in this fashion. P-192 at 22

As irreducible complexity is only a negative argument against evolution, it is refutable and accordingly testable, unlike ID, by showing that there are intermediate structures with selectable functions that could have evolved into the allegedly irreducibly complex systems. (2:15–16 [Miller]). Importantly, however, the fact that the negative argument of irreducible complexity is testable does not make testable the argument for ID. (2:15 [Miller]; 5:39 [Pennock]). Professor Behe has applied the concept of irreducible complexity to only a few select systems: (1) the bacterial flagellum; (2) the blood-clotting cascade; and (3) the immune system. Contrary to Professor Behe's assertions with respect to these few biochemical systems among the myriad existing in nature, however, Dr. Miller presented evidence, based upon peer-reviewed studies, that they are not in fact irreducibly complex.

First, with regard to the bacterial flagellum, Dr. Miller pointed to peer-reviewed studies that identified a possible precursor to the bacterial flagellum, a subsystem that was fully functional, namely, the Type-III Secretory System. (2:820 [Miller]; P-854.23–854.32). Moreover, defense expert Professor Minnich admited that there is serious scientific research on the question of whether the bacterial flagellum evolved into the Type-III Secretary System, the Type-III Secretory System into the bacterial flagellum, or whether they both evolved from a common ancestor. (38:12–16 [Minnich]). None of this research or thinking involves ID. (38:12–16 [Minnich]). In fact, Professor Minnich testified about his research as follows: "we're looking at the function of these systems and how they could have been derived one from the other. And it's a legitimate scientific inquiry." (38:16 [Minnich]).

Second, with regard to the blood-clotting cascade, Dr. Miller demonstrated that the alleged irreducible complexity of the blood-clotting cascade has been disproven by peer-reviewed studies dating back to 1969, which show that dolphins' and whales' blood clots despite missing a part of the cascade, a study that was confirmed by molecular testing in 1998. (1:122–29 [Miller]; P-854.17–854.22). Additionally and more recently, scientists published studies showing that in puffer fish, blood clots despite the cascade missing not only one, but three parts. (1:128–29 [Miller]). Accordingly, scientists in peer-reviewed publications have refuted Professor Behe's predication about the alleged irreducible complexity of the blood-clotting cascade. Moreover, cross-examination revealed that Professor Behe's redefinition of the blood-clotting system was likely designed to avoid peer-reviewed scientific evidence that falsifies his argument, as it was not a scientifically warranted redefinition. (20:26–28, 22:112–25 [Behe]).

The immune system is the third system to which Professor Behe has applied the definition of irreducible complexity. Although in *Darwin's Black Box*, Professor Behe wrote that not only were there no natural explanations for the immune system at the time, but that natural explanations were impossible regarding its origin. (P-647 at 139; 2:26–27 [Miller]). However, Dr. Miller presented peer-reviewed studies refuting Professor Behe's claim that the immune system was irreducibly complex. Between 1996 and 2002, various studies confirmed each element of the evolutionary hypothesis explaining the origin of the immune system. (2:31 [Miller]). In fact, on cross-examination, Professor Behe was questioned concerning his 1996 claim that science would never find an evolutionary explanation for the immune system. He was presented with fifty-eight peer-reviewed publications, nine books, and several immunology textbook chapters about the evolution of the immune system; however, he simply insisted that this was still not sufficient evidence of evolution, and that it was not "good enough." (23:19 [Behe]).

We find that such evidence demonstrates that the ID argument is dependent upon setting a scientifically unreasonable burden of proof for the theory of evolution. As a further example, the test for ID proposed by both Professors Behe and Minnich is to grow the bacterial flagellum in the laboratory; however, no one inside or outside of the IDM, including those who propose the test, has conducted it. (P-718; 18:125–27 [Behe]; 22:102–106 [Behe]). Professor Behe conceded that the proposed test could not approximate real-world conditions and even if it could, Professor Minnich admitted that it would merely be a test of evolution, not design. (22:107–10 [Behe]; 2:15 [Miller]; 38:82 [Minnich]).

We therefore find that Professor Behe's claim for irreducible complexity has been refuted in peer-reviewed research papers and has been rejected by the scientific community at large. (17:45–46 [Padian]; 3:99 [Miller]). Additionally, even if irreducible complexity had not been rejected, it still does not support ID as it is merely a test for evolution, not design. (2:15, 2:35–40 [Miller]; 28:63–66 [Fuller]).

We will now consider the purportedly "positive argument" for design encompassed in the phrase used numerous times by Professors Behe and Minnich throughout their expert testimony, which is the "purposeful arrangement of parts." Professor Behe summarized the argument as follows: We infer design when we see parts that appear to be arranged for a purpose. The strength of the inference is quantitative; the more parts that are arranged, the more intricately they interact, the stronger is our confidence in design. The appearance of design in aspects of biology is overwhelming. Since nothing other than an intelligent cause has been demonstrated to be able to yield such a strong appearance of design, Darwinian claims notwithstanding, the conclusion that the design seen in life is real design is rationally justified. (18:90–91, 18:109–10 [Behe]; 37:50 [Minnich]). As previously indicated, this argument is merely a restatement of the Reverend William Paley's argument applied at the cell level. Minnich, Behe, and Paley reach the same conclusion, that complex organisms must have been designed using the same reasoning, except that Professors Behe and Minnich refuse to identify the designer, whereas Paley inferred from the presence of design that it was God. (1:67 [Miller]; 38:44, 57 [Minnich]). Expert testimony revealed that this inductive argument is not scientific and as admitted by Professor Behe, can never be ruled out. (2:40 [Miller]; 22:101 [Behe]; 3:99 [Miller]).

Indeed, the assertion that design of biological systems can be inferred from the "purposeful arrangement of parts" is based upon an analogy to human design. Because we are able to recognize design of artifacts and objects, according to Professor Behe, that same reasoning can be employed to determine biological design. (18:116–17, 23:50 [Behe]). Professor Behe testified that the strength of the analogy depends upon the degree of similarity entailed in the two propositions; however, if this is the test, ID completely fails.

Unlike biological systems, human artifacts do not live and reproduce over time. They are nonreplicable, they do not undergo genetic recombination, and they are not driven by natural selection. (1:131–33 [Miller]; 23:57–59 [Behe]). For human artifacts, we know the designer's identity, human, and the mechanism of design, as we have experience based upon empirical evidence that humans can make such things, as well as many other attributes including the designer's abilities, needs, and desires. (D-251 at 176; 1:131–33 [Miller]; 23:63 [Behe]; 5:55–58 [Pennock]). With ID, proponents assert that they refuse to propose hypotheses on the designer's identity, do not propose a mechanism, and the designer, he/she/it/they, has never been seen. In that vein, defense expert Professor Minnich agreed that in the case of human artifacts and objects, we know the identity and capacities of the human designer, but we do not know any of those attributes for the designer of biological life. (38:44–47

[Minnich]). In addition, Professor Behe agreed that for the design of human artifacts, we know the designer and its attributes and we have a baseline for human design that does not exist for design of biological systems. (23:61–73 [Behe]). Professor Behe's only response to these seemingly insurmountable points of disanalogy was that the inference still works in science fiction movies. (23:73 [Behe]).

It is readily apparent to the Court that the only attribute of design that biological systems appear to share with human artifacts is their complex appearance, that is, if it looks complex or designed, it must have been designed. (23:73 [Behe]). This inference to design based upon the appearance of a "purposeful arrangement of parts" is a completely subjective proposition, determined in the eye of each beholder and his/her viewpoint concerning the complexity of a system. Although both Professors Behe and Minnich assert that there is a quantitative aspect to the inference, on cross-examination they admitted that there is no quantitative criteria for determining the degree of complexity or number of parts that bespeak design, rather than a natural process. (23:50 [Behe]; 38:59 [Minnich]). As the plaintiffs aptly submit to the Court, throughout the entire trial only one piece of evidence generated by the defendants addressed the strength of the ID inference: the argument is less plausible to those for whom God's existence is in question, and is much less plausible for those who deny God's existence. (P-718 at 705).

Accordingly, the purported positive argument for ID does not satisfy the ground rules of science, which require testable hypotheses based upon natural explanations. (3:101–103 [Miller]). ID is reliant upon forces acting outside of the natural world, forces that we cannot see, replicate, control, or test, which have produced changes in this world. While we take no position on whether such forces exist, they are simply not testable by scientific means and therefore cannot qualify as part of the scientific process or as a scientific theory. (3:101–102 [Miller]).

It is appropriate at this juncture to address ID's claims against evolution. ID proponents support their assertion that evolutionary theory cannot account for life's complexity by pointing to real gaps in scientific knowledge, which indisputably exist in all scientific theories, but also by misrepresenting well-established scientific propositions. (1:112, 1:122, 1:136–37 [Miller]; 16:74–79, 17:45–46 [Padian]).

Before discussing the defendants' claims about evolution, we initially note that an overwhelming number of scientists, as reflected by every scientific association that has spoken on the matter, have rejected the ID proponents' challenge to evolution. Moreover, the plaintiffs' expert in biology, Dr. Miller, a widely recognized biology professor at Brown University who has written university-level and high school biology textbooks used prominently throughout the nation, provided unrebutted testimony that evolution, including common descent and natural selection, is "overwhelmingly accepted" by the scientific community and that every major scientific association agrees. (1:94–100 [Miller]). As the court in *Selman* explained, "evolution is more than a theory of origin in the context of science. To the contrary, evolution is the dominant scientific theory of origin accepted by the majority of scientists." *Selman*, 390 F. Supp. 2d at 1309 (emphasis in original). Despite the scientific community's overwhelming support for evolution, the defendants and ID proponents insist that evolution is unsupported by empirical evidence. The plaintiffs' science experts, Drs. Miller and Padian, clearly explained how ID proponents generally, and *Pandas* specifically, distort and misrepresent scientific knowledge in making their antievolution argument.

In analyzing such distortion, we turn again to *Pandas*, the book to which students are expressly referred in the disclaimer. The defendants hold out *Pandas* as representative of ID and the plaintiffs' experts agree in that regard. (16:83 [Padian]; 1:107–108 [Miller]). A series of arguments against evolutionary theory found in *Pandas* involve paleontology, which studies the life of the past and the fossil record. The plaintiffs' expert Professor Padian was the only testifying expert witness with any expertise in paleontology.[4] His testimony therefore remains unrebutted. Dr. Padian's demonstrative slides, prepared on the basis of peer-reviewing scientific literature, illustrate how *Pandas* systematically distorts and misrepresents established, important evolutionary principles.

We will provide several representative examples of this distortion. First, *Pandas* misrepresents the "dominant form of understanding relationships" between organisms, namely, the tree of life, represented by classification determined via the method of cladistics. (16:87–97 [Padian]; P-855.6–855.19). Second, *Pandas* misrepresents "homology," the "central concept of comparative biology," that allowed scientists to evaluate comparable parts among organisms for classification purposes for hundreds of years. (17:27–40 [Padian]; P-

855.83–855.102). Third, *Pandas* fails to address the well-established biological concept of exaptation, which involves a structure changing function, such as fish fins evolving fingers and bones to become legs for weight-bearing land animals. (16:146–48 [Padian]). Dr. Padian testified that ID proponents fail to address exaptation because they deny that organisms change function, which is a view necessary to support abrupt appearance. Ibid. Finally, Dr. Padian's unrebutted testimony demonstrates that *Pandas* distorts and misrepresents evidence in the fossil record about pre-Cambrian-era fossils, the evolution of fish to amphibians, the evolution of small carnivorous dinosaurs into birds, the evolution of the mammalian middle ear, and the evolution of whales from land animals. (16:107–17, 16:117–31, 16:131–45, 17:6–9, 17:17–27 [Padian]; P-855.25–855.33, P-855.34–855.45, P-855.46– 855.55, P-855.56–855.63, P-855.64–855.82).

In addition to Dr. Padian, Dr. Miller also testified that *Pandas* presents discredited science. Dr. Miller testified that *Pandas*'s treatment of biochemical similarities between organisms is "inaccurate and downright false" and explained how *Pandas* misrepresents basic molecular biology concepts to advance design theory through a series of demonstrative slides. (1:112 [Miller]). Consider, for example, that he testified as to how *Pandas* misinforms readers on the standard evolutionary relationships between different types of animals, a distortion which Professor Behe, a "critical reviewer" of *Pandas* who wrote a section within the book, affirmed. (1:113–17 [Miller]; P-854.9–854.16; 23:35–36 [Behe]).[5] In addition, Dr. Miller refuted *Pandas*'s claim that evolution cannot account for new genetic information and pointed to more than three dozen peer-reviewed scientific publications showing the origin of new genetic information by evolutionary processes. (1:133–36 [Miller]; P-245). In summary, Dr. Miller testified that *Pandas* misrepresents molecular biology and genetic principles as well as the current state of scientific knowledge in those areas in order to teach readers that common descent and natural selection are not scientifically sound. (1:139–42 [Miller]).

Accordingly, the one textbook to which the Dover ID Policy directs students contains outdated concepts and badly flawed science, as recognized by even the defense experts in this case.

A final indicator of how ID has failed to demonstrate scientific warrant is the complete absence of peer-reviewed publications supporting the theory. Expert testimony revealed that the peer-review process is "exquisitely important" in the scientific process. It is a way for scientists to write up their empirical research and to share the work with fellow experts in the field, opening up the hypotheses to study, testing, and criticism. (1:66–69 [Miller]). In fact, defense expert Professor Behe recognizes the importance of the peer-review process and has written that science must "publish or perish." (22:19–25 [Behe]). Peer review helps to ensure that research papers are scientifically accurate, meet the standards of the scientific method, and are relevant to other scientists in the field. (1:39–40 [Miller]). Moreover, peer review involves scientists submitting a manuscript to a scientific journal in the field, journal editors soliciting critical reviews from other experts in the field and deciding whether the scientist has followed proper research procedures, employed up-to-date methods, considered and cited relevant literature, and generally, whether the researcher has employed sound science.

The evidence presented in this case demonstrates that ID is not supported by any peer-reviewed research, data, or publications. Both Drs. Padian and Forrest testified that recent literature reviews of scientific and medical-electronic databases disclosed no studies supporting a biological concept of ID. (17:42–43 [Padian]; 11:32–33 [Forrest]). On cross-examination, Professor Behe admitted: "There are no peer-reviewed articles by anyone advocating for intelligent design supported by pertinent experiments or calculations which provide detailed rigorous accounts of how intelligent design of any biological system occurred." (22:22–23 [Behe]). Additionally, Professor Behe conceded that there are no peer-reviewed papers supporting his claims that complex molecular systems, like the bacterial flagellum, the blood-clotting cascade, and the immune system, were intelligently designed. (21:61–62 [complex molecular systems], 23:4–5 [immune system], and 22:124–25 [blood-clotting cascade] [Behe]). In that regard, there are no peer-reviewed articles supporting Professor Behe's argument that certain complex molecular structures are "irreducibly complex."[6] (21:62, 22:124–25 [Behe]). In addition to failing to produce papers in peer-reviewed journals, ID also features no scientific research or testing. (28:114–15 [Fuller]; 18:22–23, 105–106 [Behe]).

After this searching and careful review of ID as espoused by its proponents, as elaborated upon in submissions to the Court, and as scrutinized over a six-week trial, we find that ID is not science and cannot be adjudged a valid, accepted scientific theory as it has failed to publish in peer-reviewed journals, engage in research and testing, and gain acceptance in the scientific community. ID, as noted, is grounded in theology, not science. Accepting for the sake of argument its proponents', as well as the defendants' argument that to introduce ID to students will encourage critical thinking, it still has utterly no place in a science curriculum. Moreover, ID's backers have sought to avoid the scientific scrutiny which we have now determined that it cannot withstand by advocating that the controversy, but not ID itself, should be taught in science class. This tactic is at best disingenuous, and at worst a canard. The goal of the IDM is not to encourage critical thought, but to foment a revolution which would supplant evolutionary theory with ID.

To conclude and reiterate, we express no opinion on the ultimate veracity of ID as a supernatural explanation. However, we commend to the attention of those who are inclined to superficially consider ID to be a true "scientific" alternative to evolution without a true understanding of the concept the foregoing detailed analysis. It is our view that a reasonable, objective observer would, after reviewing both the voluminous record in this case and our narrative, reach the inescapable conclusion that ID is an interesting theological argument, but that it is not science.

...

H. Conclusion

The proper application of both the endorsement and *Lemon* tests to the facts of this case makes it abundantly clear that the board's ID Policy violates the Establishment Clause. In making this determination, we have addressed the seminal question of whether ID is science. We have concluded that it is not, and moreover that ID cannot uncouple itself from its creationist, and thus religious, antecedents.

Both the defendants and many of the leading proponents of ID make a bedrock assumption which is utterly false. Their presupposition is that evolutionary theory is antithetical to a belief in the existence of a supreme being and to religion in general. Repeatedly in this trial, the plaintiffs' scientific experts testified that the theory of evolution represents good science, is overwhelmingly accepted by the scientific community, and that it in no way conflicts with, nor does it deny, the existence of a divine creator.

To be sure, Darwin's theory of evolution is imperfect. However, the fact that a scientific theory cannot yet render an explanation on every point should not be used as a pretext to thrust an untestable alternative hypothesis grounded in religion into the science classroom or to misrepresent well-established scientific propositions.

The citizens of the Dover area were poorly served by the members of the board who voted for the ID Policy. It is ironic that several of these individuals, who so staunchly and proudly touted their religious convictions in public, would time and again lie to cover their tracks and disguise the real purpose behind the ID Policy.

With that said, we do not question that many of the leading advocates of ID have bona fide and deeply held beliefs which drive their scholarly endeavors. Nor do we controvert that ID should continue to be studied, debated, and discussed. As stated, our conclusion today is that it is unconstitutional to teach ID as an alternative to evolution in a public school science classroom. Those who disagree with our holding will likely mark it as the product of an activist judge. If so, they will have erred, as this is manifestly not an activist Court. Rather, this case came to us as the result of the activism of an ill-informed faction on a school board, aided by a national public interest law firm eager to find a constitutional test case on ID, who in combination drove the board to adopt an imprudent and ultimately unconstitutional policy. The breathtaking inanity of the board's decision is evident when considered against the factual backdrop which has now been fully revealed through this trial. The students, parents, and teachers of the Dover Area School District deserved better than to be dragged into this legal maelstrom, with its resulting utter waste of monetary and personal resources.

NOTES

1. The Court has received numerous letters, amicus briefs, and other forms of correspondence pertaining to this case. The only documents submitted by third parties the Court has considered, however, are those that have become an official part of the record. Consistent with the foregoing, the Court has taken under consideration the following: (1) Brief of Amici Curiae Biologists and Other Scientists in Support of

Defendants (doc. 245); (2) Revised Brief of Amicus Curiae, the Discovery Institute (doc. 301); (3) Brief of Amicus Curiae the Foundation for Thought and Ethics (doc. 309); and (4) Brief for Amicus Curiae Scipolicy Journal of Science and Health Policy (doc. 312).

The Court accordingly grants the outstanding Motions for Leave to File Amicus Briefs, namely, the Motion for Leave to File a Revised Amicus Brief by the Discovery Institute (doc. 301), the Motion for Leave to File Amicus Brief by the Foundation for Thought and Ethics (doc. 309), and the Petition for Leave to File Amicus Curiae Brief by Scipolicy Journal of Science and Health Policy (doc. 312).

2. Defendants contend that the Court should ignore all evidence of ID's lineage and religious character because the board members do not personally know Jon Buell, president of the Foundation for Thought and Ethics (hereinafter "FTE"), the publisher of *Pandas*, or Phillip Johnson, nor are they familiar with the Wedge Document or the drafting history of *Pandas*. The defendants' argument lacks merit legally and logically.

The evidence that the defendants are asking this Court to ignore is exactly the sort that the court in *McLean* considered and found dispositive concerning the question of whether creation science was a scientific view that could be taught in public schools, or a religious one that could not. The *McLean* court considered writings and statements by creation science advocates like Henry Morris and Duane Gish, as well as the activities and mission statements of creationist think tanks like the Bible Science Association, the Institution for Creation Research, and the Creation Science Research Center. *McLean*, 529 F. Supp. at 1259–60. The court did not make the relevance of such evidence conditional on whether the Arkansas Board of Education knew the information. Instead, the court treated the evidence as speaking directly to the threshold question of what creation science was. Moreover, in *Edwards*, the Supreme Court adopted *McLean*'s analysis of such evidence without reservation, and without any discussion of which details about creation science the defendant school board actually knew. *Edwards*, 482 U.S. at 590 n9.

3. Further support for this proposition is found in the Wedge Strategy, which is composed of three phases: Phase I is scientific research, writing, and publicity; Phase II is publicity and opinion making; and Phase III is cultural confrontation and renewal. (P-140 at 3). In the "Five Year Strategic Plan Summary," the Wedge Document explains that the social consequences of materialism have been "devastating" and that it is necessary to broaden the wedge with a positive scientific alternative to materialistic scientific theories, which has come to be called the theory of ID. "Design theory promises to reverse the stifling dominance of the materialist worldview, and to replace it with a science consonant with Christian and theistic convictions." Ibid. at 6.

Phase I of the Wedge Strategy is an essential component and directly references "scientific revolutions." Phase II explains that alongside a focus on influential opinion-makers, "we also seek to build up a popular base of support among our natural constituency, namely, Christians. We will do this primarily through apologetics seminars. We intend these to encourage and equip believers with new scientific evidence that support the faith, as well as to 'popularize' our ideas in the broader culture." Ibid. Finally, Phase III includes pursuing possible legal assistance "in response to resistance to the integration of design theory into public school science curricula." Ibid. at 7.

4. Moreover, the Court has been presented with no evidence that either the defendants' testifying experts or any other ID proponents, including *Pandas*'s authors, have such paleontology expertise as we have been presented with no evidence that they have published peer-reviewed literature or presented such information at scientific conferences on paleontology or the fossil record. (17:15–16 [Padian]).

5. Additionally, testimony provided by Professor Behe revealed an increasing gap between his portrayal of ID theory and how it is presented in *Pandas*. Although he is a "critical reviewer" of the work, he disagrees with language provided in the text, including but not limited to the text's very definition of ID. (P-11 at 99–100).

6. The one article referenced by both Professors Behe and Minnich as supporting ID is an article written by Behe and Snoke titled "Simulating Evolution by Gene Duplication of Protein Features That Require Multiple Amino Acid Residues." (P-721). A review of the article indicates that it does not mention either irreducible complexity or ID. In fact, Professor Behe admitted that the study which forms the basis for the article did not rule out many known evolutionary mechanisms and that the research actually might support evolutionary pathways if a biologically realistic population size were used. (22:41–45 [Behe]; P-756).

CHAPTER 28

CAN'T PHILOSOPHERS TELL THE DIFFERENCE BETWEEN SCIENCE AND RELIGION? DEMARCATION REVISITED

ROBERT T. PENNOCK

> [W]e have addressed the seminal question of whether ID is science. We have concluded that it is not, and moreover that ID cannot uncouple itself from its creationist, and thus religious, antecedents.
>
> —*Kitzmiller v. Dover* (2005, 136)

INTRODUCTION

Intelligent design, the latest version of creationism to try to wedge its way into science classes, suffered a legal death blow in a federal district court in the 2005 *Kitzmiller et al. v. Dover Area School District* case. After hearing twenty-one days of testimony over a forty-day period in which intelligent design (ID) proponents and their critics presented their best evidence and arguments regarding the purported scientific and educational merits of "design theory," Judge John E. Jones III ruled that ID was not science but disguised sectarian religion and thus that teaching it in the public schools is illegal, a violation of the United States Constitution.

With nary a sentence in the ruling that granted even the least element of

Original essay. Reprinted by permission of the author.

their claims, ID proponents howled that the judge "got on his soapbox to offer his own views of science, religion, and evolution" and had overstepped his authority in ruling on the question of whether ID was science, calling him "an activist judge who has delusions of grandeur" (Discovery Institute spokesman John West, quoted in Anonymous 2006). Such ad hominem denunciations were an about-face from some ID pretrial writings, which had lauded him as a Bush-appointed good ol' boy with impeccable conservative credentials. Discussing in a recent interview the question of whether he should have ruled about whether ID qualified as science, Judge Jones noted that this was not only a legally relevant question in the trial, but that both sides in the case had asked him to rule on just this point (*Philadelphia Inquirer* 2006). The defense's pretrial memorandum, for instance, stated that "the evidence will show that IDT [ID theory] is a scientific argument, advanced by scientist [*sic*] relying on evidence and technical knowledge proper to their specialties," and that ID's reliance on supernatural explanations "does not place [it] beyond the bounds of 'science.' Quite the contrary, IDT's refusal to rule out this possibility represents the essence of scientific inquiry" (Thomas More Law Center 2005, 10–11). It is disingenuous, to say the least, for ID proponents to call for a ruling on this issue and then, when the ruling did not go their way, to complain that the judge overstepped.

Indeed, ID leaders had been hoping for a test case that would rule on this from the moment law professor Phillip Johnson brokered a truce between young-earth and old-earth creationists and united them under the banner of ID as a way of wedging their "theistic science," as he called it, through the wall of separation between church and state. To this end, ID leaders produced law review articles and legal guides promoting the legality of teaching ID and planned for direct legal assistance to public schools (DeWolf 1999; DeWolf, Meyer, et al. 1999; Discovery Institute 1999; DeWolf, Meyer, et al. 2000). *Kitzmiller* gave them the opportunity but not the outcome they had sought, finding that ID was "not science" but rather "creationism relabeled" and "a religious alternative masquerading as a scientific theory." A few ID opponents argued that it should have been sufficient for the plaintiffs to show that ID is religion without asking the court to also rule on whether or not ID is science, but *Kitzmiller* attorney Richard B. Katskee gives a detailed explanation of why this was a central legal question in this historic test case (2006).

The *Kitzmiller* case involved a policy that had been instituted in 2004 in the public schools of Dover, Pennsylvania, by the school district's board of directors, which was then dominated by creationists. The policy spoke of purported gaps and problems with what it called "Darwin's theory" and changed the science curriculum to allow inclusion of intelligent design as an alternative theory. Biology teachers were directed to read a statement that warned students that Darwin's theory "is not a fact" and told them about ID as a differing explanation. The ID textbook *Of Pandas and People* was made available for students to gain an understanding of what ID actually involves.

Eleven parents filed a suit against the district, charging that allowing ID in the schools was unconstitutional. The school district and the board were defended by the Thomas More Law Center (TMLC), which calls itself "The Sword and Shield for People of Faith." TMLC drew primarily from Fellows of the Discovery Institute (DI), the ID think tank, in its initial list of expert witnesses—key players and leaders of the ID movement, including Michael Behe, Scott Minnich, William Dembski, Stephen C. Meyer, and John Angus Campbell. (The last three of these abruptly withdrew from the case at the last minute before their depositions. Two other ID experts, Dick Carpenter and Warren Nord, also withdrew. Steve Fuller was added to the ID roster and did testify.)

The *Kitzmiller* case was widely described as a twenty-first-century replay of the *Scopes* monkey trial, but it was in many more ways a replay of the 1981 *McLean v. Arkansas* trial. The *McLean* case involved a state bill (Act 590) that had mandated that so-called creation science be given "balanced treatment" with evolution in public school science classes. As in the ID case, creation science proponents had claimed that they were offering a scientific alternative theory devoid of religious commitments. As ID speaks of an unspecified "designer," creation science spoke generically of a "creator" and did not identify it as God or make explicit reference to the Bible. There were numerous other parallels, including the final judgment: the judge in the *McLean* decision, William Overton, ruled that creation science was not science but religion, and thus that teaching it in the public schools is unconstitutional.[1]

An important part of Overton's decision relied upon expert testimony of philosopher of science Michael Ruse, who offered five criteria to distinguish science from nonscience, namely:

(1) It is guided by natural law;
(2) It has to be explanatory by reference to natural law;
(3) It is testable against the empirical world;
(4) Its conclusions are tentative, i.e., are not necessarily the final word; and
(5) It is falsifiable. (Overton 1982, §IV[C])

Two philosophers, Larry Laudan and Philip Quinn, subsequently took issue in print with Overton's decision and with Ruse's role in it. Ruse reprinted their articles in his original anthology *But Is It Science?* about this philosophical question in the *McLean* trial. He gave brief rebuttals, but then let his critics have the last word (Ruse 1988). Unfortunately, this generous editorial gesture left the impression that the critics could not be answered, and tens of thousands of students have had no further exposure to the issue than this limited and misleading exchange. Creationists have exploited this misleading impression ever since.

ID creationists (IDCs) cite Laudan and to a lesser extent Quinn, who mostly makes the same points, in almost everything they have written that discusses the question of whether ID is science. Lauding Laudan for insight and honesty, they proclaim *McLean* a hollow victory based upon an irresponsible misrepresentation of the nature of scientific demarcation. There is no way to legitimately rule out ID as science, they claim, as Laudan showed that there is no way to distinguish science from nonscience. J. P. Moreland, for instance, a philosopher and ID advocate at Biola University (previously known as the Bible Institute of Los Angeles), which houses the ID movement's model teaching program, makes this claim and appeals to Laudan in an article (Moreland 1994) in which he still used creation science terminology. The Discovery Institute administrator and core ID leader Stephen C. Meyer (who has previously taught philosophy at Whitman College and Christian Apolgetics at Palm Beach Atlantic University) does the same in an article in which he argued that it was legitimate to advance a supernatural "Theory of Creation" as methodologically equivalent to the theory of evolution (Meyer 1994). IDCs have especially relied upon Laudan in articles claiming the legality of teaching ID, such as a legal guidebook they published on how to include ID in public school science curricula (DeWolf 1999).

They continued to do so following the *Kitzmiller* case. Criticizing my own expert testimony on the questions of whether ID is science and whether it is religion, Discovery Institute staffer Casey Luskin quoted Laudan's and Quinn's criticisms of Ruse and went on to claim, bizarrely, that Ruse himself recanted his testimony (Luskin 2005). IDCs trumpeted an online preprint by the philosopher Bradley Monton, who claimed to have no sympathy for ID, but who echoed their criticisms of the court's decision, mostly repeating their exact arguments and their appeal to Laudan (Dembski 2006; Monton 2006; Wirth 2006). As Laudan and Quinn had questioned Ruse's testimony, so did Luskin and Monton question mine, on the same grounds—we were purportedly misrepresenting philosophy of science not only by appealing to outmoded demarcation criteria, but by not recognizing that the demarcation problem was dead.

In this chapter I hope to correct some of the common errors these commentators have made and to offer a more reasonable approach to how to think about distinguishing science from pseudoscience in general and religion in particular. Why revisit this? Because Laudan's and Quinn's discussions of demarcation, which can only be described as histrionic and ill considered, and those of their careless imitators continue to muddy the waters to the detriment of both science and philosophy of science.

RUMORS OF DEMISE

Commenting on the *McLean* case, Laudan chided Judge Overton, and indirectly Ruse, for basing his decision against creation science on what he called "a false stereotype of what science is and how it works" (Laudan 1982, 355). He opined that *McLean* was an "anachronistic effort to revive a variety of discredited criteria for distinguishing between the scientific and the non-scientific" and that for the scientific community to leave it unchallenged would "raise grave doubts about that community's intellectual integrity" (ibid.). (It is worth pointing out that Laudan's critique focuses almost entirely on Ruse's criteria, which was only one part of what Overton took into account in ruling that creation science was not science. Overton also discussed how the activities of so-called creation scientists differed so markedly from that of real scientists. He took into account the dearth of peer-reviewed publications to establish an evidential basis for creation science and the absence of appropriate educational materials. And he explicitly discussed the nature of religion and the ways that creation science was religious. None of this should have been ignored.) In another article—"The Demise of Demarcation"—Laudan went further and charged that the problem of demarcation was itself dead. He wrote: "The problem of demarcation between science and non-science is a pseudo-problem (at least as far as philosophy is concerned)" (Laudan 1983a, 348).

However, even if Laudan had been correct that philosophers viewed demarcation as a pseudoproblem, that would not mean that it is a pseudoproblem in other settings or for scholars with other interests. The relevant context in *McLean* is the legal arena, and deciding constitutional questions regarding the establishment of religion is hardly a pseudoproblem for plaintiffs, defendants, attorneys, and courts.[2] Neither does it mean that philosophy could not set aside its peculiarly abstract and rarified interests and make a useful contribution in these other contexts for their more practical purposes. However, we need not develop this avenue of reply because the premise of the objection is false on its face. Indeed, it is hard to know what to make of the superficial scholarship that leads IDCs and others to cite Laudan as though he provided the last word on the subject and as the official coroner of philosophical interest in the issue of demarcation. Laudan's obituary of the demarcation problem was premature, to say the least.

It would have been one thing had Laudan simply been describing his own view or giving his judgment that philosophers *should* give up the demarcation problem as dead, but he wrote as though he were stating a historical fact. However, even at the time he wrote his article it was false to say that demarcation was no longer a live topic. Moreover, subsequent to Laudan's paper and up to this day, demarcation questions continue to be regularly discussed in the philosophical literature. Even a cursory search turns up well over a dozen articles and several books that directly address the demarcation question and many more deal with it or assume it indirectly.

Several of these are explicitly critical of Laudan's treatment (Ruse 1982; Gross 1983; Derksen 1993), some highly so. Barry Gross, who served as a philosophy consultant to the ACLU in the *McLean* trial, found Laudan's treatment to be almost willfully naive and misguided.

> [Laudan] not only missed the context of this inquiry and the essential features of the creationist position, but has also shown lack of comprehension of the constitutional issues and standards of proof involved, of the nature of adversary trial, of the weight of legal decision, of the dynamics of preparation for trial undertaken by a large team of attorneys, and of the nature of state and local text of decisions. (Gross 1983, 30)

He wrote, "Larry Laudan presents in his jeremiad on *McLean v. Arkansas* a perfect example of a philosopher richly deserving an exclusion from 'the conversation of mankind'" (ibid.), concluding with a stinging philosophical rebuke that "Mr. Laudan in proposing himself as the Socrates of the *Gorgias* has, instead, read us the lines of Euthyphro" (Gross 1983, 37). Many philosophers quickly reject Laudan's conclusions and proceed to defend various demarcation criteria, while several do so without even bothering to mentioning his pronouncements. If the problem of demarcation is a philosophical pseudoproblem, then there is a long list of first-rank pseudophilosophy journals (*Philosophy of Science, British Journal for the Philosophy of Science, Studies in the History and Philosophy of Science, Philosophy of Social Sciences*, and so on) and an even longer list of top-notch pseudophilosophers publishing on the question at the time and since. Among those who have continued to tackle the problem are Deborah Mayo (1996), Keith Abney (1997), George Reisch (1998), Michael Ruse (2001), and many more.

Naturally, different philosophers continue to disagree about the best way to demarcate science. Joseph Agassi and Nathaniel Laor (2000) emphasize the importance of repeatability, arguing that the "scientific method sharply characterizes facts given to scientific inquiry: all and only those facts are scientific that are given to repeatable observation." David Resnik (2000) offers a pragmatic approach for distinguishing what is scientific. James Roper (2005) argues that intelligent design creationism is not science by making use of Goodman's notion of projectability. Even those who approach the matter from a constructivist perspective, such as Charles Taylor (1996), agree that there is a difference, though perhaps a historically contingent difference, between what is and what is not science. Thomas Gieryn (1983) gives a sociological account of how scientists draw professional boundaries to distinguish science from nonscience. Thus, while there are different view about exactly how to draw a line between science and nonscience, there is widespread agreement not only that there is a real difference but that it is of philosophical interest.

If ID supporters continue to cite Laudan's pronouncements on the death of the demarcation problem, they should be recalled of Mark Twain's wry comment, slightly paraphrased, that the rumors of its demise are greatly exaggerated.

THE *KITZMILLER* PHILOSOPHY

Before going on to look in more detail at the demarcation problem and Laudan's discussion in the contemporary setting, it will be worthwhile to briefly explain our approach to this issue in *Kitzmiller* and highlight ways in which it differed from the *McLean* reasoning. This will help avoid some common misunderstandings found in commentaries about both cases.

(1) First of all, there was no attempt in *Kitzmiller* to follow Ruse's five criteria from the *McLean* case. Indeed, my recommendation to the legal team from the beginning was to avoid the philosophical problems inherent in Overton's listing of these and to revise and simplify the argument. There were indeed problems with some of the *McLean* criteria, but more than that it was overly and unnecessarily ambitious to attempt to lay out criteria that are necessary and sufficient to define science. Thus, for instance, we made no appeal to falsifiability or tentativeness as scientific litmus tests. Even when we discussed some of the same concepts, such as notions of explanation, natural law, and testability, we did so in quite different ways that reflected more current thinking in philosophy of science.

IDC critics of the *Kitzmiller* opinion, and even some critics unsympathetic to IDC like Monton, seem not only to presume that Laudan had given the last word on the demarcation problem but also that nothing else had changed in thinking about explanation, causation, confirmation, and other philosophical issues relevant to the demarcation question in the twenty-plus years of philosophical discussion since *McLean.* ID leader Stephen Meyer's defense of appeal to supernatural agency rebuttal of Ruse's demarcation criteria at the 1992 Southern Methodist University (SMU) conference that publicly launched the ID movement was already so out of date and confused in its discussion of the relations among law, cause, and scientific explanation (Meyer 1992) that it inadvertently undermined its own argument. Meyer's subsequent papers have not corrected the problems (Pennock 2004). By 2005, philosophy of science had progressed well beyond many of the old philosophical debates of the middle part of the century (some of which will be discussed in the penultimate section) that had still lingered in the *McLean* debate. The *Kitzmiller* argument was able to draw upon the lessons of the last three decades and avoid earlier confusions.

(2) Moreover, even the overlapping concepts were not used as demarcation criteria in the sense for which Laudan criticized the *McLean* ruling. Neither did we substitute an alternative set. Indeed, we made no attempt to give a list of criteria to strictly define science. It was not necessary to do so. The relevant demarcation problem is far simpler than Laudan would have us believe. The task was not to demarcate science by pinning down its precise borders in the formal sense of giving a set of necessary and sufficient conditions that are shared without exception among all and only sciences, both historical and contemporary. I am as skeptical as Laudan that such a clear bright pinline border could ever be discovered, though more for general reasons about the nature of classification than for anything special about the case of science and pseudoscience. However, contra Laudan, I would argue that this standard of demarcation is not only unrealistic but fundamentally misguided. It is certainly an inappropriate standard in this context. What is needed is not a historical formal definition but something more pragmatic and down to earth—what might be called a *ballpark* demarcation that simply identifies a position as violating a basic value, or *ground rule*, inherent in the practice. One need not be able to list all the rules that distinguish baseball from softball or stickball to be able to say that someone who wants to use immaterial balls and bats and call in a supernatural pinch hitter is playing a totally different game. Showing that creationism is not science requires no more complicated notion of demarcation than that—it violates a scientific ground rule and is not even in the ballpark.[3] Indeed, for the constitutional case the problem is simpler still because the contrast classes are not even science and pseudoscience, but rather science and religion. Laudan's entire critique of demarcation, which expects a precise line that can unambiguously rule any possible theory in or out of science, addresses quite a different question than was at issue in

Kitzmiller, which required as its first part only showing that a particular thing—"ID theory"—was not science. (This chapter has space to deal only briefly with the second part of showing that ID is religion.)

(3) *Kitzmiller* articulated a simple ballpark approach in ruling out creationism, identifying methodological naturalism (MN) as a ground rule of science that ID and other forms of creationism violate. MN holds that as a principle of research we should regard the universe as a structured place that is ordered by uniform natural processes, and that scientists may not appeal to miracles or other supernatural interventions that break this presumed order. Science does not hold to MN dogmatically, but because of reasons having to do with the nature of empirical evidence. I initially laid out the arguments for this (1996a) and elaborated upon them (1996b; 1998; 1999) and will not rehearse them here. Neither did we rehearse them in any detail in court, but tried to illustrate points with examples and to put the arguments in terms that were as simple as possible without sacrificing accuracy. As one illustrative example of methodological naturalism, I noted in my testimony that we cite the Hippocratic corpus as at least protoscientific precisely because it begins to reject supernatural explanations; epilepsy is not to be thought of as a "sacred disease" but one for which we seek an explanation and cure in terms of ordinary natural causes. Hippocrates even begins to offer some good methodological reasons for this: "Men think epilepsy divine, merely because they do not understand it. But if they called everything divine which they do not understand, why, there would be no end of divine things" (Hippocratic Corpus).

It is worth noting that expert witnesses are advised to testify in as simple and succinct a manner as is possible, and are cautioned against going into detailed, technical explanations unless called upon to do so. Usually such questions come during cross-examination as the opposing attorneys attempt to challenge an expert's opinion. These in-court challenges, in turn, are typically based upon detailed questioning that occurred previously during the pretrial deposition. In my case, a Thomas More attorney spent nearly nine hours probing every argument and claim I had made in my written opinion, and on key points also questioned me about my published articles I had based it upon, hoping to find holes or weaknesses to exploit.[4] I mention this aspect of the legal process to highlight the critical fact that a judge's written opinion in a case is a distillation of in-court testimony and documentation which is itself a distillation of prior oral and written testimony and other evidence, which itself is often based upon a previously published body of material. It is thus an embarrassment that many philosophers have felt free to opine on *McLean* (and a few on *Kitzmiller* as well) based on no more than a superficial review of the final opinion and in near-total ignorance of the documentation and justificatory process that stood behind it. Criticisms, for example, that Ruse assumed a naive Popperian falsificationism without regard to issues of Duhemian holism are plausible only if one stops with a bare reading of Overton's list. ID creationists and a few philosophers have similarly misread *Kitzmiller* by assuming that methodological naturalism was being offered as a replacement a priori demarcation criterion in the same sense that Laudan took Ruse to be doing with his five criteria. Even a cursory reading of my publications would have prevented such an impression.[5]

Put simply, the argument was that as a point of method science does not countenance appeals to the supernatural. Again, we did not claim *only* science requires this ground rule. Such appeals are disallowed in court as well, for instance; MN is tacitly assumed in legal reasoning just as it is in science and should be so for the same sorts of reasons (Pennock 1999, 294–300). Suffice to say that no judge would take seriously a plaintiff who sought damages against someone for laying a curse upon his car or a defendant who pleaded innocent on the grounds that the crime had actually been committed by a ghost. A lawyer would be laughed out of court who argued that judges and juries should consider "alternative theories" that a crime was committed by a supernatural intelligence. The IDC's call for a "theistic science" is similarly unworkable (Pennock 1998).

Methodological naturalism is such a basic assumption that it is mostly taken for granted even among those who disagree about criteria of demarcation. Many philosophers do mention a basic prohibition against appeals to the supernatural in discussions of scientific demarcation. Reisch, for instance, emphasizing the unity of science, writes, "In the case of creation science, statements about immaterial agencies and the creation of things through supernatural processes . . . would render it 'isolated' from the existing network of science" (1998, 345–46). Agassi and Laor focus on repeatability in large part because they say that ignoring it leads to vague metaphysical "magic" whereas "science is the search for natural laws" (2000, 556). Many more do not mention the proscription explicitly but nevertheless can be seen to take it for granted in the possibilities they do or do not take seriously, or in the way they tend to misunderstand those who seek to overturn it. Even critics of demarcation often seem to assume it. For instance, just a week after the *Kitzmiller* ruling, Alexander George wrote an op-ed criticizing it along Laudanian lines for trying to draw a line between science and non-science. But in the same breath that he concludes that philosophers should dismiss the demarcation question he gives his own "liberal" definition of science: "Let's abandon this struggle to demarcate and instead let's liberally apply the label 'science' to any collection of assertions about the workings of the natural world" (2005). Presumably even such a liberal definition would not apply the label to anything at all, so George has not really abandoned the demarcation struggle. And the limit that he seems to presume is at least very like the ballpark notion of MN in restricting science to claims about the natural world. As we will see later, there is even reason to think that this is true of Laudan.

Equally important, especially in the context of the case, even creationists grant that naturalism is a ground rule of science (though they typically confuse or conflate methodological and metaphysical naturalism). One of the reasons I advocated focusing on MN in the *Kitzmiller* case rather than other characteristic scientific values was that all of the major ID creationists had explicitly acknowledged it as a ground rule of science—though a ground rule they think should be overturned—often using those very words. In my written and oral testimony in court I gave a sample of representative quotations from ID leaders to show this (Pennock 2005). Indeed, as discussed previously, the primary goal of the ID movement from the beginning has been to change the definition of science to allow appeal to the supernatural in a revolutionary new theistic science.

In light of this fact, it is interesting that *Kitzmiller* ID witness Steve Fuller gives an odd explanation of his testimony against methodological naturalism. Fuller argues that MN improperly conflates "the source of hypotheses and the conditions under which they are testable" (2006, 829). He defends the role of the supernatural as a legitimate *inspirational heuristic* in the context of discovery, and then points out how this role "comes to be erased" once the previously mysterious (he gives two examples, gravity and genes) has been given experimental proof in the context of justification. He writes how the separation of these "explains the studious neutrality that philosophers of the scientific method have tended to adopt toward 'metaphysics,' in which both naturalism and its opposite, supernaturalism, are normally included: neither metaphysics offers a royal road to scientific validity but both have had significant heuristic value" (ibid.) and argues that MN is a Whiggish reading back of where-science-ended-up into where it emerged. But this defense is of no help to the ID movement, which does not advocate supernatural design as a mere heuristic that may later be discarded in the context of justification, but as a metaphysical truth that should be substantively incorporated into the scientific picture of the world. It is always their substantive claim, not their inspiration, that is at issue.

Ironically, Fuller uses one of the same examples I have used in print and in my oral testimony to illustrate the point about naturalizing the supernatural (which will be discussed further below); the scientific investigation of gravity is a perfect example of how MN views the world in terms of natural regularities. Again, Fuller's example provides no comfort to creationists, whose view of Newton's thinking about gravity is that he was right to call upon God to keep the planets orbiting regularly when his laws of gravitation seemed to fall short of a full explanation. No scientist today would take seriously such appeal to divine nudges to explain the heavens. Fuller writes that he is "not a known advocate of—or expert in" intelligent design, so it is possible that he was unaware of the core philosophical commitments of the ID movement, which may explain why he often said things in his testimony that better supported the claims of the plaintiffs than the defendants. His point above about the studious neutrality toward both supernaturalism and metaphysical naturalism in the context of justification fits quite well with the view that we articulated—and that ID creationists adamantly opposed—in *Kitzmiller.*

Rejecting the naturalism of science is not a peripheral issue but is the central point of the ID movement, which aims to serve as a "wedge" to break apart the naturalism they see as having driven God from the public square.[6] IDCs are often cagey about this core commitment, sometimes even seeming to deny that theirs is a supernaturalist view (occasionally even by attempting to redefine nature), but we were able to provide ample documentation of their view in court. To give just one additional example here, in his paper at the SMU ID symposium William Dembski claims that scientific naturalism (which he links to atheism, materialism, scientism, and secular humanism) is incomplete because it excludes appeal to the supernatural, and he tries to argue "that it is legitimate within scientific discourse to entertain questions about supernatural design" (1994).

IDCs will also sometimes say that they could accept "evolution" but by this they have in mind change where every increase in biological complexity necessarily is purposefully directed by a supernatural entity. We will again limit ourselves to just one example from Dembski who once wrote that, "the design theorists' beef is not with evolutionary change per se, but with the claim by Darwinists that all such change is driven by purely naturalistic processes which are devoid of purpose" (1995). This quotation also illustrates a second, related matter, namely, that IDCs illegitimately build atheism into their definitions of evolution (here "Darwinism") and scientific naturalism. In fact, neither of these are inherently atheistic or theistic but are neutral with regard to metaphysical views about possible transcendent purposes. IDCs regularly try to claim that naturalism in science simply defines away what may be a real possibility about the world that it is created supernaturally (for example, Plantinga 1991, 345) But it is IDCs who employ idiosyncratic definitions. Nothing in our account of science defines away anything about such metaphysical possibilities. On the contrary, it remains scrupulously neutral regarding the existence or nonexistence of God or other transcendent beings. Judge Jones made this quite explicit in his ruling; the ID belief in a supernatural designer may indeed be true, but it is just not science.

(4) Consistent with this view, we did not claim, as creationists often charge (for example, Witt 2005), both that ID is unfalsifiable and that it has been tested and found false. I have already mentioned that we did not appeal to a falsifiability demarcation criterion, but there are two other common misunderstandings regarding this issue that are worth mentioning.

The first arises in relation to the conclusion in the *Kitzmiller* opinion that ID fails as science not just for the primary reason that it violates the ground rules of science, but for two additional reasons, namely, that "the argument from irreducible complexity, central to ID, employs the same flawed and illogical contrived dualism that doomed creation science in the 1980s" and also that "ID's negative attacks on evolution have been refuted by the scientific community" (2005). Neither of these points contradicts the claim that the central positive thesis of creationism (that the complexity of life is the result of the purposeful action of a transcendent designer) is untestable. As I pointed out in my oral testimony, irreducible complexity (IC) and the other common creationist arguments do not offer evidence for their positive thesis but are simply challenges to evolution. As such *they are not tests of ID but rather of evolution*, a critical point that is regularly overlooked. Moreover, they are tests that evolution passes. For instance, regarding the IC argument, I had presented a counterexample in *Tower of Babel* (Pennock 1999) that Behe later admitted undermined his definition of IC (Behe 1999). He had yet to provide a promised revision to fix the error. I also testified about experiments that colleagues and I had done using evolving computer organisms to test some of Darwin's hypotheses about the evolution of complex features (Lenski et al. 2003). Some of these experiments turned out to also be relevant to the case in that some of the traits that we observed were irreducibly complex in Behe's sense and so provided a direct observational refutation of the core ID claim that such systems cannot evolve. Biologist Ken Miller also testified in court about IC and other failed challenges, as did paleontologist Kevin Padian. Moreover, even if science did not have a ready answer to such negative arguments, such explanatory gaps would not have supported the positive ID claim, as poking holes in evolution does not prove creationism. Overton had previously identified this fallacy of the dual-model argument of the creation science arguments and Jones was correctly pointing out that ID made exactly the same mistake.

The second misunderstanding arises in a different way, with ID proponents and even some opponents (typically supporters of metaphysical naturalism), claiming that science can indeed test the supernatural. This confusion often seems to turn on an inadvertent naturalizing of the supernatural, such as treating creationist hypotheses as though they were meant in the ordinary way. For instance, both Laudan and Quinn cite the young-earth creationist view that God created the earth 6,000 to 10,000 years ago as a hypothesis that is testable and found to be false. But this and other examples that are offered to show the possibility of tests of the supernatural invariably build in naturalistic assumptions that creationists do not share. Confronted with the empirical evidence for an ancient earth, creation scientists dismiss the relevance of any such observations on the ground that God simply made the earth *appear* to be old (or "mature"). Some think of this as a test of faith so that one learns to accept the authority of the Bible over that of one's (mere) senses. The point here is that we cannot overlook or ignore, as Laudan and company regularly do, the fact that creationists have a fundamentally different notion from science of what constitutes proper evidential grounds for warranted belief. The young-earth view is certainly disconfirmed if we are considering matters under MN, but if one takes the supernatural aspect of the claim seriously, then one loses any ground upon which to test the claim.

The "design" hypothesis is another common case in point; IDCs regularly conflate natural notions of design such as are used in archeology and forensics (notions that are unproblematic scientifically) with supernatural notions that science cannot countenance. One must carefully distinguish between the truly supernatural and what is only apparently so (Pennock 1999, chap. 6, esp. 301–308). This distinction is exemplified historically in the difference between supernatural magic, which called upon demons or angels, and "natural magic," which might seem mysterious to the uninitiated but was assumed to rely upon natural, albeit esoteric, cause-effect relationships. Sometimes it is relatively easy to identify a truly supernatural claim and see why it is not testable, such as is the case with the religious explanation of how communion wine can be said to change to blood metaphysically without modification of any of its mundane properties (Ruse 1982; Pennock 2006). There are a host of familiar religious mysteries one could also cite, ranging from the view of Jesus as simultaneous God and man to the orthodox claim that God's "being" is a concept beyond being and not being. Nor is this issue limited to such overtly religious theses. That a nonreligious term is used does not tell by itself that a thesis can be understood naturalistically; for instance, it would be a mistake to think that spiritual "energy" is testable as the scientific notion of energy is (Pennock 2000). Would it even be intelligible to speak of supernatural "weight" or supernatural "color"? If these were truly meant to be different than the notions of weight and color as we understand these concepts in terms of our ordinary natural experience, then we have no ground upon which to draw any inference about them. Supernatural "design" is of a kind. As David Hume pointed out, we have no experience and thus no knowledge of divine attributes. Those who think otherwise, whether in the service of proving or disproving the divine, invariably do so by illegitimately assuming naturalized notions of the key terms or other naturalized background assumptions.

Evan Fales makes this mistake in arguing that *Kitzmiller* reached the right conclusion for the wrong reason, claiming that there is no reason that "suitably precise claims about the supernatural could not have distinctive empirical implications, and hence be testable." He faults ID "not merely because it invokes the supernatural... [but because] it refuses sufficiently to flesh out its supernatural hypotheses" (2006). Ironically, Fales's way of putting this shows the problem he misses; it is only by adding flesh to ghostly supernatural hypotheses that they become amenable to test. To specify who the designer is, what its purposes are, and how it achieves them helps make design hypotheses testable when we are speaking in natural terms, but the moment we acknowledge the supernatural element as "other-worldly" and as truly different in kind from the natural the terms lose any connection to testable reality.

Elliott Sober occasionally seems to take a similar line to Fales in suggesting that at least some claims about supernatural beings are testable if they are suitably stated, giving by way of example "the claim that an omnipotent supernatural being wanted above all that everything in nature be purple" (forthcoming). Presumably he thinks the ordinary observation that not everything in nature is purple shows that the hypothesis is false because an omnipotent being with such a desire would surely have made the world so. If we are thinking of this in terms of a naturalistic understanding of notions such as "a being," "desire," "above all," "nature," "to be," "purple," and so on, that test may be perfectly fine.[7] But what can we say when we treat the hypothesis supernaturally? Might not all of nature now indeed "be purple" in its noumenal substance, irrespective of its accidents, as wine purportedly becomes blood without observable change in the miracle of the Eucharist? Is it even possible for God (to specify the being behind this generic talk of omnipotent supernatural entities), to want such a thing "above all"? Even if one sets aside Leibnizian problems with such a notion, it is not clear how we could tell whether this is a coherent thesis. And what can we say follows from a claim of supernatural "wanting," by an omnipotent being or not, unless we treat that desire naturalistically like our own? One could easily continue, but let us not belabor the point—it is only under a tacit naturalistic reading that the testability of hypotheses containing such concepts could be thought plausible.

Admittedly, interpreting such "hypotheses" is inherently confusing because of the pervasive inconsistency of religious claims in general and creationist claims in particular. Only rarely may we proceed as though they presume MN; more often we are obliged to address these under the standard assumption that they reject it. ID creationists are intentionally vague and prevaricating when speaking of their design hypothesis to obscure the inherent supernaturalism of their views. However, the principle of charity requires that we consider the strongest version of their argument, which requires the truly supernaturalist interpretation. As far as possible I have tried always to be careful to indicate when I am treating a thesis purely scientifically and assuming MN and when I am stepping back to take a supernatural thesis seriously. For instance, I pointed out in my oral testimony that even our experimental observation of the evolution of an irreducibly complex system refutes Behe's challenge to Darwinian evolution only under the presumption of MN—if some supernatural entity is intervening in the computer core to simply make it appear that evolution happened naturally, we have no way of checking. To put the general point philosophically: it will not do to argue only with Cleanthes and ignore Demea, or to inadvertently treat Demea's mysterious God like Cleanthes' (naturalized) anthropomorphic God.

(5) Finally, we did not assume, as some creationists charged, that something is religion simply by virtue of not being science. Rather we identified a characteristic—namely, appeal to the supernatural—that by itself was sufficient to rule ID as not science and that independently was sufficient to show that it was religion for legal purposes.[8] We provided extensive documentary evidence to support this, showing not only that ID is religious, but that it is sectarian religion.

There is more that one could say about the *Kitzmiller* philosophy and ways in which it compared to or differed from *McLean*, but this brief overview is sufficient for our present purposes and puts us in a position to return to discuss of Laudan's objections to demarcation.

LAUDAN CONTRA *McLEAN* AND DEMARCATION

Laudan gave his key arguments against the demarcation problem in a triptych of articles (1982; 1983a; 1983b), the upshot of which is this strongly worded and oft-quoted conclusion.

> If we would stand up and be counted on the side of reason, we ought to drop terms like "pseudo science" and "unscientific" from our vocabulary; they are just hollow phrases which do only emotive work for us. As such, they are more suited to the rhetoric of politicians and Scottish sociologists of knowledge than to that of empirical researchers. (1983a, 349)

Laudan tries to support this conclusion through two main lines of argument.

His first argument is to appeal to the lack of unity among philosophers regarding proposals for criteria of demarcation.

> From Plato to Popper, philosophers have sought to identify those epistemic features which mark off science from other sorts of beliefs and activity. Nonetheless, it seems pretty clear that philosophy has largely failed to deliver the relevant goods. Whatever the specific strengths and deficiencies of the numerous well-known efforts at demarcation . . . it is probably fair to say that there is no demarcation line between science and non-science, or between science and pseudo-science, which would win assent from a majority of philosophers. (1996, 210)

On this point, we may here again briefly mention Quinn's commentary. ID leader William Dembski, for instance, cites Quinn's conclusion that "one bad precedent, particularly one so extensively publicized and so apt to arouse passionate feelings, is already one too many" (Quinn, quoted in Dembski 1995). Quoting this out of context, Dembski improperly makes it appear that the purportedly "bad precedent" Quinn speaks of was the *McLean* case itself, though Quinn was actually referring to what he claimed was the bad precedent of a philosopher as expert witness. Quinn wrote that "the major problem in *McLean v. Arkansas* [is that] Ruse's views do not represent a settled consensus of opinion among philosophers of science" (1984, 384). He also faults the opinion and Ruse's demarcation criteria on the same grounds as Laudan. Before dealing with the more substantive points about purportedly failed, false, and fallacious arguments, the complaint about general acceptance deserves a brief rebuttal.

While it is true that there was no consensus in 1983 among philosophers of science regarding Ruse's five criteria, the lack of a "settled consensus" should hardly be seen as a bar to engagement at the bar. The law does not require unanimity before a professional may be called as an expert witness; there would be no such testimony possible if it did. Philosophy is hardly unique in its internal professional disagreements. It is well known that economists laid end to end still all point in different directions. Of course there are probably always more devil's advocates on any given question among the ranks of philosophers, but contrarians are to be found on even well-settled issues in any profession, including science. ID advocates continue to cite astronomer Fred Hoyle's contrarian rejection of the big bang and evolution on earth. Hoyle's like-minded colleague Chandra Wickramasinghe testified on behalf of creation science in the *McLean* case. IDCs likewise have a group of "Darwinism dissenters" they regularly trot out. However, unlike the *McLean* criteria, there is good reason to think that MN is accepted by a large majority of philosophers of science and is probably as close to a settled consensus as is possible in our profession. In any case, as will be discussed in detail in the next section, there is excellent evidence that it is all-but-universally accepted as a tacit ground rule of science among scientists, which is the more relevant standard. This last fact is also relevant to the second major criticism.

Laudan's second, more substantive, approach is to ask whether *McLean* accurately captured how science works:

> The victory in the Arkansas case was hollow, for it was achieved only at the expense at perpetuating and canonizing a false stereotype of what science is and how it works. (1982, 355)

This and the first line from the quotation that headed this section are repeated endlessly by ID creationists and, sad to say, probably are the most influential sentences in Laudan's entire body of work. But what exactly is the false and "anachronistic" stereotype that Ruse supposedly perpetuated in *McLean* that calls into question the scientific community's intellectual integrity? It is what Overton called the five essential characteristics of science distilled from the testimony of Ruse and the scientific witnesses in the case. Laudan's strategy in criticizing Ruse's five criteria and presumably any other demarcation criteria was to find counterexamples from the history of science. For each of the five he offers one or another exception and so concludes that they fail as a list of necessary and sufficient conditions.

These considerations lead Laudan to conclude that there is no sensible distinction between science and pseudoscience. Applying this reasoning to creationism, he says that we should admit that it is science, but just very bad science. Creationism is testable, he opines, writing: "[To claim that] creationism is neither falsifiable nor testable is to assert that creationism makes no empirical assertions whatever. That is surely false" (1982, 352). Laudan then goes on to give a list of what he says are testable creationist assertions drawn from the *McLean* decision itself, such as that the earth is of very recent origin. Indeed, he chides Judge Overton for mentioning these "apparently without seeing the implications." But Laudan and others who make this sort of statement are wide of the mark.

First of all, Laudan's statement that a claim that is neither falsifiable nor testable implies that it makes no empirical assertions whatever is odd unless he means to define "empirical" in terms of falsifiability and testability. That is a curious move to make given that there are nonscientific empirical matters that do not involve either. And it is certainly a strange statement for someone to make in a discussion of the definition of empirical science who has just rejected both of these criteria for just that purpose. (Of course there is a contrast between the empirical and the mathematical sciences, but no one took the debate in *McLean* to be about the latter.) If Laudan's counterexamples work against these as demarcation criteria for defining science, then they or ones like them should work equally well against them for defining the empirical. Thus it is Laudan here who apparently fails to see the implications of his arguments, not Overton, who surely would not have quibbled over whether we disqualify creationism because it isn't science or because it isn't empirical science.

Second, Laudan and company fundamentally misunderstand the nature of the creationist claims he cites as having already been disconfirmed. Take again the key example that Laudan mentions, namely, that the earth is of very recent origin. In the Arkansas balanced-treatment act the thesis was put forward with just this vague language, but the notion of "recent" creation is a standard term in creationist circles to refer to the young-earth creationist view that the earth is no more than 6,000 to 10,000 years old. Laudan does not say how he thinks scientists know that this is false, but presumably he would cite the usual sorts of scientific evidence. Indeed, if we judge the evidence in the ordinary scientific manner, then this conclusion is inescapable. But as we have seen and must continually emphasize, creationists do not view the evidence in the ordinary scientific manner.

DEFENDING DEMARCATION

Laudan and company are wrong to think that scientific demarcation is a pseudoproblem and that there is no point to maintaining a distinction between science and pseudoscience or religion. In this section I will give four reasons for philosophers to reject Laudanian anti-demarcationism and to take the task of demarcation seriously. I begin by briefly defending a weak version of the distinction that would be reasonable even if one were to grant most of Laudan's other points.

The Dustbin of History Argument

The conclusion that creation science does not qualify as science is defensible even if one were to grant Laudan's superficial view about creationist claims, namely, that "these claims are testable, they have been tested, and they have failed those tests" (1982, 352). Take the geocentric view of the world, which is still advanced by some creationists. While one may say that such a claim was historically scientific or even that it remains scientific in the abstract sense that it is testable, it would nevertheless be fair to conclude, because this claim has been decisively disconfirmed (at least under the assumptions of MN), that it is unscientific to continue to hold and teach it today. The scientific picture of the world does not include claims that have been decisively refuted and effectively relegated to the dustbin of scientific history. Creationists want their claims about the age of the earth, the universal flood, the sudden emergence of life-forms with all their features intact, or what have you, to be taught as the truth or at least as live alternatives. But this is unscientific in a perfectly straightforward sense.

Confronting a would-be biologist who intentionally or because of incompetence paid no heed to empirical evidence and what it has shown, a responsible academic advisor would have to say, I'm sorry, but you are not doing science and you have no business being in this program. Similarly, a school administrator would be irresponsible who did not say the equivalent thing and remove a teacher who was teaching creation science or ID in a science class. Putting this another way, once the two senses of the term are recognized, we may declare that sufficiently bad science is not science at all. In *Kitzmiller* we covered this base as well.

As a supplement to the primary reasons for why ID fails to qualify as science, Judge Jones also notes that "ID has failed to gain acceptance in the scientific community, it has not generated peer-reviewed publications, nor has it been the subject of testing and research" (*Kitzmiller* 2005, 64). We do not need to say anything beyond what has already been mentioned in passing about the first and last points; not only has ID failed to subject its claims to test, but it has failed even to offer any positive research program and statements from dozens of national and international professional scientific societies are unanimous in rejecting ID as science. Regarding the second point, I have elsewhere remarked upon the dearth of peer-reviewed scientific publications by ID proponents (2002), but it is worth briefly looking at a few of the claims IDCs make about their publications before we conclude, if only to highlight some of the possible pitfalls a commentator must learn to watch out for to understand how ID theory tries to masquerade as science.

One case in point is *Darwinism, Design, and Public Education*, an anthology edited by philosopher Stephen C. Meyer and rhetorician John Angus Campbell. (Both were mentioned previously as ID leaders who had originally been listed as expert witnesses for the defense in the *Kitzmiller* case but who withdrew at the last minute.) When the book appeared the Discovery Institute issued a press release, hailing it as "a peer-reviewed book from Michigan State University Press that presents a multi-faceted scientific case for the theory of intelligent design" (2004), and in a letter to the Chronicle of Higher Education Meyer held it up as "a peer-reviewed... scientific anthology" (2004). However, in a letter to the *Chronicle*, the director of the MSU Press corrected these false characterizations.

> Of concern to us is the fact some individuals now are stating that MSU Press's publication of *Darwinism, Design, and Public Education* proves that the "Intelligent Design" (ID) theories presented in the work have been subjected to a "scientific vetting," which, in turn, proves or supports their credibility. Such comments are inaccurate and wrong.

> [T]he vetting was specifically for a work that would appear in our Rhetoric & Public Affairs Series; the procedures and criteria applied to this review were fundamentally different from those applied to manuscripts we would consider "scientific" in nature. In other words, Michigan State University Press's publication of *Darwinism, Design, and Public Education* should not be construed as demonstrating that the book's contents have scientific validity.[9]

This is not an isolated incident of misrepresentation. On the stand in the *Kitzmiller* case, Behe touted his own book as a prime example of peer-reviewed science, claiming that it was actually reviewed even more stringently. His claims were directly impeached during cross-examination and subsequent information further undermined them.[10] The handful of other publications that IDCs cite that are peer-reviewed give no evidence for ID. For instance, articles mentioned of Discovery Institute fellow Douglas Axe (2000; 2004), who now works for the DI-funded Biologic Institute, provide no positive evidence for ID whatsoever and simply try to show problems that evolution faces in producing functional biomolecules. As with Behe's and Dembski's publications and those of every other creationist I know of, such "research" completely fails to test their own "alternative theory" but, as previously noted, rather are tests of evolution.

Although one could disqualify creationism with this weaker notion of science, I want to argue now that we should not grant Laudan's analysis in either its specific claims about creationists' views or in its major claims about demarcation as it applies to the case at hand. Again, *Kitzmiller* did not attempt to draw a pinline border between science and anything else, but needed only to show that ID was not science but religion.

The Perversity Argument

We may begin by pointing out the prima facie absurdity of Laudan's claim that searching for demarcation criteria is a pointless pseudoproblem; philosophers distinguish fifty-seven varieties of realism and antirealism, but they can't tell science from pseudoscience or religion? What ever happened to the adage that you can always sell a philosopher a distinction? As much as philosophers delight in provoking their listeners by problematizing the commonplace, there comes a point at which this pleasure becomes a perversity. To hold that there is no sensible difference between science and pseudoscience is to abandon any claim of insight into the analysis of knowledge or questions about distinguishing the real from the deceptive. And to hold that there is no difference between science and religion is to make philosophy appear absurdly out of touch and irrelevant to scientists.

When philosophers step back from being merely provocative they do acknowledge the real difference. As a case in point, one need only mention Paul Feyerabend, whose infamous defense of epistemological anarchism and claim that in science "anything goes" earned him the reputation as perhaps the most extreme philosophical critic of scientific method. Yet even Feyerabend, despite his playful acceptance of all manner of odd views, knew he had to draw a line somewhere. He does so in giving criteria for ruling out cranks. Feyerabend writes that "the distinction between the crank and the respectable thinker lies in the research that is done once a certain point of view is adopted." As he explains this distinction,

> The crank usually is content with defending the point of view in its original, undeveloped, metaphysical form, and he is not at all prepared to test its usefulness in all those cases which seem to favor the opponent.... It is this further investigation... which distinguishes the "respectable thinker" from the crank. The original content of his theory does not. (1981, 199)

What is Feyerabend doing here if not offering a practice-based criterion of demarcation between science and crank science? Indeed, it is almost as though he meant to apply this criterion to rule out creationism as crank science. Creationists of all stripes are well known as beginning with beliefs, both metaphysical and empirical, that they hold immune from empirical test. Feyerabend only fails to see that this problem is not just a matter of attitude—metaphysical immunity to test can indeed be built into the original content of the theory. As we have seen, this is exactly what creationism does.

My purpose here is not to defend Feyerabend's or any other particular proposal to demarcate science from pseudoscience in the sense that Laudan had in mind, but rather to point out the absurdity of the view that philosophers view demarcation as a pointless pseudoproblem. Contra Laudan's blithe dismissal, pseudoscience in the ballpark sense is a useful, reasonable concept that even someone like Feyerabend had to acknowledge. To hold that there is no useful conceptual difference between science and pseudoscience is to lose touch with reality in a profound way.

The Pragmatic Argument

If philosophers really couldn't tell the difference between science and religion, and if anti-demarcationism were really taken seriously and held consistently in philosophy departments, our course listings and hiring practices would be quite different than they are. But one would be hard pressed to find a department that does not list philosophy of science and philosophy of religion as separate courses. We expect that these should cover quite different subject matter. If a philosophy of science course touches on religion at all, it would likely involve how science views religion, and vice versa. The fundamental assumptions and characteristic concepts that are subject to philosophical analysis are essentially different in these courses. No philosophy department would be taken seriously that failed to distinguish between these.

Quinn may contend that there is "no settled consensus of opinion among philosophers of science" (1984) about what criteria distinguishes science from religion, but even he (and in the same breath) acknowledges that there are philosophers of science. Presumably he also acknowledges that there are philosophers of religion, because he is one himself. His professional colleagues would surely balk if he proposed that one could not tell the difference between these when conducting a job search. Laudan would not have remained long as department chair if he started hiring philosophers of religion for philosophy of science openings or vice versa. The fact is that philosophy departments have no trouble recognizing the difference between science and religion or making practical decisions based on that difference.

Even if this were not the case, it would be more a mark against philosophy than a sign that there is no real distinction. Indeed, the basic commitment of a philosopher of science is to analyze and explicate the concepts and assumptions of science as it is practiced, so it behooves us to consider what science actually says and does with regard to this question.

The Empirical Argument

When we look empirically at what scientists and science educators themselves say science is, then we see immediately that they all ignore Laudan and clearly operate on the idea that there is a real distinction between science and non-science. Indeed, the evidence for this view is so pervasive that it is hard to see how one can take Laudan's incredible pronouncements as anything but indicating a cavalier disregard for the balance of evidence and a foolhardy disengagement from what should be the subject matter of philosophy of science. I can here only give an outline of some of what Laudan had to ignore in his anti-demarcationist screed.

Just as they were when Laudan was writing, resolutions today from professional scientific associations on this issue are in broad agreement. The National Academy of Sciences, for instance, dismisses both classical creationism and ID as unscientific:

> Creationism, intelligent design, and other claims of supernatural intervention in the origin of life or of species are not science because they are not testable by the methods of science. (1999, 25)

By a recent count, over seventy-five professional scientific organizations have issued public statements opposing ID and other forms of creationism and nearly all say explicitly that these are not science. Moreover, almost all of these statements mention in one way or other science's restriction to natural explanations as a reason for disqualifying creationism.

Professional science education organizations have issued similar statements, rejecting creation science and ID as not real science. The National Science Teachers Association (NSTA) statement on the nature of science is but one case in point. It reads, in part:

> Although no single universal step-by-step scientific method captures the complexity of doing science, a number of shared values and perspectives characterize a scientific approach to understanding nature. Among these are a demand for naturalistic explanations supported by empirical evidence that are, at least in principle, testable against the natural world. Other shared elements include observations, rational argument, inference, skepticism, peer

> review and replicability of work.... Science, by definition, is limited to naturalistic methods and explanations and, as such, is precluded from using supernatural elements in the production of scientific knowledge. (2000)

In another statement on the teaching of evolution, NSTA explicitly rejects creation science and ID on the grounds that they are not science for just such reasons (2003). One can find dozens of similar statements from both scientific and science education organizations that in more or less direct ways articulate a presumption of natural regularity and the requirement that science appeal only to naturalistic explanations.

The number and consistency of such statements is good evidence by itself that scientists see a difference between science and nonscience and that they count creationism as falling in the latter group. Moreover, the methodological restriction of science to testable, natural hypotheses is a key reason given for ruling creation science and ID out of bounds. But is this just the propaganda of scientists defending their social authority, as creationists sometimes charge? Again, the evidence does not support such a complaint. Indeed, one gets an even stronger sense of the importance of this ground rule by examining scientific practice directly, where it is simply taken for granted. The few cases of what some might superficially take to be exceptions to the ground rule, such as attempts to weigh the soul or to test the efficacy of petitionary prayer, actually turn out to be confirming examples when one examines them more carefully, for these all work by naturalizing the relevant concepts.

To find plausible counterexamples, one usually has to look to the early history of science. By far the most common counterexample cited is the one we noted previously, namely, Isaac Newton and his appeal to such a being in the General Scholium of the *Principia*, but one occasionally hears mention of William Whewell, Charles Lyell, and even Darwin, all of whom left open the door to some degree for interventions by the Creator. This is not the place to examine in detail such historical examples, which are not always as clear as they are purported to be. Many do not actually make use of supernatural intervention but remain properly agnostic and simply acknowledge it as a possibility (which is consistent with MN); some reserve it for ultimate explanations in a way that is more philosophical than scientific; some seem to hold it by inertia as an inconsistent holdover from a prescientific way of thinking;

and a few seem to be merely pious lip service.

But we do not need to explain or explain away every purported counterexample; it is no surprise that there are some cases to be found, especially in the early history of science. Again, we are not proposing a conception of science that ignores how conceptions have changed or may continue to change over time. The point is that science as it is currently understood would not countenance their supernatural explanations. Whether we view them with indulgence or embarrassment, Newton's and others' appeals to miracles are mostly simply ignored now and are not taken to be part of what was scientific about these scientists' pioneering work. Contemporary scientists who opine for or against the supernatural mostly confine their speculations to popular "philosophical" writings, for they have no place in the scientific journals.

Nor is there any sign that this requirement of MN as a scientific ground rule is changing. As part of my research for my expert opinion in the *Kitzmiller* case, I did a systematic search of major indices of scientific journals to see whether there was any evidence that appeals to the supernatural were being countenanced in scientific studies (Pennock 2005). In databases that covered tens of thousands of peer-reviewed scientific articles there were only a tiny number that even mentioned the supernatural and these mostly dealt with medical studies about how to deal with patients' belief in the supernatural. A single article I found that did seem to take supernatural possibilities seriously was by an advocate of prayer in alternative medicine; yet even that author did not take exception to the ground rule of methodological naturalism but explicitly acknowledged that considering such nonnatural possibilities took one outside of science (Levin 1996).

The Philosopher's Task

I have emphasized how MN is assumed in statements and resolutions about the nature of science, but it is important to point out that such statements also make reference to other scientific values, including ones in Overton's list. Someone might say that these statements are an incoherent hodgepodge. But even if they are, this is where the job of the philosopher of science begins. To reiterate, the basic task of philosophy of science is to explicate scientific reasoning and practice in the Carnapian sense of giving a rational reconstruction of the relevant concepts and their interrelationships.

My own account has been to explicate scientific naturalism as a methodological commitment, not an a priori metaphysical one, and to rationally reconstruct it as arising from a basic value in science, namely, to the idea of testability or, more precisely, to science's epistemic value commitment to the authority of empirical evidence. MN is not dogma; it continues to be accepted in part because of its success—it works. Moreover, we do not necessarily rule out modifying the ground rule if someone were to find a workable method of finding evidence for supernatural hypotheses. On my analysis of the relevant concepts I find it hard to even imagine what such an alternative method would look like and I have seen no proposal that comes close to being conceptually coherent (certainly IDCs do not have such a method), but I remain open to being shown wrong. Such an attitude usually goes without saying in philosophy, but in this context one must mention it explicitly because IDCs regularly try to tar defenders of evolution and scientific naturalism as closed-minded ideologues.

Finally, my explication of the ground rule is obviously not the only possibility. A few scientists do appear to take science's naturalism in a metaphysical sense. Biologist Richard Lewontin is probably the clearest example (1997), though in his case this view of science likely stems from his Marxism. (Creationists nevertheless endlessly quote Lewontin on this point as though he represented all of science. They also regularly cite a comment Ruse once made in a talk about the *McLean* case in which he seemed to say that science assumes metaphysical naturalism.)[11] Moreover, whether one holds to a metaphysical or to a methodological form of naturalism, a philosopher of science could explicate its justification and its relationship to other scientific concepts in different ways. What one may not do is ignore or lightly dismiss such a pervasive and fundamental ground rule.

Similarly, while it is certainly a philosophical option to argue that the commonly understood distinction between science and pseudoscience (or science and religion) is a pseudoproblem, it should be acknowledged as an extreme view for a philosopher of science to take because it departs so radically from the actual scientific norms and practices that are the subject matter of our analysis. In light of such evidence, it is hard to think that Laudanians are taking our subject seriously. How could Laudan have gone so wrong?

Diagnosing and Rehabilitating Laudan

One reason for Laudan's errors seems to have been that he was ignorant of many important aspects of creationists' real claims and of the epistemological assumptions that they do not share with science. Unfortunately, it is a common problem for many philosophers and scientists that because they do not take creationism seriously—it being merely a public controversy—they fail to do their homework before opining on the subject. Though they may mean well, probably more harm has been done to the defense of science education by such ill-prepared and politically naive commentators than could have ever been done by creationists unaided by their ill-considered remarks.

A second reason involves the nature of classification; most of the problems with Laudan's analysis arise because of the way he frames the demarcation problem. At the very least, the perverse insistence that demarcation requires finding a set of exceptionless necessary and sufficient conditions is making perfection the enemy of the good. We should not expect a sharp, bright pinline of demarcation. I suspect that Laudan viewed the demarcation problem this way because he was thinking about it primarily in reaction against Karl Popper's treatment of the question. It was Popper who set up demarcation as an issue in philosophy of science in the latter half of the twentieth century and who proposed his own deductive notion of falsification as the criterion to mark the border. Laudan's article was written near the end of an era of widespread "Popper bashing" within analytic philosophy, and it should really be seen as an attack on Popper by way of an attack on Overton and Ruse, who unfairly seem to be taken as his surrogates.

This would also explain the anti-demarcationists' emphasis on problems with the falsification criterion. Quinn, appealing to the work of Duhem, objects that statements are not testable or falsifiable in isolation. Then, like Laudan, he goes on to cite the young-earth creationist view as something that "has been repeatedly tested and is so highly disconfirmed that, for all practical purposes it has been falsified" (1984). Setting Quinn's inconsistencies aside, these points would be unproblematic if directed against a naive falsificationism and if creationists' claims about the age of the earth were understood under the ground rule of MN. However, one sees that they are wide of the mark once one moves beyond a superficial reading of Overton's opinion and becomes more familiar with what creationists actually hold.

In any case, as discussed previously, *Kitzmiller* did not appeal to falsification as a demarcation criterion.[12] However, we might now ask whether Laudan's other arguments would put him at odds with our ballpark demarcation approach that judges creationism as unscientific because it violates methodological naturalism. Interestingly, Laudan gives Overton's criterion about the "natural law" (which is related though not identical to MN) only a very cursory mention (Laudan 1982) compared to the material about falsifiability. Moreover, he criticizes just the limited bit about "law," noting, correctly, that we can study phenomena without having the laws. Again, we need to read Laudan in light of issues of interest to the logical empiricism of the period; he is probably here just making a passing gesture to a well-known problem with Hempel's Deductive-Nomological (D-N) model of scientific explanation. It is likely that he assumed that Ruse's second criterion was referring to the D-N model (specifically to its requirement that the *explanans* of a scientific explanation contain a law) and, of course, he was quite correct to call that requirement into question as it had already been shown that one may have legitimate explanations even without being able to specify the relevant covering law. But, again, this technical issue does not bear on and is no criticism of the more general issue of the rule of MN, which does not depend at all upon a D-N account of the relation between explanation and laws. Quinn also devotes only a brief paragraph to Ruse's condition about natural law and scientific explanation (1984), and his counterexamples are of the same sort as Laudan's and similarly do not bear on MN.

Although Laudan's discussion of creationism and demarcation has been a boon to ID proponents and other creationists, we ought to at least briefly consider whether Laudan is being used or misused. Unfortunately, it seems that Laudan's usual good sense did abandon him in this instance; he takes himself too seriously in these pieces and his attack on Ruse is too personal. As noted above, it seems as though Laudan was insufficiently familiar with creationists' real views and was oblivious to much of what was at issue in the *McLean* case. However, although Laudan has only himself to blame for the trouble he caused following *McLean*, it remains possible to interpret him in a manner that would put him in line with the ballpark demarcation argument of *Kitzmiller*. Indeed, I think that a fair case can be made that Laudan actually takes for granted the ground rule of naturalism in something close to the sense I have advocated it.

For instance, even in his response to Ruse's reply to his criticism of the *McLean* decision, Laudan does note that Ruse advanced the thesis of transubstantiation as one example of a nonscience that doesn't fit Overton's definition of science and, for what it's worth, he does not take issue with the description of transubstantiation as nonscience. Another bit of evidence for this interpretation has already been touched upon indirectly, namely, in the way that Laudan treated creationist claims as if they had already been tested and refuted, and failed to take into account their real supernatural content. Laudan either does not recognize the distinctive religious aspect of views or else simply does not take them seriously. Either way, it appears that he tacitly, perhaps unconsciously, assumes that science should treat them naturalistically.[13] And more explicitly, although he objects to the thesis that a claim is unscientific until we have found the laws upon which the phenomenon depends (as do I), Laudan clearly endorses what he acknowledges is the ultimate goal of explaining phenomena "in a lawlike way" (1982, 354). Without putting too fine a point on it, that is essentially what the regulative ground rule of MN requires.[14]

CONCLUSION

Barry Gross thought that Laudan's basic mistake was a disastrous application of inappropriate standards. Laudan, he wrote, "has confused the outlines of a Constitutional conflict with a colloquium in philosophy" and in doing so neglected his own wise pragmatic advice about the need to pay attention to the relevant context of inquiry and to the actual course of the evolution of science (1983, 30). Gross was certainly right about this, but as we have seen there is more to the problem, for Laudan and others who have echoed him fail in philosophically more serious ways.

Laudan's broad claim that philosophy regards demarcation as a dead pseudoproblem was and remains inaccurate and it is shoddy scholarship, to say the least, for creationists or others to cite his pronouncement as authoritative. This is not to say that Laudan was wrong on all counts. Many of his criticisms of Ruse's five criteria were correct if the demarcation task is taken to require the identification of a historically exceptionless set of necessary and sufficient criteria to mark a pinline border between science and non-science. But this is an unrealistic and inappropriate standard. In any case, I did not advocate such a list in *Kitzmiller* or elsewhere. Nor did I simply substitute methodological naturalism as an alternative or attempt to draw an a priori sharp line to demarcate all and only science; rather I explicated MN as a basic ground rule that one finds as an all-but-universally-accepted assumption of scientific practice and that is well justified on epistemological grounds as a rational basis for empirical research. Only such a ballpark demarcation judgment is needed to determine that intelligent design or some other form of creationism is not science. Furthermore, I showed that ID creationists themselves recognize naturalism as a scientific ground rule and that their revolutionary aim is to redefine science in the interest of using its authority to support not just a general religious view, but a narrow sectarian one.

As we have seen, Laudan and those who echo his views are completely out of step with the theory and practice of actual scientists. If Laudan's view were indeed the norm in philosophy of science, then it is little wonder that some say philosophy is irrelevant to any matters of practical consequence. Is philosophy going to be so removed from the realities of the world that it has nothing of value to say even on topics that ostensibly are its core concerns? It would be a sad commentary on our profession if philosophers could not recognize the difference between real science and a sectarian religious view masquerading as science. When squinting philosophers like Laudan, Quinn, and their imitators such as Monton and George purport that there is no way to distinguish between science and pseudoscience or religion they bring to mind Hume's observation that "generally speaking, the errors in religion are dangerous; those in philosophy only ridiculous" (1978 [1739], book I, part iv, sec. vii). Unfortunately, in giving succor, inadvertently or not, to creation science and now to ID, such philosophers compound the error, making the ridiculous dangerous.

Judging creationism in the ballpark sense requires doing one's homework to learn what creationists actually hold and it requires some philosophical care to frame one's critique within a sound epistemological framework and to avoid various pitfalls, but it is not a matter of controversial conceptual hair-splitting. This is not like the umpire calling a player out who attempts a dusty slide to home plate but more like distinguishing a real ball game at Fenway Park from the "baseball movie" *Field of Dreams*. Critics like Laudan first demand a precise line of demarcation for any possible case and then, failing to find one, petulantly declare that there is no difference and try to take away the ball and make everyone go home. But demarcation, properly understood, is not dead and ID does not just miss the line by a hair; the rational judgment here is that creationism does not even belong in the stadium, that it is playing a different game entirely—Sudoku perhaps. We do not need to precisely delimit the boundaries of science any more than we need the precise boundaries of a pin to conclude that it is not science to ask how many angels can dance on its head. Fortunately, a wise judge understood that, even if a few myopic philosophers still do not.

NOTES

1. The *McLean* case, which was decided in early 1982, was not appealed. A second case, *Edwards v. Aguillard*, which ruled against a parallel Louisiana law, was appealed and made its way to the US Supreme Court, which in 1987 reached the same conclusion that teaching creation science was unconstitutional. One of the revelations of the *Kitzmiller* trial was documentary evidence showing how, immediately following the *Edwards* decision, creationists had simply replaced the term "creation" with the term "design" and "creationist" with "design proponent" in the textbook for the public schools (originally titled *Creation Biology* but eventually published as *Of Pandas and People*) that they were producing without changing the substance of the material (Matke 2005a, 2005b).

2. I will not take the time to address those who have claimed that the courts have no business saying anything about what is or is not science. Such complaints are myopic to the point of absurdity. Determining what is and is not science is absolutely critical in legal settings, quite apart from the issue of the status of creationism. The courts have to determine for all sorts of cases what to admit as scientific expert testimony and what to exclude. Crystal ball readers are not recognized as scientific experts, nor would someone who claimed that God told him that the butler did it. Through various legal precedents, the courts have laid out ways to make such determinations, such as the Frye rule and the Daubert criteria. As we shall see, courts countenance neither suits nor defenses that appeal to nonscientific, supernatural "alternative theories."

3. The notion of ground rules has its origin in baseball as the rules governing

play on a particular field but has come to refer to any basic rule(s) of procedure and behavior to be taken for granted. It is the latter, stronger sense that have I have in mind here—scientific ground rules that are tacitly understood as being so fundamental that they usually do not even need to be listed among the official rules—but I will certainly not take issue with a Red Sox fan who insists that the peculiar Fenway Park ground rule that a ball that rolls under the tarp in Canvas Alley counts as two bases should be taken no less seriously. (I have elsewhere spoken metaphorically of creationism as attempting to sneak into the "stadium of science" [Pennock 1999, 215], so the first sense of ground rule could similarly apply metaphorically if one wanted to retain the historical usage.)

4. Failing to find the hoped-for weaknesses, their cross-examination in court mostly took a different approach, asking me many questions about more or less obscure persons and facts (mostly from the history of science) in what I learned was a standard tactic used to try to make an expert appear ignorant. I thank my professors from graduate seminars many years ago at Pittsburgh for serendipitously preparing me for many of these questions.

5. Even some philosophers who have written on the right side of the issue have sometimes made similar mistakes. Neil Tennant, for instance, says that what he calls "Pennock's positive account of testability" is satisfactory for a lay reader but "will not pass muster for the logician" (Tennant, in press). But he bases this assessment on a single sentence from my testimony, improperly reading into it a notion of *logical* consequence where I was careful to avoid just that term and meaning. Indeed, he attributes to me a view of testability that I have explicitly argued against in publications over fifteen years, and offers by way of rebuttal a logical counterexample of the very sort I have used against that view myself. I predict that some ID creationist will try to undermine *Kitzmiller* by quoting Tennant against me for a view I have actively rejected.

6. For accounts of the ID "Wedge" strategy and the leaked ID Wedge document manifesto that became an important exhibit in the *Kitzmiller* trial see especially the accounts by *Kitzmiller* expert witness Barbara Forrest (Forrest 2001; Forrest and Gross 2003).

7. The example is not elaborated, so I set aside possible complications that could make the test problematic even in the naturalized case. For instance, we are not given the purported observation evidence or the confirmation relation, so it is not clear whether the hypothesized agent is relevantly tested or whether it is illegitimately "confirmed," say, as an irrelevant conjunction.

8. This should be obvious but, for the record, neither did we assume that the supernatural is the defining characteristic of all religions.

9. Personal communication. Bohm copied his e-mail to the *Chronicle* to me, as well as to the dean of the College of Natural Sciences and several other campus administrators. Unfortunately, it appears that the *Chronicle* never published his rebuttal.

10. For instance, Michael Atchison, who identified himself as the reviewer who the publisher said was the deciding factor to publish Behe's book, revealed that he gave his recommendation based simply on a description of it in a ten-minute phone call, attributing his critical role to divine intervention by which the Lord placed Behe's book in the hands of an editor whose wife just happened to be a student in a class in which Atchison identified himself as a Christian (Atchison 2004).

11. Creationists posted a transcript of a talk Ruse gave in February 1993 at an AAAS symposium "The New Antievolutionism" in which he said: "But those of us who are academics, or for other reasons pulling back and trying to think about these things, I think that we should recognize, both historically and perhaps philosophically, certainly that the science side has certain metaphysical assumptions built into doing science, which—it may not be a good thing to admit in a court of law—but I think that in honesty that we should recognize, and that we should be thinking about some of these sorts of things," www.arn.org/docs/orpages/or151/mr93tran.htm. More recently, however, Ruse has defended methodological naturalism (Ruse 2002).

12. The notion of testability that we did make use of does, of course, involve the possibility of disconfirmation as well as confirmation, but this inductive notion is quite different from Popperian falsificationism.

13. This is not enough to tell whether Laudan is presuming just methodological naturalism or the stronger metaphysical notion, but we need not get into that interpretive issue here. In his *Beyond Positivism and Relativism* (1997) Laudan does endorse what he calls Methodological Naturalism. However, in his use of the term there he means naturalism about methodology—the view that methodology is an empirical discipline. He holds that a sound methodology is one that leads to success in achieving our goals and so whether a methodology is sound is an empirical matter. This is not the sense of MN that we have been discussing here, though I suspect that Laudan's notion would not make sense without it, so this may be another reason to think he does presume it.

14. Again, this does not mean that scientific explanations must explicitly cite a law as a premise or even that a specific law be known; we are well past discussions of the D-N model (Pennock 1995).

REFERENCES

Abney, K. 1997. "Naturalism and Nonteleological Science: A Way to Resolve the Demarcation Problem between Science and Nonscience." *Perspectives on Science and Christian Faith* 49.

Agassi, J., and N. Laor. 2000. "How Ignoring Repeatability Leads to Magic." *Philosophy of Social Sciences* 30, no. 4: 528–86.

Anonymous. "Intelligent-Design Backers Downplay *Dover*." WorldNetDaily, http://www.worldnetdaily.com/news/article.asp?ARTICLE_ID=48057 (accessed January 15, 2006).

Atchison, M. "Mustard Seeds." Leadership U, http://www.leaderu.com/real/ri9902/atchison.html (accessed October 21, 2005).

Axe, D. D. 2000. "Extreme Functional Sensitivity to Conservative Amino Acid Changes on Enzyme Exteriors." *Journal of Molecular Biology* 301, no. 3: 585–95.

———. 2004. "Estimating the Prevalence of Protein Sequences Adopting Functional Enzyme Folds." *Journal of Molecular Biology* 341, no. 5: 1295–1315.

Behe, M. 1999. "The God of Science: The Case for Intelligent Design." *Weekly Standard*, p. 35.

Dembski, W. A. 1994. "The Incompleteness of Scientific Naturalism." In *Darwinism: Science or Philosophy*, ed. J. Buell and V. Hearn, 79–98. Richardson, TX: Foundation for Thought and Ethics.

———. 1995. "What Every Theologian Should Know about Creation, Evolution, and Design." *Center for Interdisciplinary Studies Transactions* 3, no. 2: 1–8.

———. "Bradley Monton—Important Article on *Dover*." Uncommon Descent: The Intelligent Design Weblog of Bill Bembski & Friends, http://www.uncommondescent.com/index.php/archives/611 (accessed January 7, 2006).

Derksen, A. A. 1993. "The Seven Sins of Pseudo-Science." *Journal for General Philosophy of Science* 21, no. 1: 17–42.

DeWolf, D. K. "Teaching the Origins Controversy: A Guide for the Perplexed." Discovery Institute, http://www.discovery.org/crsc/articles/article6.html (accessed September 1999).

DeWolf, D. K., and S. C. Meyer, et al. 1999. *Intelligent Design in Public School Science Curricula: A Legal Guidebook.* Richardson, TX: Foundation for Thought and Ethics.

———. 2000. "Teaching the Origins Controversy: Science, or Religion, or Speech?" *Utah Law Review* 39, no. 1: 39–110.

Discovery Institute. "The Wedge Strategy," http://www.stephenjaygould.org/ctrl/archive/thomas_wedge.html (accessed May 1999).

———. "Darwin, Design, and Public Education—New Book Examines the Scientific Evidence for Intelligent Design and Advocates Teaching Both Darwinism and Design to Improve Science Education," http://www.discovery.org/scripts/viewDB/index.php?command=view&id=1694&program=News-CSC&callingPage=discoMainPage (accessed January 8, 2004).

Fales, E. 2006. "Dover Judge Makes the Right Ruling Using the Wrong Premise." Commentary, *Science & Theology News*, March 10.

Feyerabend, P. 1981. *Realism and Instrumentalism: Comments on the Logic of Factual Support. Realism, Rationalism & Scientific Method: Philosophical Papers*, vol. 1, pp. 176–202. Cambridge: Cambridge University Press.

Forrest, B. 2001. "The Wedge at Work: How Intelligent Design Creationism Is Wedging Its Way into the Cultural and Academic Mainstream." In *Intelligent Design Creationism and Its Critics: Philosophical, Theological and Scientific Perspectives*, ed. R. T. Pennock, 5–53. Cambridge, MA: MIT Press.

Forrest, B., and P. R. Gross. 2003. *Creationism's Trojan Horse: The Wedge of Intelligent Design.* New York: Oxford University Press.

Fuller, S. 2006. "A Step Toward the Legalization of Science Studies." *Social Studies of Science* 36, no. 6: 827–34.

George, A. 2005. "What's Wrong with Intelligent Design, and with Its Critics." *Christian Science Monitor.*

Gieryn, T. F. 1983. "Boundary-Work and the Demarcation of Science from Non-science: Strains and Interests in Professional Ideologies of Scientists." *American Sociological Review* 48, no. 6: 781–95.

Gross, B. R. 1983. "Commentary: Philosophers at the Bar—Some Reasons for Restraint." *Science, Technology & Human Values* 8, no. 4: 30–38.

Hume, D. 1978. *A Treatise on Human Nature.* Oxford: Clarendon Press. Originally published in 1939.

Katskee, R. B. 2006. "Why It Mattered to *Dover* That Intelligent Design Isn't Science." *First Amendment Law Review* 5 (Fall): 112–61.

Laudan, L. 1982. "Science at the Bar—Causes for Concern." *Science, Technology & Human Values* 7, no. 41: 16–19. Reprinted in *But Is It Science?* ed. Michael Ruse, 351–55. Amherst, NY: Prometheus Books, 1996.

———. 1983a. "More on Creationism." *Science, Technology & Human Values* 8, no. 42: 36–38. Reprinted in *But Is It Science?* ed. Michael Ruse, 363–66. Amherst, NY: Prometheus Books, 1996.

———. 1983b. "The Demise of the Demarcation Problem." In R. S. Cohen and L. Laudan, eds., *Physics, Philosophy, and Psychoanalysis*, 111–27. Dordrecht: Reidel.

Reprinted in *But Is It Science?* ed. Michael Ruse, 337–50. Amherst, NY: Prometheus Books, 1996.

———. 1996. *Beyond Positivism and Relativism: Theory, Method, and Evidence.* Boulder, CO: Westview Press.

Lenski, R. E., and C. Ofria, C. Adami, and R. T. Pennock. 2003. "The Evolutionary Origin of Complex Features." *Nature* 423: 139–44.

Levin, J. S. 1996. "How Prayer Heals: A Theoretical Model." *Alternative Theories of Health Medicine* 2, no. 1: 66–73.

Lewontin, R. 1997. "Billions and Billions of Demons." *New York Review of Books.*

Luskin, C. "Will Robert Pennock Become the Next Michael Ruse?" *Evolution News and Views,* http://www.evolutionnews.org/index.php?title=will_robert_pennock_become_the_next_mich&more=1&c=1&tb=1&pb=1 (accessed October 27, 2005).

Matzke, N. 2005a. "I Guess ID Really Was 'Creationism's Trojan Horse' After All." The Panda's Thumb. http://www.pandasthumb.org/archives/2005/10/i_guess_id_real.html (accessed October 13, 2005).

———. 2005b. "Missing Link Discovered!" *Kitzmiller et al. v. Dover Area School District.* Legal documents, trial materials, updates. http://www2.ncseweb.org/wp/?p=80 (accessed November 7, 2005).

Mayo, D. G. 1996. "Ducks, Rabbits, and Normal Science: Recasting the Kuhn's-Eye View of Popper's Demarcation of Science." *British Journal for the Philosophy of Science* 47: 271–90.

Meyer, S. C. 1992. "Laws, Causes, and Facts: Response to Michael Ruse." In *Darwinism: Science or Philosophy.* Richardson, TX: Foundation for Thought and Ethics.

———. 1994. "The Methodological Equivalence of Design & Descent: Can There Be a 'Theory of Creation'?" In J. P. Moreland, *The Creation Hypothesis,* 66–112. Downers Grove, IL: InterVarsity Press.

———. "Meyer Responds to Errors in *Chronicle of Higher Education* Article." http://www.discovery.org/scripts/viewDB/index.php?command=view&id=2207 (accessed September 13, 2004).

Monton, B. 2006. "Is Intelligent Design Science? Dissecting the Dover Decision." Online manuscript draft.

Moreland, J. P. 1994. "Scientific Creationism, Science, and Conceptual Problems." *Perspectives on Science and Christian Faith* 46: 2–13.

National Academy of Sciences. 1999. *Science and Creationism: A View from the National Academy of Sciences,* 2nd ed.

National Science Teachers Association. "The Nature of Science." http://www.nsta.org/about/positions/natureofscience.aspx (accessed July 13, 2006).

———. "The Teaching of Evolution." http://www.nsta.org/about/positions/evolution.aspx (accessed July 13, 2006).

Overton, W. R. "United States District Court Opinion: *McLean v. Arkansas.*" In *But Is It Science? The Philosophical Question in the Creation/ Evolution Controversy*, ed. Michael Ruse, 307–31. Amherst, NY: Prometheus Books, 1982.

Pennock, R. T. 1995. "Epistemic and Ontic Theories of Explanation and Confirmation." *Philosophy of Science*, Japan 28: 31–45.

———. 1996a. "Naturalism, Evidence and Creationism: The Case of Phillip Johnson." *Biology and Philosophy* 11, no. 4: 543–59.

———. 1996b. "Reply: Johnson's Reason in the Balance." *Biology and Philosophy* 11, no. 4: 565–68.

———. 1998. "The Prospects for a Theistic Science." *Perspectives on Science and Christian Faith* 50, no. 3.

———. 1999. *Tower of Babel: The Evidence against the New Creationism.* Cambridge, MA: MIT Press.

———. 2000. "The Wizards of ID: Reply to Dembski." The Global Spiral METANEXUS 089. http://www.metanexus.net/magazine/ArticleDetail/tabid/68/id/2645/Default.aspx.

———. 2002. "Intelligent Design & Peer Review: What If They Gave a War and Nobody Came." *Research News & Opportunities in Science & Technology* 2, nos. 11–12.

———. 2004. "DNA by Design?: Stephen Meyer and the Return of the God Hypothesis." In *Debating Design*, ed. M. Ruse and W. Dembski, 130–148. New York: Cambridge University Press, 2004.

———. 2005. *Kitzmiller v. Dover Area School District.* Expert Report.

———. 2006. "God of the Gaps: The Argument from Ignorance and the Limits of Methodological Naturalism." In *Scientists Confront Creationism: Creation Science, Intelligent Design and Beyond*, ed. A. J. Petto and L. R. Godfrey. New York: W. W. Norton.

Philadelphia Inquirer. 2006. "The Opinion Speaks for Itself."

Plantinga, A. 1991. "When Faith and Reason Clash: Evolution and the Bible." *Christian Scholars Review* 21, no. 1: 8–32.

Quinn, P. L. 1984. "The Philosopher of Science as Expert Witness." In *Science and Reality: Recent Work in the Philosophy of Science*, ed. J. T. Cushing, C. F. Delaney, and G. M. Gutting. South Bend, IN: Notre Dame University Press.

Reisch, G. A. 1998. "Pluralism, Logical Empiricism, and the Problem of Pseudoscience." *Philosophy of Science* 65 (June): 333–48.

Resnik, D. B. 2000. "A Pragmatic Approach to the Demarcation Problem." *Studies in History and Philosophy of Science* 31, no. 2: 249–67.

Roper, J. 2005. "Should We Teach Both Evolution and 'Creationism' in Science Classes?" In *The Midwest Philosophy of Education Society Proceedings for 2001–2003*, ed. O. Jagusah, D. Smith, and A. Makedon, 485–504. Bloomington, IN: Author House.

Ruse, M. 1982. "Pro Judice." *Science, Technology & Human Values* 7, no. 4: 19–23.

———, ed. 1998. *But Is It Science? The Philosophical Question in the Creation/Evolution Controversy.* Amherst, NY: Prometheus Books.

———. 2001. "Methodological Naturalism under Attack." In *Intelligent Design Creationism and Its Critics: Philosophical, Theological and Scientific Perspectives*, ed. R. T. Pennock, 363–85. Cambridge, MA: MIT Press.

Sober, E. Forthcoming. "Intelligent Design Theory and the Supernatural—The 'God or Extra-Terrestrials' Reply." *Faith and Philosophy.*

Tammy Kitzmiller, et al. v. Dover Area School District, et al. 2005. Judge Jones, United States District Court for the Middle District of Pennsylvania.

Taylor, C. A. 1996. *Defining Science: A Rhetoric of Demarcation.* Madison: University of Wisconsin Press.

Tennant, N. Forthcoming. "What Might Logic and Methodology Have Offered the Dover School Board, Had They Been Willing to Listen?" *Public Affairs Quarterly.*

Thomas More Law Center. 2005. Defendants' Pretrial Memorandum.

Wirth, K. "The Grinch Opinion in *Kitzmiller v Dover.*" http://www.kevs-korner.com/CREVO/ (accessed January 25, 2006).

Witt, J. 2005. "Miller on Witness Stand: ID Isn't Falsifiable, So It Isn't Science: Plus, We've Already Falsified It." *Evolution News & Views.* http://www.evolutionnews.org/2005/09/title_43.html.